EXS 83

# Environmental Stress, Adaptation and Evolution

Edited by R. Bijlsma
V. Loeschcke

**Birkhäuser Verlag**
**Basel · Boston · Berlin**

Editors:

Dr. R. Bijlsma
Department of Genetics
University of Groningen
Kerklaan 30
NL-9751 NN Haren
The Netherlands

*QP*
*82*
*.E625*
*1997*

Dr. V. Loeschcke
Department of Ecology and Genetics
University of Aarhus
Ny Munkegade
Building 540
DK-8000 Aarhus
Denmark

**Library of Congress Cataloging-in-Publication Data**
**Environmental stress, adaptation and evolution :** / edited by
  R. Bijlsma, V.Loeschcke
    p.   cm. -- (EXS ; 83)
    Includes bibliographical references and index.
    ISBN 3-7643-5695-2 (hardcover : alk. paper). -- ISBN 0-8176-5695-2 (hardcover : alk. paper)
    1. Adaptation (Physiology)   2. Adaptation (Biology)   3. Stress (Physiology)
    4. Evolution.   I. Bijlsma, R. (Rudolf), 1945–   ,
  II. Loeschcke, V. (Volker), 1950–   .   III. Series.
  QP82.E825   1997
  576.8`5--dc21

**Deutsche Bibliothek Cataloging-in-Publication Data**
**Environmental stress, adaptation and evolution :**  / ed. by R. Bijlsma ;
V. Loeschcke. - Basel ; Boston ; Berlin : Birkhäuser, 1997
    (EXS: 83)
    ISBN 3-7643-5695-2 (Basel …)
    ISBN 0-8176-5695-2 (Boston)
**EXS.** - Basel ; Boston ; Berlin : Birkhäuser
    Früher Schriftenreihe
    Fortlaufende Beil. zu: Experientia

© 1997 Birkhäuser Verlag, P.O. Box 133, CH- 4010 Basel, Switzerland
Cover design: Els Meeles, Groningen, The Netherlands
Printed on acid-free paper produced from chlorine-free pulp
Printed in Germany
ISBN 3-7643-5695-2
ISBN 0-8176-5695-2

9 8 7 6 5 4 3 2 1

# Contents

**Stress, selection and extinction**

**Evolution and stress**

# List of Contributors

Albert F. Bennett, Department of Ecology and Evolutionary Biology, University of California, Irvine, CA 92696, USA.

R. Bijlsma, Department of Genetics, University of Groningen, Kerklaan 30, NL-9751 Haren, The Netherlands.

Anneke C. Boerema, Department of Genetics, University of Groningen, Kerklaan 30, NL-9751 Haren, The Netherlands.

Paul M. Brakefield, Institute of Evolutionary and Ecological Sciences, Leiden University, P.O. Box 9516, NL-2300 RA Leiden, The Netherlands.

Jørgen Bundgaard, Department of Ecology and Genetics, University of Aarhus, Ny Munkegade, Building 540, DK-8000 Aarhus C, Denmark.

Reinhard Bürger, Institute of Mathematics, University of Vienna, Strudlhofgasse 4, A-1090 Wien, Austria.

Peter Calow, Department of Animal and Plant Sciences, The University of Sheffield, Sheffield S10 2UQ, UK.

Andrew G. Clark, Department of Biology, 208 Mueller Laboratory, Pennsylvania State University, University Park, PA 16802, USA.

Jesper Dahlgaard, Department of Ecology and Genetics, University of Aarhus, Ny Munkegade, Building 540, DK-8000 Aarhus C, Denmark.

Martin E. Feder, Department of Organismal Biology and Anatomy, The Commitee on Evolutionary Biology, and The College, The University of Chicago, 1027 East 57th Street, Chicago, IL 60637, USA.

Valery E. Forbes, Department of Life Sciences and Chemistry, Roskilde University, P.O. Box 260, DK-4000 Roskilde, Denmark.

Ary A. Hoffmann, School of Genetics and Human Variation, La Trobe University, Bundoora, Victoria 3083, Australia.

Päivi K. Hurme, Department of Biology, University of Oulu, P.O. Box 333, FIN-90571 Oulu, Finland.

Nicole L. Jenkins, School of Genetics and Human Variation, La Trobe University, Bundoora, Victoria 3083, Australia.

Albert Kamping, Department of Genetics, University of Groningen, Kerklaan 30, NL-9751 Haren, The Netherlands.

Robert A. Krebs, Department of Organismal Biology and Anatomy, The Commitee on Evolutionary Biology, The University of Chicago, 1027 East 57th Street, Chicago, IL 60637, USA.

Richard E. Lenski, Center for Microbial Ecology, Michigan State University, East Lansing, MI 48824, USA.

Volker Loeschcke, Department of Ecology and Genetics, University of Aarhus, Ny Munkegade, Building 540, DK-8000 Aarhus C, Denmark.

Michael Lynch, Department of Biology, University of Oregon, Eugene, Oregon 97403, USA.

Mark R. Macnair, Hatherly Laboratories, University of Exeter, Department of Biological Sciences, Prince of Wales Rd, Exeter EX4 4PS, UK.

Ivan Matić, Laboratoire de Mutagenèse, Institut Jacques Monod, 2 Place Jussieu, F-75251 Paris, France.

Pawel Michalak, Department of Ecology and Genetics, University of Aarhus, Ny Munkegade, Building 540, DK-8000 Aarhus C, Denmark.

Anders P. Møller, Laboratoire d'Ecologie, CNRS URA 258, Université Pierre et Marie Curie, Bât. A, 7ème étage, 7 quai St. Bernard, Case 237, F-75252 Paris Cedex 5, France.

Peter A. Parsons, School of Genetics and Human Variation, La Trobe University, Bundoora, Victoria 3083, Australia.
For correspondence: 21 Avenue Road, Glebe, NSW 2037, Australia.

Miroslav Radman, Laboratoire de Mutagenèse, Institut Jacques Monod, 2 Place Jussieu, F-75251 Paris, France.

Outi A. Savolainen, Department of Biology, University of Oulu, P.O. Box 333, FIN-90571 Oulu, Finland.

Carla M. Sgrò, School of Genetics and Human Variation, La Trobe University, Bundoora, Victoria 3083, Australia.

Peter R. Sheldon, Department of Earth Sciences, The Open University, Milton Keynes MK7 6AA, UK.
e-mail: P.R.Sheldon@open.ac.uk

François Taddei, Laboratoire de Mutagenèse, Institut Jacques Monod, 2 Place Jussieu, F-75251 Paris, France.

Wilke van Delden, Department of Genetics, University of Groningen, Kerklaan 30, NL-9751 Haren, The Netherlands.

Welam F. Van Putten, Department of Genetics, University of Groningen, Kerklaan 30, NL-9751 Haren, The Netherlands.

Marin Vulić, Laboratoire de Mutagenèse, Institut Jacques Monod, 2 Place Jussieu, F-75251 Paris, France.

Lev A. Zhivotovsky, Vavilov Institute of General Genetics, Russian Academy of Sciences, 3 Gubkin St., Moscow 117809, Russia.

# Preface

Considering environmental stress from an evolutionary viewpoint almost naturally leads to a definition of stress as an environmental factor that impairs Darwinian fitness. Even if the occurrence of stress is only a rare event not necessarily experienced by each individual in its lifetime, or if stress can be avoided behaviourally most of the time, stressful conditions may have a huge impact on evolutionary change if stress-resistant organisms are the only survivors on these rare occasions. Environmental stress, e.g. in the form of extreme temperatures or humidity, may also set limits to species distribution and abundance and therefore have a decisive role in shaping the composition of biological communities. All organisms have apparently evolved mechanisms to cope with stress or to reduce its negative impact on fitness. For example, in response to extreme temperatures (and many other stresses), organisms express a suite of proteins called heat shock proteins or stress proteins that prevent the damaging impact of thermal stress.

We invited as authors scientists whose emphases reflect their varied approaches to the study of environmental stress – from molecules and proteins to individuals, populations and ecosystems – with the aim of exploring how organisms adapt to extreme environments, how stress changes genetic structure and affects life histories, how organisms cope with thermal stress through acclimation, and how environmental and genetic stress induce fluctuating asymmetry, shape selection pressure and cause extinction of populations. Finally, we asked the authors to discuss the role of stress in evolutionary change, from stress-induced mutations and selection to speciation and evolution at the geological timescale. The authors were asked to review their field but also to include their latest and often not yet otherwise published research and to identify lines of future research activities. The book therefore contains reviews as well as novel scientific results on the subject and will be of interest to both researchers and graduate students and may also serve as a text for graduate courses. Some of the papers were presented at a symposium on stress and evolution held at the Fifth International Congress of Evolutionary Biology and Systematics, Budapest, Hungary, in August 1996. We are grateful to those who contributed for sharing their goals and perspectives with us, and to the reviewers of the manuscripts for their prompt and helpful advice.

*Kuke Bijlsma and Volker Loeschcke*

Groningen, The Netherlands, and Aarhus, Denmark, January 1997.

Environmental Stress, Adaptation and Evolution
ed. by R. Bijlsma and V. Loeschcke
© 1997 Birkhäuser Verlag Basel/Switzerland

# Introductory remarks: Environmental stress, adaptation and evolution

Kuke Bijlsma[1] and Volker Loeschcke[2]

[1] *Department of Genetics, University of Groningen, Kerklaan 30, NL-9751 NN Haren, The Netherlands*
[2] *Department of Ecology and Genetics, University of Aarhus, Ny Munkegade, Building 540, DK-8000 Aarhus C, Denmark*

There is a growing awareness that environmental stress may play and may have played a significant role in the evolution of biological systems, from the level of the gene to that of ecosystems. Historically, environmental stress has not been considered important in the development of evolutionary theories. Although the adaptation of an individual to its natural physical and biotic environment is central to Darwin's theory of evolution as expressed in *On the Origin of Species by Means of Natural Selection* (1859), he thought intra- and interspecific competition to be much more significant than environmental stress. A comprehensive overview and the historical role of stress in evolutionary thinking can be found in Hoffmann and Parsons (1991). Since the 1940s, for Drosophilists starting with the publication of Timofeeff-Ressovsky (1940), there has been increasing interest in linking environmental stress with natural selection for stress resistance and adaptation. Recently, an accumulating number of data do suggest that environmental stress may have a considerable impact on the evolutionary and ecological processes that affect and shape the genetic structure and evolution of populations (Calow and Berry, 1989; Hoffmann and Parsons, 1991), and may even play a significant role in the process of speciation (Parsons, this volume; Sheldon, this volume).

Most biologists are familiar with the word "stress", but it is used in many different ways and different contexts for several reasons. First, as pointed out by Hoffmann and Parsons (1991), two components are involved in dealing with stress, "the external and internal forces that are applied to organisms or other biological systems, and changes in biological systems that occur as a consequence of these forces". Both are clearly interdependent, and the degree of stress caused by an environment can only be valued in relation to the organism or population experiencing this environment (see e.g. Zhivotovsky, this volume). Although the term "stress" is often used to designate either the environmental component or the biological component, from an evolutionary perspective the environmental force and the biological response should be viewed as integrative.

Second, stress is level-dependent: it can be viewed at different biological levels, e.g. at the molecular, physiological, organismal and populational level. Responses at one level do not necessarily have to become manifest at another. For examle, phenotypic plasticity may prevent biochemical changes from being revealed at the organismal level, although this process itself may be costly and may be revealed as a change in fitness.

Third, the term "stress" is often associated with the intensity of stress. Often the environment is considered to be stressful only if the response it causes exceeds an arbitrary threshold, e.g. when more than a certain fraction of the population is affected. Others consider the intensity of stress to be continuous, including zero.

Given the foregoing problems, it is not surprising that many different definitions of stress have been formulated. Biological definitions, generally, fall into two classes. The first class of definitions considers stress in a physiological context. These definitions focus mostly on the physiological effects of stress on the individual. Selye (1973), for example, defined stress as a syndrome of physiological responses to environmental stresses that affect the well-being of individuals. In this view, these stresses are non-specific, and various stresses can cause a similar syndrome at the physiological level.

The second class of definitions considers stress in an evolutionary context. Most of these definitions focus on both environmental forces and the specific effect of stress on the biological system. For example, Sibly and Calow (1989) defined stress as "an environmental condition that, when first applied, impairs Darwinian fitness". Koehn and Bayne (1989) defined it as "any environmental change that acts to reduce the fitness of an organism". Definitions of this type and many more can be found elsewhere in this volume, and have in common that they emphasize the reduction in fitness caused by the environmental factor. As such, the organism or population may respond to stress phenotypically and/or genetically, and evolve adaptive mechanisms to overcome it. All contributors to this book, either explicitly or implicitly, consider stress in this manner, and aim to understand the impact of stress on biological systems from an evolutionary perspective.

The growing interest in environmental stress has been stimulated by recent developments in molecular genetics. These have not only made it possible to study stress responses in more detail, but have also revealed that most organisms have evolved sophisticated mechanisms to cope with different environmental stresses, such as heat shock proteins to counteract thermal and other stresses, mixed function oxidases to degrade xenobiotics, and the major histocompatibility complex to fight biotic attacks. Additionally, molecular techniques have made it possible to study responses to stress at the molecular level. The finding that thermal stress induces the production of a suite of proteins (heat shock proteins) to prevent the damaging impact of high temperatures in most organisms, from bacteria to humans, has evoked a significant increase in stress-related research.

Moreover, the impact of the human population in the last century on the biosphere at a global scale is unprecedented. This has caused and will increasingly cause major environmental changes, such as climatic shifts, chemical pollution and habitat destruction. The size and rate of these changes form an increasing threat for the existence of life on this planet. Global warming may exert thermal stress, the consequences of chemical pollution are yet incalculable and the destruction of habitats has caused an accelerating rate of species extinction. Understanding the nature and consequences of these stresses at a global level from an ecological and evolutionary perspective is of paramount importance for the development and evaluation of countermeasures.

*About the book*

Clearly most if not all organisms and populations have to cope with hostile environments that threaten their existence. Their ability to respond phenotypically and genetically to these challenges and to evolve adaptive mechanisms is, therefore, crucial. In this book we focus on understanding, from an evolutionary perspective, the impact of stress on biological systems.

The first part of the book is concerned with the adaptation of species to extreme stresses in their "natural" environment. Macnair reviews the current information on adaptation to metal-contaminated soils, and discusses both the genetic and physiological mechanisms involved. Interestingly, he shows that only those species that have tolerance genes present in low frequencies in normal populations prior to the selective agent being imposed seem to be able to evolve metal tolerance. Forbes and Calow focus on stress associated with exposure to chemical pollutants of marine and freshwater aquatic systems. They emphasize that the study of the response of biological systems to novel pollutants can provide significant information for addressing fundamental ecological and evolutionary issues, but, conversely, that ecological and evolutionary understanding is crucial for effective development of ecological tests and risk assessment models. Savolainen and Hurme discuss the genetic and evolutionary aspects of adaptation of conifers to the harsh environmental conditions of the northern boreal environments of Finland that are characterized by severe cold stress and a short growing season.

The second part of the book is concerned with the effect of stress on the structure and maintenance of genetic variation. Brakefield describes the striking seasonal polyphenism of the African butterfly *Bicyclus anynana* that reflects alternative evolutionary responses to alternating seasons, one of which is favorable, the other a stress environment. Evolution of phenotypic plasticity, mediated by hormonal mechanisms, has lead to genetic and physiological coupling of fundamental life history traits and plasticity in wing pattern. Jenkins et al. study the effect of environmental extremes on the expression of genetic variation of life history and other relevant traits in *Drosophila*. Their results, among others, suggest that under extreme conditions the pattern of genetic variation may be much more complex than under constant laboratory conditions, and that extreme conditions may markedly affect the heritability of important life history traits. Van Delden and Kamping discuss the role of environmental stress, such as toxic concentrations of ethanol and high temperature, for explaining the worldwide latitudinal cline observed for polymorphism at the *Adh* locus in *Drosophila melanogaster*. They also discuss the mechanisms by which the polymorphism and the cline are maintained. Clark studies the effect of various stresses on a set of metabolic characters in genetically defined lines of *D. melanogaster*, and describes the observed magnitude of environmental effects and genotye × environment interactions.

The third part of the book focuses on the evolutionary response to thermal stress and the role of acclimation, a special case of phenotypic plasticity. Bennett and Lenski report the evolutionary adaptation of *Escherichia coli* to thermal stress during thousands of generations. Their results show that fitness significantly increased in adapted populations compared with the ancestral clone, but that this improvement was achieved by several distinct pathways for the different replicates. Contrary to expectations, adaptation did not result in a change in the thermal niche, and acclimation to heat stress seemed not to increase, but rather to decrease, fitness. Feder and Krebs present a combined molecular and evolutionary approach to study the adaptive significance of acclimation and the concurrent production of heat shock proteins (Hsps) for the *hsp70* gene in *Drosophila*. Their results suggest that increasing the number of copies of this gene results in elevated levels of heat shock proteins after acclimation, and consequently in increased thermotolerance. Loeschcke et al. review the evolution of thermal resistance in *Drosophila* and discuss the contribution of acclimation and heat shock proteins to this process. Their experimental results suggest that several other mechanisms may also play a significant role in the evolution of stress tolerance.

The fourth part of the book focuses on the importance of the level and structure of genetic variation within populations for adaptation to changing environmental conditions and their impact on population extinction. Furthermore, the role of different models of selection in the process of adaptation is discussed. Bijlsma et al. study the consequences of genetic stress, i.e. loss of fitness due to an increase in homozygosity caused by genetic drift and/or inbreeding, for the fitness and persistence of small *D. melanogaster* populations under both optimal and stressful conditions. Their results suggest that genetic stress and environmental stress both increase the extinction risk of small populations and, more important, that both stresses are not independent but can act synergistically. Bürger and Lynch review theoretical models concerning the vulnerability of small populations to demographic stochasticity and directional changes in the environment, and study the effect of mutation and genetic variability for the persistence of populations in relation to the rate of environmental change. Zhivotovsky studies theoretically the evolution of the level of stress experienced by a population evolved in a homogeneous "home" environment, when it is exposed to a "foreign" environment, and he analyses the factors involved in this process. One of the interesting results of his model is the observation that a population will not adapt to a foreign environment when the frequency of occurrence of this environment is rare. Møller discusses how sexual selection may give rise to an increase in the general stress levels experienced by individuals of both animal and plant species, and the role of secondary characters and developmental stability in this process. Both Møller and Brakefield discuss the possible importance of fluctuating asymmetry for assessing the level of stress experienced by individuals or populations.

The final part of the book is concerned with the role of environmental stress in the process of speciation and evolution. Taddei et al. present evidence from bacterial systems that stress can induce greatly elevated mutation rates by activating a mutagenic response and by inhibiting anti-mutagenic mechanisms like the mismatch repair system. They argue that such stress-induced mutations not only accelerate the adaptation process but also may lead to a burst in speciation. Parsons argues that organisms have to cope with abiotic stresses far more than with biotic variables, and discusses the role of abiotic stresses in shaping species life histories and in speciation. Sheldon discusses the finding that morphological stasis seems to be the usual response to widely fluctuating physical stresses over a geological time scale. His explanation of this observation is that fluctuating environments could have given rise to generalist species that could survive these fluctuations for millions of years. On the other hand, more stable environments may be characterized by specialization and gradualistic evolution.

# References

Calow, P. and Berry, R.J. (1989) *Evolution, Ecology and Environmental Stress*. Academic Press, London.

Darwin, C. (1859) *On the Origin of Species by Means of Natural Selection*. Murray, London.

Hoffmann, A.A. and Parsons, P.A. (1991) *Evolutionary Genetics and Environmental Stress*. Oxford University Press, Oxford.

Koehn, R.K. and Bayne, B.L. (1989) Towards a physiological and genetical understanding of the energetics of the stress response. *Biol. J. Linn. Soc.* 37:157–171.

Selye, H. (1973) The evolution of the stress concept. *Amer. Scientist* 61:629–699.

Sibly R.M. and Calow, P. (1989) A life cycle theory of response to stress. *Biol. J. Linn. Soc.* 37:101–116.

Timofeeff-Ressovsky, N.W. (1940) Mutations and geographical variation. *In*: J. Huxley (ed.): *The New Systematics*. Clarendon Press, Oxford, pp. 73–136.

# Extreme environments and adaptation

Environmental Stress, Adaptation and Evolution
ed. by R. Bijlsma and V. Loeschcke
© 1997 Birkhäuser Verlag Basel/Switzerland

# The evolution of plants in metal-contaminated environments

Mark R. Macnair

*Hatherly Laboratories, University of Exeter, Department of Biological Sciences, Prince of Wales Rd, Exeter, EX4 4PS, UK*

*Summary.* There are many areas of the world where the soil is naturally, or through anthropogenic activity, contaminated with metals. Metals are toxic to plants in excess, and the consequence has been that many species found on normal soils are excluded from these areas. The plants that can grow on these soils can normally be shown to have evolved tolerance to the metals in excess, though there are some species which may be constitutively able to tolerate high metal levels, and there is some evidence for environmentally induced tolerance in some species. Tolerance is generally under major gene control, though the degree of tolerance shown by a plant will be affected by minor genes as well, at least some of which act hypostatically to the major tolerance locus. The major tolerance loci generally are specific, so that where plants show tolerance to more than one metal, it is because they have evolved independent tolerances to more than one metal. Co-tolerance, where one gene gives pleiotropic tolerance to more than one metal, is probably rare. The mechanism of tolerance is in most cases unknown, and the problems of studying this phenomenon are discussed. There is circumstantial evidence that tolerance involves a cost, in that tolerant plants are at a selective disadvantage in an uncontaminated environment. It has not, however, been possible to establish the reason or basis of this cost. The ability of a species to evolve tolerance seems to depend on the presence of tolerance genes at low frequency in normal populations prior to the selective agent being imposed. In areas which have been naturally contaminated for very long periods, endemic species restricted to the toxic environment are found (edaphic endemics). The evolutionary processes leading to these, and the difference between an endemic and an ecotype, are discussed.

## Introduction

There are many areas of the world that are contaminated to a greater or lesser extent by metals or metalloids, either naturally or as a result of the activities of humans. Examples of natural contamination include soils formed above metal rich rocks, such as serpentine soils or the soils formed on the copper mountains of Zaire. Anthropogenic contamination often results from the mining or smelting of metal ores, but also has followed many industrial and domestic activities: roadside verges tend to be contaminated from vehicle exhausts, for instance, and the use of sewage sludges as fertiliser, and copper and arsenic as pesticides, has resulted in significant contamination of agricultural soils.

All the heavy metals, and metalloids such as arsenic and selenium, are toxic to plants in greater than trace quantities. In localities with elevated soil metal concentrations, the toxicity therefore causes stress to all or most species. This stress can cause some species to go locally extinct, while other species can evolve to increase their tolerance of the stress. The evolution can result in locally adapted ecotypes, but in long-exposed naturally contaminated sites it has also produced edaphic endemics-species restricted to soils showing a particular suite of chemical and physical properties. This chapter will examine these evolutionary responses to metal-induced stress in plants.

## Metal toxicity in soils

The toxicity of metals to plants depends on a number of factors, including the innate properties of the element, the speciation of the metal, and the availability of metal to the plant in the soil solution. Thus some elements are inherently more damaging to plants than others: for instance, copper, nickel and silver are all much more toxic than lead, manganese or aluminium, by reference to the toxic concentration in solution culture. Metal speciation affects the reactivity of the element, and thus its toxicity. For instance, the As(III) ion (arsenite) is much more phytotoxic than As(V) (arsenate), and the toxicity of aluminium is very much determined by its speciation, which in turn depends on the pH and other cations present in the solution (Kochian, 1995).

Much the most important factor affecting metal toxicity to plants, however, is the availability to the plant. The chemistry of some metals renders them much less likely to be soluble in soils than others, and therefore less toxic. Thus some metals, e.g. chromium and tin, which can be shown to be highly toxic to plants in solution culture, appear to be rarely toxic in soils because they form highly insoluble complexes in the soil. The amount of organic matter in the soil will also markedly affect the availability of metals, since metals will bind effectively to the organic ligands present in decaying vegetation. Soil pH is another critical parameter to metal toxicity: many metals are effectively only available to plants at low pH. For instance, aluminium toxicity is the primary factor limiting crop productivity on acid soils but is not often a significant problem in more normal environments (Kochian, 1995), while zinc is also much less toxic at pHs greater than about 5.

Many published methods have tried to estimate the actual amount of metal available to a plant, either by chemical extraction of the soil or by growing specific sensitive plants in a soil and determining growth of the plant or concentration of metal in the plant. None are universally accepted or applicable in all (any?) circumstances. The consequence is that we have no objective way of measuring or defining the degree of stress that a particular metal is putting on a particular ecosystem or species. We generally can only recognise that the metal in question is toxic in the particular soil we are studying because of its effects, genetic or ecological, on the species growing (or not growing) in it. This can be unsatisfactory when we wish to

determine whether it is the metal or some other factor, e.g. pH, that is causing the effect, or when we find that some species does not appear to have
been affected by the soil. It is that this species is highly tolerant of stress or
is simply not experiencing any?

## Metal tolerance in plants

Prat (1934) was the first to show that a plant species had evolved metal
tolerance. He compared the growth of a normal population and a copper
mine population of *Silene vulgaris* on soil from the copper mine, and showed that the seeds from the copper mine were better able to survive the
metal-contaminated soil. Bradshaw (1952) performed a similar experiment
with *Agrostis capillaris* from the Goginan lead mine in Wales. Both these
early studies suffered from the problem that the contaminated soil was
being treated as a black box, and the actual amount of metal available to the
plant was unknown. It was the development of the rooting test for tolerance
by Jowett and Wilkins (see Wilkins, 1978) that enabled the tolerance of
plants to specific metals in solution culture to be studied. In this test, cuttings of a species are rooted in both a control solution and a similar solution to which the metal under investigation has been added; the relative root
growth of the cuttings in the two solutions can be used as a measure of the
tolerance of the plant to the metal. There are a number of variations on this
basic theme that have been used by various authors (see Wilkins (1978);
Macnair (1993) for discussion of their advantages and disadvantages), but
all depend on the fact that root growth is more swiftly inhibited by metals
than are other parts of the plant.

   Using this test, it is possible to demonstrate that individuals and populations of a species differ in their ability to tolerate metals, and to test this
characteristic independently of any other variation. Generally, when a contaminated environment is studied, it can be shown that the plants surviving
in the stressful environment are tolerant of at least the metals present
(e.g. Antonovics et al., 1971). The more rigorous studies have put the plants
through a seed cycle and demonstrated that tolerance is heritable. The
phenomenon is so common that I suspect that new observations are un-
publishable. There is a bias towards studies involving herbs and grasses
because of the ease with which they can be investigated by the rooting
test, but woody perennials and trees have also been shown to be tolerant
(e.g. Brown and Wilkins, 1985).

   More interesting are reports of species that can grow on contaminated
land without evolving genetic tolerance. There are a number of different
phenomena:

   *Constitutive tolerance:* Tolerance is generally manifested as a difference
between plants found on contaminated soil and those in normal soil. It is
the difference which defines tolerance. A species that could grow on metal-

contaminated land without evolving tolerance, while others in the same habitat did show tolerance, would be a candidate for this class of plant. McNaughton et al. (1974), investigating *Typha latifolia*, and Gibson and Risser (1982), studying *Andropogon virginicus*, failed to detect tolerance in plants collected from sites heavily contaminated with zinc, lead and cadmium. Neither study however, positively demonstrated, that the metals were actually at phytotoxic concentrations in the soil solution.

Species do appear to differ in their degree of sensitivity to phytotoxic metals. At least in solution culture, the tolerance index of normal populations of different species at low levels of a metal may differ considerably (Baker and Proctor, 1990; Murphy and Taiz, 1995). There is an indication that species with the lowest sensitivity may evolve full tolerance more easily than those with greater inhibition. It is likely that full tolerance for these species involves a smaller "step" and thus is more easily achieved.

*Inducible tolerance:* There are a number of reports that cadmium tolerance can be induced by growing the plants on Cd-contaminated soil. Thus Baker et al. (1986) found that both tolerant and non-tolerant clones of a variety of grass species showed a greater degree of Cd tolerance in the standard rooting test if they were given a pretreatment with the metal. Outridge and Hutchinson (1991) were also able to induce Cd tolerance in the fern *Salvinia minima*. Unfortunately, none of these species has been crossed or seed produced to investigate whether the induced tolerance is heritable.

Dickinson and his colleagues have suggested that trees may grow on metal-contaminated sites without evolving true genetic tolerance. Turner and Dickinson (1993a) studied sycamore (*Acer pseudoplatanus*) collected from copper- and zinc-contaminated sites. They could find no differences in metal tolerance, as measured by rooting tests, between seedlings from contaminated and control sites. The same authors did find evidence of stable tolerance in callus tissue derived from a mature tree from a copper refinery site (Turner and Dickinson, 1993b). They were also able to induce tolerance in a non-tolerant callus from an uncontaminated site. This latter result might suggest that tolerance can evolve in somatic tissue as a result of selection acting in the meristem or cell culture. However, somatic selection in plants should still be heritable, and the lack of heritable tolerance (Turner and Dickinson, 1993a) has resulted in the hypothesis that trees have an innately flexible genome that can acclimate in a stable but non-heritable manner (Dickinson et al., 1992). However, this hypothesis does not really explain what is happening in trees on mines. Trees seem just as affected by contaminated soil as other species, and it is easy to find tree seedlings dying from a lack of ability to tolerate the metals on mines. Evolved tolerance has been demonstrated in birch (Brown and Wilkins, 1985; Denny and Wilkins, 1987a). Further research on how trees become established on mines, and the roles of mycorrhizae (Denny and Wilkins, 1987b; Wilkinson and Dickinson, 1995) and soil heterogeneity (contaminated sites are never uniformly contaminated) is required to establish whether trees are really different in any way from the herbaceous species on which most work on the evolutionary dynamics of metal tolerance has been hitherto carried out.

## Genetics of tolerance

Initial studies on the inheritance of metal tolerance suggested that this was a classic polygenic character (Antonovics et al., 1971). In this respect metal tolerance differed from many other classic cases of rapid evolution in respect to anthropogenic change to the environment (e.g. insecticide resistance or industrial melanism) in which dominant major genes were commonly found. However, a number of recent studies (reviewed in Macnair, 1993) have demonstrated major genes for metal tolerance, and it seems likely that metal tolerance is actually much more similar to insecticide resistance or melanism than previously thought. The systems studied in most detail are copper tolerance in *Mimulus guttatus* (Macnair, 1983; Smith and Macnair, in preparation) and in *Silene vulgaris* (Schat and ten Bookum, 1992a; Schat et al., 1993). In both species, a single major gene has been found, with evidence for hypostatic modifiers that enhance tolerance and generate the variation found in degree of tolerance between individuals from contaminated sites, all of which are tolerant, and possess the major gene.

For instance, Smith and Macnair (in preparation) investigated the difference in degree of copper tolerance between two lines, both initially homozygous for tolerance (*TT*) that had been selected for increased and decreased degree of tolerance. Both lines were crossed to a single non-tolerant individual (*tt*). The F1s were selfed, and both F2s segregated tolerants:non-tolerants in a 3:1 ratio (confirming that both lines were homozygous for the major tolerance gene, *TT*). The non-tolerant progeny were crossed to a single tester plant, a low-tolerance plant also of genotype *TT*. They were also selfed, and their progeny (all non-tolerant) were again crossed to the tester plant. Some typical results are shown in Figure 1. Plants 1/20 and 13/14 are non-tolerant progeny from two different F2 families. Figure 1, (a) and (b), shows the results of crossing these plants to the tester plant. Clearly 13/14 enhances the tolerance of the tester more than does 1/20. The two plants were selfed, and five offspring again crossed to the tester plant. The tolerance of the resulting progeny is given in Figure 1, (c) and (d). Again, the progeny derived from 13/14 are more tolerant than are those from 1/20, showing that the differences observed in Figure 1, (a) and (b), are heritable.

I argued (Macnair, 1991) that major genes are expected to evolve where anthropogenic change has produced a large change in the environment, such that a big shift in mean (relative to the underlying genetic variation in

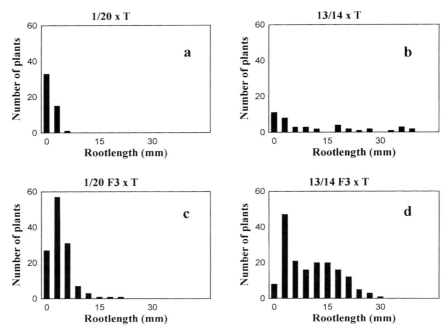

Figure 1. Hypostatic modifiers of copper tolerance in *M. guttatus*. (a) and (b): Tolerance of the progeny produced by crossing a single NT plant from a cross between a highly tolerant plant and a plant homozygous for the non-tolerant gene (*tt*) to a tester plant that was homozygous for the tolerance gene (*TT*); (a) plant 1/20; (b) plant 13/14. (c) and (d): Tolerance of the progeny produced by crossing five of the selfed progeny of 1/20 (c) and 13/14 (d) to the same tester plant. Data is the length of roots produced by plants grown in a 1.0-ppm solution of Cu.

the base population) is required to produce an individual able even to survive in the contaminated environment. I showed that selection acting on polygenic variation simply cannot produce a sufficiently big shift in the mean, but a single gene of large effect can. Once the major gene has enabled a population to establish in the novel environment, albeit of low absolute fitness, minor (polygenic) genes can spread that enhance the effect of the major gene and increase the ability of the species to survive and reproduce in the novel habitat.

Another issue that has caused some concern in the literature is the question of dominance. Generally, tolerance has been found to be dominant, either completely or partially. This is to be expected, since genes whose phenotype is detectable in the heterozygote (i.e. at least partially dominant genes) evolve so much more readily than do recessive ones in outbreeding species. However, it must be recognised that dominance is not a property of genes but a character like any other, dependent for its expression on the environment, both genomic and physical. It is therefore possible for the

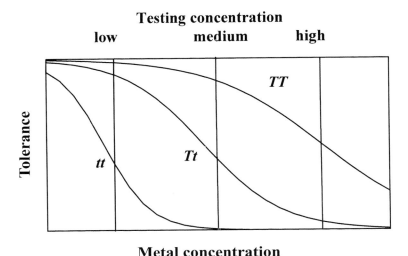

Figure 2. Hypothetical growth curves of different genotypes showing the dependence of dominance on testing concentration. At low concentrations tolerance is dominant, but at high it is recessive.

dominance of tolerance to alter both with the testing environment and with the genetic background in which the gene is found. Figure 2 illustrates one effect of the testing environment. If the dose-response curve of the hetero-zygote is intermediate between those of the two homozygotes, then it will follow that at low testing concentrations tolerance will be dominant, while at high concentrations it will be recessive. This was seen for copper toler-ance in *M. guttatus* by Allen and Sheppard (1971). Another effect of the testing environment was found by Lefèbvre (1974), who found that the dominance of zinc tolerance in *Armeria maritima* was reduced when the background solution in the tolerance test included calcium nitrate. What is important of course is the dominance of the genes at the metal concen-trations actually found in the field, and this unfortunately cannot be determined. The effect of the background genotype on dominance has been shown by Strange and Macnair (1991). In *M. guttatus* we have isolated more-or-less isogenic lines in which the tolerant gene has been inserted into a non-tolerant genetic background by repeated backcrossing. In these lines, the dose-response curves of homozygote and heterozygote are very similar, in contrast to that of mine plants as shown by Allen and Sheppard (1971). This suggests that the intermediate dose-response curve of mine plants may be a property of the modifiers rather than of the major tolerance gene.

## Specificity of tolerance

The early work on metal tolerance suggested that tolerance to different metals was achieved independently, and that where plants showed tolerance to more than one metal, this was because multiple contamination had selected for tolerance to more than one metal (e.g. Gregory and Bradshaw, 1965; Antonovics et al., 1971). Challenging this paradign have been a growing number of papers which have claimed that plants tolerant to metal X are also tolerant to metal Y, even though metal Y is not present at elevated levels in the soil. One of the first was the study of Allen and Sheppard (1971) of copper tolerance in *M. guttatus* at the large mine at Copperopolis. They found that, while the levels of zinc, lead and nickel were no greater at Copperopolis than in control sites, the level of tolerance to these metals shown by the copper-tolerant Copperopolis population was significantly greater. Schat and ten Bookum (1992b) have studied a number of populations of *S. vulgaris* from northern Europe and related the tolerance of each population to the metals found at the site. While each population is tolerant to the metals found in excess at that mine, many populations are also tolerant to metals that are only found in trace quantities. They suggested that selection for copper tolerance would also increase tolerance to cadmium, zinc, cobalt and nickel, while selection for tolerance to zinc, cadmium and lead also produces tolerance to cobalt and nickel. A number of authors have suggested that while high level tolerances may be specific, many tolerances may produce low-level tolerance to other metals (Symeonidis et al., 1985; Baker and Walker, 1989).

Two phenomena will produce plants which show correlated tolerances to more than one metal. The first is co-tolerance, where genes for tolerance to metal X also give tolerance to Y. The second is multiple tolerance, where metal-tolerant populations tend to acquire genes giving tolerance to both X and Y. Technically, this is the distinction between pleiotropy and linkage disequilibrium. Multiple tolerance can have a number of causes, but two are particularly important. First, mines are generally very heterogeneous environments, with distinct areas displaying very different concentrations of various metals. It is difficult to exclude the possibility that one area of the mine is selecting for tolerance to one metal, and a second for another. Second, multiple tolerances can evolve by migration between mines. It is not unlikely that many tolerant populations arise by migration from nearby mines, possibly mediated by the movement of miners between sites, rather than by *de novo* evolution *in situ*.

A purely phenotypic analysis cannot distinguish between pleiotropy and linkage disequilibrium. It is essential to conduct a proper genetic analysis to show whether the correlation persists through a breeding programme. This is not straightforward, and has rarely been attempted. Humphreys and Nicholls (1984) analysed copper and lead tolerance in *A. capillaris* and found that the two tolerances were independently inherited. In my laboratory, we have recently been studying the classic case of multiple tolerances in the Copperopolis population of *M. guttatus* first investigated by Allen and Sheppard (1971). We have used two principal techniques to look for a genetic correlation between metals. First, we have studied a series of lines selected for increased or decreased copper tolerance, testing whether they show any correlated response to other metals. Second, we have crossed plants tolerant to metal X to a non-tolerant, and investigated whether there is any correlation between tolerance to X and other metals in backcross and F2 families. The results have excluded a number of co-tolerances, particularly copper and cadmium, and copper and nickel (Tilstone and Macnair, 1997; Tilstone et al., 1997), and zinc and nickel (Tilstone, 1996). We have no evidence for co-tolerance in this benchmark species. That does not mean, of course, that in other species, and other metal pairs, there is not co-tolerance, but we are not aware of any well-documented cases of co-tolerance (but see Schat and ten Bookum, 1992b, who report that some of the minor copper tolerance genes may give zinc tolerance).

## The mechanism of tolerance

There are intuitively a large number of ways in which plants could protect themselves from the toxic effects of metals (see Tab. 1). Before the metal enters the plant root, it could be chelated or precipitated by exudates from the root. It could be complexed within the cell wall, preventing or reducing access to the cell. The rate of uptake into the cell could be reduced, or the rate of efflux increased. Once the metal is in the cell, it could be detoxified, or transported into the vacuole and detoxified. Alternatively, the particularly sensitive components of the cell machinery (membranes or proteins) could be altered to render them less sensitive to the toxic effects of the metal. These mechanisms can obviously operate singly or in combination, and all have been proposed or shown to operate in some organism or metal. Baker (1987) has suggested that there is likely to be a "syndrome of

Table 1. Postulated mechanisms of metal tolerance in higher plants

A. Extracellular
   1. Exudation of chelators to bind metal
   2. Exudation of substances that change rhizosphere pH and thus metal speciation
   3. Production of ion-exchange sites on cell walls to bind metal ions

B. Intracellular
   1. Alteration of cell membrane or other structural protein to reduce attack by metal
   2. Alteration of sensitive enzymes to prevent inhibition by metal
   3. Alteration of influx/efflux of metal ions to reduce rate of increase of metal concentration in cell
   4. Production of substances that bind metal in cell and render it non-toxic
   5. Transport of metal to vacuole, where detoxification takes place

tolerance" whereby several mechanisms may operate to different extents in different species.

However, it is important to distinguish between primary and secondary mechanisms of tolerance (Macnair and Baker, 1994; Meharg, 1994). Primary mechanisms are those that actually cause the difference between tolerant and non-tolerant plants, and are directly related to the product(s) of the gene(s) responsible for tolerance. Secondary mechanisms are those that both tolerant and non-tolerant plants possess, and are essential components of the whole tolerance syndrome, but they do not differ qualitatively or quantitatively between the two phenotypes. Meharg (1994) calls these adaptive and constitutive mechanisms, respectively. The secondary mechanisms are of interest in understanding the whole plant response to metals, but they have nothing to do with the phenomenon of tolerance *per se*.

The problem is that it is not necessarily easy to distinguish between primary and secondary mechanisms. Thus, if two plants differing in tolerance are exposed to a high concentration of a metal, one may be able to detect a difference in the concentration of some protein or substance produced by the two plants. But it is likely that in practice this experiment is comparing the products of live, functioning cells with those of dead or dysfunctional cells. It is likely that the two classes of cells will have a very different metabolism, but it does not follow that all or indeed any of the differences detected have anything to do with the primary mechanism of tolerance. Further problems arise when the experiment compares mine plants with non-mine plants, since these plants may differ by many other genes adapting them to the prevailing environment in addition to those conferring tolerance.

The problem may be illustrated by the story of phytochelatins and tolerance. It is well known that a class of proteins called metallothioneins is involved in metal tolerance in animals (Hamer, 1986). Rauser and Curvetto (1980), studying copper tolerant plants of *A. gigantea*, found a compound that they suggested might be a plant metallothionein. Later work showed that this was not a metallothionein (which is gene encoded) but a class of peptides called phytochelatins, which are synthesised from glutathione by an enzyme $\lambda$-glutamylcysteine dipeptidyl transpeptidase (or phytochelatin synthase, for short). The enzyme is produced constitutively in plant cells, and is activated by heavy metals (Loeffler et al., 1989). It appears to be found ubiquitously in higher plants, and Grill et al. (1987) suggested that phytochelatins were functionally equivalent to animal metallothioneins. However, while it was possible to show that tolerant plants exposed to metal had a higher phytochelatin production (e.g. Salt et al., 1989), it was not clear that these were responsible for the primary tolerance mechanism. For a start, phytochelatins are induced by a wide range of metals, and thus could not explain the specificity of tolerance (see above). Detailed work on the relationship between copper uptake, copper tolerance and phytochelatin synthesis in *S. vulgaris* showed conclusively that phytochelatin production had nothing to do with the primary mechanism of copper tolerance in this species (Schat and Kalff, 1992), though it remained possible that they were an essential part of the secondary detoxification mechanism in the live cells produced by the primary tolerance mechanism. However, even that has now been cast in doubt by the demonstration that *Arabidopsis thaliana* plants carrying the *cad-1* mutation, which knocks out the phyto-chelatin (PC) synthase gene, are highly sensitive to cadmium, but not to copper (Howden et al., 1995).

A genetic approach in which a putative tolerance mechanism has been shown to co-segregate with a tolerance gene has proved to be the most effective at disentangling primary mechanisms from secondary mecha-nisms, and adaptations to other features of the environment. For instance, we showed that arsenic-tolerant plants of *Holcus lanatus* had an altered uptake system for the two chemically related ions arsenate and phosphate (Meharg and Macnair, 1990). Normal plants of most species possess two uptake systems: a high-affinity system operational at low external concen-trations, and a low-affinity system that is effective at high concentrations. The high-affinity system is highly inducible, but the low-affinity system is constitutive. We found that arsenate-tolerant plants appeared to have lost the high-affinity system (Meharg and Macnair, 1990) and thus took up arsenate more slowly at physiologically important concentrations than did non-tolerants. Proof that this was really the mechanism of tolerance, and not a subsidiary adaptation to, for instance, low nutrients in mine soils, was provided when we showed that arsenate tolerance and low arsenate uptake co-segregated in F2s of crosses between mine and non-mine plants (Meharg and Macnair, 1992). Such evidence has also supported a role for an altered cell membrane to be the primary mechanism of copper tolerance in *M. guttatus* (Strange and Macnair, 1991) and *S. vulgaris* (de Vos et al., 1989), and for the efflux of malate as the cause of aluminium tolerance in wheat (Delhaize et al., 1993).

Despite these few examples, we still know frustratingly little about the true mechanisms of tolerance in plants. Real progress is likely to follow the cloning and molecular characterisation of a tolerance locus. Unfortunately, searching for mutants in *Arabidopsis* is unlikely to be helpful in this case, since it is not known to evolve true tolerance, though small differences in susceptibility can be found (Murphy and Taiz, 1995), and none of the wild plants showing real tolerance have been sufficiently well studied molec-ularly that any strategy for the cloning of a tolerance gene suggests itself.

## The cost of tolerance

It is a fundamental assumption of evolutionary theory that the evolution of a novel adaptation should involve some cost or trade-off. Were it not so, there would be no reason that all individuals or species should not have the character in question. The demonstration of these costs, however, is not uncontroversial.

In the case of metal tolerance, there is substantial circumstantial evidence that tolerance is disadvantageous in the absence of the selective metal. First, the genes for tolerance are generally at very low frequency in populations growing on normal soils, and there are very steep clines for tolerance at the edge of mines (Hickey and McNeilly, 1975). McNeilly (1968) found that the seed of the wind-pollinated grass *A. capillaris* growing downwind of a mine was more tolerant than its maternal parent, suggesting that selection acted against the gene flow coming from the mine. In the case of *M. guttatus*, the clines are steeper for the modifiers of tolerance than for tolerance itself, suggesting that selection may act more on the degree of tolerance than on the possession of tolerance *per se* in this species (Macnair et al., 1993). The only exception to the rule that tolerance genes are usually at low frequency is the case of arsenic tolerance in *H. lanatus*, where nationwide surveys have found that this character is polymorphic in all populations studied (Meharg and Macnair, 1993; Naylor et al., 1996). The second line of evidence comes from comparing the growth or competitiveness of plants collected from mines with plants of the same species from uncontaminated environments. Slower growth rate of tolerant ecotypes has been reported for *S. vulgaris* (Ernst et al., 1990), for *A. capillaris* (Wilson, 1988) and for barnyard grass (Morishima and Oka, 1977). Hickey and McNeilly (1975) found that four species of tolerant plant were less competitive than their non-tolerant conspecifics when competed against two races of rye grass (*Lolium perenne*). A third line of evidence for a cost is the frequent demonstration that mine plants appear to have a greater requirement for the metal to which they are tolerant. Many root-growth experiments have found an apparent stimulation of root growth at low levels of the metal in tolerant plants. There have also been reports that some tolerant plants (particularly zinc-tolerant ones) require exogenous metal in order to grow successfully in normal potting compost (Antonovics et al., 1971). Schat and ten Bookum (1992b) suggest that plants from one very highly copper tolerant population of *S. vulgaris* show symptoms of copper deficiency when grown in soil from a lead/zinc mine.

However, all this evidence cannot be said to show a cost for tolerance *per se*, i.e. for the tolerance genes themselves. Mine plants differ from normal plants in many ways apart from just being tolerant. The whole ecotype may be disadvantageous, but that does not mean that any one of the genes contributing to the whole phenotype need be so. In particular, many mines are free draining and have low nutrient status, and it is not unlikely that adaptations to these features of mines may include a slow growth rate and reduced competitiveness in a richer environment (Chapin, 1991). The problem here is precisely the same as that discussed above concerning the specificity of tolerance. How does one distinguish between a cost associated with the tolerance gene (= pleiotropy) from one associated with another gene(s) (= linkage disequilibrium)? Three studies have explicitly tried to separate the tolerance gene and test its effect directly, and none of them detected a cost: Nicholls and McNeilly (1985) and Walker (1990) compared tolerant and non-tolerant plants selected from a non-tolerant population, while Macnair and Watkins (1983) compared copper-tolerant hetero-zygotes and non-tolerant homozygotes of *M. guttatus* from segregating backcross families.

In this laboratory we have recently been examining the cost of copper tolerance in *M. guttatus* in more detail. We have explicitly tested the two principal hypotheses for how the cost could arise: (1) that the mechanism of tolerance requires energy/resources for its maintenance that are diverted from other activities, reducing growth rate and/or competitiveness; and (2) that the mechanism of tolerance reduces the intracellular concentration of copper available to normal metabolism, thus creating a greater requirement for this essential micronutrient. We investigated the effect of the tolerance gene using isogenic lines in which the tolerance gene has been separated from other genes by backcrossing to a recurrent non-tolerant population. However, the evidence from the clines at the edge of the mine suggested that selection might act more on degree of tolerance (i.e. the hypostatic modifier loci) than on the tolerance gene itself (Macnair et al., 1993). To investigate the effect of having more or fewer modifier loci, we produced five replicate pairs of selection lines in which one line had been selected for increased tolerance and the other for reduced tolerance (but ensuring that it was still tolerant, so all lines were homozygous for the major toler-ance locus). A cost associated with the modifiers would be manifested by a consistent fitness difference between all five replicate lines.

No differences between the isogenic lines were found in growth or cop-per utilisation. The first and last generations of all five selection lines were grown together in a randomised block experiment, and the plants were measured for various growth and fitness characteristics (Harper et al., 1997). While the plants were very variable, almost no consistent effects were detected. All five upward lines produced a greater number of aerial roots than did the low lines. This is almost certainly an artefact of the method of selection, whereby plants are selected on the basis of their root length in nutrient solution: it is likely that the increase in aerial roots re-presents an increase in the innate ability to form roots. It would be unwise to interpret this as a cost, but it acts as a positive control to show that the method should be able to detect any consistent differences in growth or fit-ness characters. The upward lines also showed a consistent change in sexual resource allocation with increasing tolerance. The more tolerant plants appeared to have a greater allocation to later reproduction and female func-tion (ovaries, seeds) and less to male function (pollen). They also had a greater allocation to attractive structures (corolla). These differences are not large and are difficult to interpret in functional terms, but it is clear that a change in the effective gender of a plant might well not be neutral.

Many experiments were carried out looking at the effect on the plants' growth and fitness in suboptimal copper concentrations (Harper, 1996). There is absolutely no evidence that highly tolerant plants require more copper than less tolerant ones. This is an interesting result, if disappointing in terms of identifying the cost of tolerance. It shows that the evolution of tolerance has increased the range of environments in which the plant is buffered in terms of its copper nutrition.

Overall, this work has shown that if there is a cost to tolerance, it is either very small or is manifested through a mechanism different than those which have hitherto been suggested. It also adds weight to the suggestion that the disadvantage of tolerance in mine ecotypes arises through the evolution of the whole co-adapted gene complex, rather than the individual costs associated with the spread of particular genes.


## Evolution of tolerance *de novo*

Only a minority of species have evolved metal-tolerant ecotypes. Even in species which have evolved one or more tolerant races, tolerance may only have evolved to one or a few metals. The question of why species do not evolve an adaptation can be just as fundamental as why they do, and this aspect of microevolution has been illuminated by the study of the initial evolution of metal tolerance.

Walley, Khan and Bradshaw (1974) showed it was normally possible to select tolerant individuals directly out of non-tolerant populations of the grass *A. capillaris*. They sowed large numbers of seeds onto ameliorated mine soil and obtained a small number of survivors (ca. 0.1%). It is normally possible to find this species, or another species in the genus *Agrostis* on almost any mine site in Europe. Gartside and McNeilly (1974) and Ingram (1988) extended this work and studied a number of species, some of which were found on mines and some of which were not. They found that it was possible to select tolerant individuals from non-tolerant races of species that were found on mines, but not from those that were not (Tab. 2). Bradshaw (1991) has argued that species that are unable to evolve because of the lack of appropriate genetic variability are constrained by what he has called *genostasis*.

Since metal tolerance appears to be generally controlled by major genes (see above), these results suggest that in normal populations tolerant gene frequencies are low, of the order of 0.001. At these frequencies, observed gene frequencies are going to be highly susceptible to stochastic forces, and one would expect considerable between-population variation, with some populations having a frequency of zero, even in species which can normally evolve metal tolerance.

Al-Hiyaly et al. (1988, 1993) have investigated this aspect of the within species variability in the possession for genetic variance associated with

Table 2. The occurrence of copper-tolerant individuals in normal populations of various grass species studied by Ingram (1988). Whether or not these species have been found in nature on mines is indicated

| Species | Occurrence of tolerant individuals (%) | Presence of species on copper mines |
|---------|----------------------------------------|-------------------------------------|
| Holcus lanatus | 0.16 | + |
| Agrostis capillaris | 0.13 | + |
| Festuca ovina | 0.07 | − |
| Dactylis glomerata | 0.05 | + |
| Deschamsia flexuosa | 0.03 | + |
| Anthoxanthum odoratum | 0.02 | − |
| Festuca rubra | 0.01 | + |
| Lolium perenne | 0.005 | − |
| Poa Pratense | 0.0 | − |
| Poa trivialis | 0.0 | − |
| Phleum pretense | 0.0 | − |
| Cynosurus cristatus | 0.0 | − |
| Alopecurus pratensis | 0.0 | − |
| Bromus mollis | 0.0 | − |
| Arrhenatherum elatius | 0.0 | − |

the evolution of tolerance. They made use of the replicated toxic environment created by the erection of zinc-galvanised electricity pylons on acid soils. Pylons are erected every 200 m or so in an otherwise unpolluted environment. Rain washes off the zinc, and it contaminates the soil under the pylon, reducing species diversity and selecting for tolerant individuals. Al-Hiyaly et al. (1988) studied a number of pylons and found that the populations under them differed in degree of tolerance; some had developed a reasonable degree of tolerance, others had essentially not evolved at all. The authors then (Al-Hiyaly et al., 1993) examined the non-tolerant populations surrounding the pylons using the methods of Ingram (1988) (see above). They found that these subpopulations differed: some possessed the genetic variance for tolerance, but some did not. Populations surrounding pylons possessing tolerant populations tended to have more variance than those surrounding pylons with low-tolerance populations.

The ability to evolve metal tolerance is only part of the evolution of a metal-tolerant ecotype. Mines differ from pastures in many ways apart from mere metal contamination: the soil structure is normally worse, and the organic content less, so that the soils dry out quickly; the soils are frequently very deficient in nitrogen and phosphorus and other essential elements; and wind and water erosion may mean that seedling establishment is very difficult (Baker and Proctor, 1990). After the gene(s) for metal tolerance have spread, it would be expceted that selection would act to improve the species' ability to cope with the other physical and biological features of the environment. The ecotype that finally develops will differ by many genes in addition to the initial metal-tolerance mutation. Ingram

(1988) found some species had the apparent ability to evolve copper tolerance but have never been found on mines (Tab. 2). It is likely that these species are excluded by an inability to tolerate some other feature of the inhospitable environment.

## The evolution of edaphic endemics

When we look at the flora of mines, we find that they are populated with species that are also found in uncontaminated environments, though not necessarily locally. It is rare to find locally endemic species, though specific status has been accorded to *Viola calameria*, a common species of zinc mines in Belgium and Germany, and Macnair (1989a) described *Mimulus cupriphilus*, a species apparently recently evolved on a small copper mine in California. The situation is different in ancient metal-contaminated soils where species endemic to the unusual geology ("edaphic endemics") are very common. One of the most dramatic environments in which edaphic endemics are found is serpentine. Serpentinisation is a common process associated with the hydration of ultramaphic (ultrabasic) rocks. They are found throughout the world, but they are particularly associated with orogenesis (Malpas, 1991). They are fast weathering, producing basic soils that are free draining, have a high heavy metal content (particularly nickel, chromium and cobalt) and also tend to have an abnormal calcium to magnesium ratio. Normal soils tend to have more Ca than Mg, while serpentine soils typically have an Mg:Ca ratio that is substantially greater than 1. Serpentine soils are typically nutrient-poor. These characteristics mean that serpentine soils tend to have low productivity and to be toxic to many plant species. Particularly in ancient serpentine soils, the level of endemism is high: for instance, in New Caledonia, two monotypic families, over 30 genera and more than 60% of the island's flora are restricted to serpentine soils (Jaffré, 1981).

Kruckeberg (1984) reviewed the flora of Californian serpentines. He found that 215 taxa were endemic to serpentine soils; though serpentine makes up less than 1% of Californian soils, more than 10% of the plant species endemic to California are found on this substrate. About 1300 species were able to grow on both serpentine and normal soils. Where these species occur in the same area, they are known as "bodenwag" species. But the majority of Californian species are unable to tolerate this environment (excluded species).

So what is the difference between the three classes of species, endemics, bodenwag and excluded? The study of the evolution of metal-tolerant races suggests that the difference between the excluded and bodenwag species is the same as the difference between those that can and cannot evolve tolerant races. This model would predict that bodenwag species have generally evolved their serpentine populations by the formation of ecotypes, and this appears to be generally true (Kruckeberg, 1984; Macnair and Gardner, 1997). The more interesting question is, What is the difference between an endemic and an ecotype?

Traditionally, endemics have been classified into palaeoendemics and neoendemics. Palaeoendemics are ancient species where the population on the serpentine is a relict of a previously much wider distribution. In this case, one can envisage an originally bodenwag species where the normal populations are eliminated following a change in the environment, either physical or biological, but the serpentine population is able to persist because of a different competitive environment. The evolution of this population will then obviously follow the classic allopatric speciation model. Neoendemics, on the other hand, are species in which the progenitor population still persists locally and may well even still occur on the serpentine as an ecotype. For instance, in the *M. guttatus* complex, *M. guttatus* is a bodenwag species, occurring in California ubiquitously, both on and off the serpentine. *M. nudatus* and *M. pardalis* are local serpentine endemics, in the coastal mountains and the Sierras, respectively. Both endemic species coexist sympatrically with their presumed progenitor, *M. guttatus.*

Speciation involves two processes. First, two populations must evolve sufficient differences in ecology that they can coexist without suffering competitive exclusion. Second, they must evolve sufficient intrinsic barriers to gene exchange that the two gene pools can evolve independently. Kruckeberg (1986) proposed a model for the evolution of a serpentine endemic in which he visualised a number of steps:

0. Some preadaptation for serpentine tolerance exists in normal populations of a species. This provides the conditions which will permit the evolution of a serpentine-tolerant ecotype.
1. Disruptive selection causes separation of the species into serpentine-tolerant and -intolerant gene pools. This is the formation of an ecotype.
2. Further genetic divergence in structural and functional traits occurs within the ecotype.
3. Isolation between the two races becomes genetically fixed, so that the two gene pools are unable to exchange genes.
4. Further divergence of the incipient species occurs, "put in motion by the initial genetic discontinuity" (Kruckeberg, 1986).

We have already seen the factors leading to steps 0–2. It is the step from 2 to 3 that is difficult to envisage without a period in allopatry. I (Macnair, 1989b; Macnair and Gardner, 1997) have argued that the difference between an ecotype (which stalls at stage 2) and an endemic (stages 3 and 4) may be the particular adaptations that evolve in the formation of an individual ecotype population. If the particular adaptations are purely physiological, and leave the fundamental ecology and breeding system of the species unchanged, then the progenitor and derived populations will continue to exchange genes, and the derived population will remain an ecotype. If, on the other hand, the characters adapting the ecotype to the novel environment also happen to change the breeding system, or alter in some other way the ability of the populations to exchange genes, then the derived population can diverge swiftly and produce an independent gene pool.

Our studies with the *M. guttatus* complex illustrate how this could have happened. *M. guttatus* has formed copper- and serpentine-tolerant ecotypes, but these differ little in ecology from the normal *M. guttatus* which is a common hydrophilic species in western North America. The ecotypes tend to grow along the streams and in damp patches, and are outcrossing, pollinated by bumblebees. The plants may also grow as small and unproductive plants in the drier areas. The derived species. *M. nudatus* and *M. pardalis* (serpentine endemics), and *M. cupriphilus* (a copper mine endemic, Macnair, 1989a) are early flowering and better adapted to grow on the drier areas of the mine/serpentine by being able to produce more seeds as small plants (Macnair et al., 1989; Macnair and Gardner, 1997). Early flowering is a potent barrier to gene flow, and this adaptation to contaminated environments has been seen in metal-tolerant ecotypes (McNeilly and Antonovics, 1968). *M. pardalis* and *M. cupriphilus* are also isolated from *M. guttatus* by a change in breeding system: they have evolved self-fertilisation, which will largely prevent gene flow from *M. guttatus*. The evolution of self-fertilisation is one of the commonest evolutionary events in plants (Stebbins, 1970) and has a number of potential selective forces favouring it (Jain, 1976). However, in this context, the most likely selective advantage is reproductive assurance: by flowering early *M. cupriphilus* and *M. pardalis* flower before the bumblebees that pollinate *M. guttatus* are very active. The evolution of autogamy is obviously adaptive in this situation but also produces the reproductive isolation that allows further independent evolution.

*M. nudatus* is allogamous, but a change in flower shape has enabled it to partially switch pollinator to a small sweat bee. However, the primary barrier to gene exchange between *M. nudatus* and *M. guttatus* is a strong postmating barrier that leads to almost complete inviability in hybrids. It is likely that the genes responsible for this barrier have spread in the evolution of the derived species, *M. nudatus* (Macnair and Gardner, 1997). Macnair and Christie (1983) showed how the copper tolerance gene that spread in Copperopolis also acts as a postmating reproductive isolating gene when crossed to particular populations. This showed how the spread of adaptive genes could produce a postmating barrier as a pleiotropic effect of the adaptation. We have not yet been able to test whether the genes causing the barrier between *M. nudatus* and *M. guttatus* are associated with any of the physiological or physical adaptations to the serpentine, but the copper tolerance gene illustrates how it could have happened.

Thus the essential difference between an ecotype and an endemic may be that an endemic is an ecotype in which adaptations were selected that produced, as a by-product, sufficient reproductive isolation to allow the two populations to evolve independently; where the adaptations do not alter the ability of progenitor and derivative to cross, the ecotype remains an ecotype.

## Conclusions

For over 30 years, the study of metal tolerance has proved to be one of the clearest examples of microevolution and an excellent system in which to study the relationship between adaptation and ecology. It provides a model for the evolution of ecotypes and edaphic endemics. Many questions remain to be answered, and we can expect this system to continue to provide insights into how plants evolve to meet abiotic stress for many years to come.

## References

Al-Hiyaly, S.A., McNeilly, T. and Bradshaw, A.D. (1988) The effects of zinc contamination from electricity pylons: Evolution in a replicated situation. *New Phytol.* 110:571–580.

Al-Hiyaly, S.A., McNeilly, T., Bradshaw, A.D. and Mortimer, A.M. (1993) The effect of zinc contamination from electricity pylons: Genetic constraints on selection for zinc tolerance. *Heredity* 70:22–32.

Allen, W.R. and Sheppard, P.M. (1971) Copper tolerance in some Californian populations of the monkey flower *Mimulus guttatus*. *Proc. Soc. Lond. B* 177:177–196.

Antonovics, J., Bradshaw, A.D. and Turner, R.G. (1971) Heavy metal tolerance in plants. *Adv. Ecol. Res.* 7:1–85.

Baker, A.J.M. (1987) Metal tolerance. *New Phytol.* 106 (Suppl.):93–111.

Baker, A.J.M. and Proctor, J. (1990) The influence of cadmium, copper, lead and zinc on the distribution and evolution of metallophytes in the British Isles. *Plant Syst. Evol.* 173:91–108.

Baker, A.J.M. and Walker, P.L. (1989) Physiological responses of plants to heavy metals and the quantification of tolerance and toxicity. *Chem. Spec. Bioavail.* 1:7–17.

Baker, A.J.M., Grant, C.J., Martin, M.H., Shaw, S.C. and Whitebrook, J. (1986) Induction and loss of cadmium tolerance in *Holcus lanatus* L. and other grasses. *New Phytol.* 102:575–587.

Bradshaw, A.D. (1952) Populations of *Agrostis tenuis* resistant to lead and zinc poisoning. *Nature* 169:1098.

Bradshaw, A.D. (1991) Genostasis and the limits to evolution. *Philos. Trans. R. Soc. Lond. B* 333:289–305.

Brown, M.T. and Wilkins, D.A. (1985) Zinc tolerance of mycorrhizal *Betula* spp. *New Phytol.* 99:101–106.

Chapin, F.S. III. (1991) Integrated responses of plants to stress. *BioScience* 41:29–36.

de Vos, C.H.R., Schat, H., Vooijs, R. and Ernst, W.H.O. (1989) Copper-induced damage to the permeability barrier in roots of *Silene cucubalus*. *J. Plant Physiol.* 135:164–169.

Delhaize, E., Ryan, P.R. and Randall, P.J. (1993) Aluminium tolerance in wheat (*Triticum aestivum* L.). II. Aluminium-stimulated excretion of malic acid from root apices. *Plant Physiol.* 103:695–702.

Denny, H.J. and Wilkins, D.A. (1987a) Zinc tolerance in *Betula* spp. I. Effect of external concentration of zinc on growth and uptake. *New Phytol.* 106:517–524.

Denny, H.J. and Wilkins, D.A. (1987b) Zinc tolerance in *Betula* spp. IV. The mechanism of ectomycorrhizal amelioration of zinc toxicity. *New Phytol.* 106:545–553.

Dickinson, N.M., Turner, A.P., Watmough, S.A. and Lepp, N.W. (1992) Acclimation of trees to pollution stress: Cellular metal tolerance traits. *Ann. Bot.* 70:569–572.

Ernst, W.H.O., Schat, H. and Verkleij, J.A.C. (1990) Evolutionary biology of metal resistance in *Silene vulgaris*. *Evol. Trends Plants* 4:45–50.

Gartside, D.W. and McNeilly, T. (1974) The potential for evolution of heavy metal tolerance in plants. II. Copper tolerance in normal populations of different plant species. *Heredity* 32:335–348.

Gibson, D.J. and Risser, P.G. (1982) Evidence for the absence of ecotypic development in *Andropogon virginicus* L. on metalliferous mine wastes. *New Phytol.* 92:589–599.

Gregory, R.P.G. and Bradshaw, A.D. (1965) Heavy metal tolerance in populations of *Agrostis tenuis* Sibth. and other grasses. *New Phytol.* 64:131–143.

Grill. E., Winnacker, E.-L. and Zenk, M.H. (1987) Phytochelatins, a class of heavy metal-binding peptides from plants, are functionally analogous to metallothioneins. *PNAS* 84:439–443.

Hamer, D.A. (1986) Metallothionien. *Ann. Rev. Biochem.* 55:913–951.

Harper, F.A. (1996) The cost of copper tolerance in *Mimulus guttatus*. Ph.D. dissertation, University of Exeter.

Harper, F.A., Smith, S.E. and Macnair, M.R. (1997) Where is the cost in copper tolerance in *Mimulus guttatus*? Testing the trade-off hypothesis. *Funct. Ecol.; in press.*

Hickey, D.A. and McNeilly, T. (1975) Competition between metal tolerant and normal plant populations: A field experiment. *Evolution* 29:458–464.

Howden, R., Goldsburgh, P.B., Anderson, C.R. and Cobbett, C.S. (1995) Cadmium-sensitive, *cad-1* mutants of *Arabidopsis thaliana* are phytochelatin deficient. *Plant Physiol.* 107:1059–1066.

Humphreys, M.O. and Nicholls, M.K. (1984) Relationships between tolerance to heavy metals in *Agrostis capillaris* (=*Agrostis tenius* Sibth.) *New Phytol.* 98:177–190.

Ingram, C. (1988) *The evolutionary basis of ecological amplitude of plant species.* Ph.D. dissertation, Liverpool University.

Jaffré, T. (1981) *Etude écologique du peuplement végétal des sols dérivés des roches ultrabasiques en Nouvelle Calédonie.* Office Rech. Sci. Technol. Outre Mer: Paris.

Jain, S.K. (1976) The evolution of inbreeding in plants: *Ann Rev. Ecol. Syst.* 7:469–495.

Kochian, L.V. (1995) Cellular mechanisms of aluminum toxicity and resistance in plants. *Annu. Rev. Plant Physiol. Plant Mol. Biol.* 46:237–260.

Kruckeberg, A.R. (1984) California serpentines: Flora, vegetation, geology, soils and management problems. *Univ. Calif. Publ. Bot.* 78:1–180.

Kruckeberg, A.R. (1986) An essay: The stimulus of unusual geologies for plant speciation. *Syst. Bot.* 11:455–463.

Lefèbvre, C. (1974) Note sur la génétique de la tolerance au zinc chez *Armeria maritima*. *Bull. Soc. Roy. Bot. Belg.* 107:217–222.

Loeffler, S., Hochberger, A., Grill, E., Winnacker, E.-L. and Zenk, M.H. (1989) Termination of the phytochelatin synthase reaction through sequestration of heavy metals by the reaction product. *FEBS Lett.* 258:42–46.

Macnair, M.R. (1983) The genetic control of copper tolerance in the yellow monkey flower, *Mimulus guttatus*. *Heredity* 50:283–359.

Macnair, M.R. (1989a) A new species of *Mimulus* endemic to copper mines in California. *Bot. J. Linn. Soc.* 100:1–14.

Macnair, M.R. (1989b) The potential for rapid speciation in plants. *Genome* 31:203–210.

Macnair, M.R. (1991) Why the evolution of resistance to anthropogenic toxins normally involves major gene changes: The limits to natural selection. *Genetica* 84:213–219.

Macnair, M.R. (1993) Tansley Review No. 49: The genetics of metal tolerance in vascular plants. *New Phytol.* 124:541–559.

Macnair, M.R. and Baker, A.J.M. (1994) Metal tolerance in plants: Evolutionary Aspects. *In:* M.E. Farago (ed.): *Plants and the Chemical Elements.* VCH, Weinheim, pp. 68–86.

Macnair, M.R. and Christie, P. (1983) Reproductive isolation as a pleiotropic effect of copper tolerance in *Mimulus guttatus*? *Heredity* 50:295–302.

Macnair, M.R. and Gardner, M. (1997) The evolution of edaphic endemics. *In.* D.J. Howard and S.H. Berlocher (eds): *Endless Forms: Species and Speciation; in press.*

Macnair, M.R. and Watkins, A.D. (1983) The fitness of the copper tolerance gene in *Mimulus guttatus* in uncontaminated soil. *New Phytol.* 95:133–137.

Macnair, M.R., Macnair, V.E. and Martin, B.E. (1989) Adaptive speciation in *Mimulus*: An ecological comparison of *M. cupriphilus* with its presumed progenitor, *M. guttatus*. *New Phytol.* 112:268–279.

Macnair, M.R., Cumbes, Q.J. and Smith, S. (1993) The heritability and distribution of variation in degree of copper tolerance in *Mimulus guttatus* on a copper mine at Copperopolis, California. *Heredity* 71:445–455.

McNeilly, T. (1968) Evolution in closely adjacent populations. III. *Agrostis tenuis* on a small copper mine. *Heredity* 23:99–108.

McNeilly, T. and Antonovics, J. (1968) Evolution in closely adjacent plant populations. IV. Barriers to gene flow. *Heredity* 23:205–218.

McNaughton, S.J., Folsom, T.C., Lee, T., Park, F., Price, C., Roeder, D., Schmitz, J. and Stockwell, C. (1974) Heavy metal tolerance in *Typha latifolia* without the evolution of tolerant races. *Ecology* 55:1163–1165.

Malpas, J. (1991) Serpentine and the geology of serpentinized rocks. *In:* B.A. Roberts and J. Proctor (eds) *The Ecology of Areas with Serpentinized Rocks: A World View.* Kluwer Academic Publishers, Dordrecht, Netherlands, pp. 7–30.

Meharg, A.A. (1994) Integrated tolerance mechanisms: Constitutive and adaptive plant responses to elevated metal concentrations in the environment. *Plant, Cell and Environment* 17:989–993.

Meharg, A.A. and Macnair, M.R. (1990) An altered phosphate uptake system in arsenate tolerant *Holcus lanatus* L. *New Phytol.* 116:29–35.

Meharg, A.A. and Macnair, M.R. (1992) Genetic correlation between arsenic tolerance and the rate of uptake of arsenate and phosphate in *Holcus lanatus. Heredity* 69:336–341.

Meharg, A.A. and Macnair, M.R. (1993) Pre-adaptation of Yorkshire fog, *Holcus lanatus* L., to arsenate tolerance. *Evolution* 47:313–316.

Morishima, H. and Oka, H.I. (1977) The impact of copper pollutionon barnyard grass populations. *Jap. J. Genetics* 52:357–372.

Murphy, A. and Taiz, L. (1995) A new vertical mesh transfer technique for metal-tolerance studies in *Arabidopsis*: Ecotypic variation and copper sensitive mutants. *Plant Physiol.* 108: 29–38.

Naylor, J., Macnair, M.R., Williams, E.N.D. and Poulton, P.R. (1996) A polymorphism for phosphate uptake/arsenate tolerance in *Holcus lanatus* L.: Is there a correlation with edaphic or environmental factors? *Heredity* 77:509–517.

Nicholls, M.K. and McNeilly, T. (1985) The performance of *Agrostis capillaris* L. genotypes, differing in copper tolerance, in ryegrass swards on normal soil. *New Phytol.* 101: 207–217.

Outridge, P.M. and Hutchinson, T.C. (1991) Induction of cadmium tolerance by acclimation transferred between ramets of the clonal fern *Salvinia minima* Baker. *New Phytol.* 117: 597–605.

Prat, S. (1934) Die Erblichkeit der Resistenz gegen Kupfer. *Berichte der Deutschen Botanischen Gesellschaft* 102:65–67.

Rauser, W.E. and Curvetto, N.R. (1980) Metallothionein occurs in roots of *Agrostis* tolerant to excess copper. *Nature* 287:563–564.

Salt, D.E., Thurman, D.A., Tomsett, A.B. and Sewell, A.K. (1989) Copper phytochelatins of *Mimulus guttatus. Proc. R. Soc. Lond. B* 236:79–89.

Schat, H. and Kalff, M.M.A. (1992) Are phytochelatins involved in differential metal tolerance or do they merely reflect metal-imposed strain? *Plant Physiol.* 99:1475–1480.

Schat, H. and ten Bookum, W.M. (1992a) Genetic control of coper tolerance in *Silene vulgaris. Heredity* 68:219–229.

Schat, H. and ten Bookum, W.M. (1992b) Metal specificity of metal tolerance syndromes in higher plants. *In:* J. Proctor, A.J.M. Baker, R.D. Reeves (eds): *The Ecology of Ultramafic (Serpentine) Soils.* Intercept, Andover, pp. 337–352.

Schat, H., Kuiper, E., ten Bookum, W.M. and Vooijs, R. (1993) A general model for the genetic control of copper tolerance in *Silene vulgaris*: Evidence from crosses between plants from different tolerant populations. *Heredity* 70:142–147.

Stebbins, G.L. (1970) Adaptive radiation in Angiosperms. I. Pollination mechanisms. *Ann. Rev. Ecol. Syst.* 1:307–326.

Strange, J. and Macnair, M.R. (1991) Evidence for a role for the cello membrane in copper tolerance of *Mimulus guttatus. New Phytol.* 119:383–388.

Symeonidis, L., McNeilly, T. and Bradshaw, A.D. (1985) Differential tolerance of three cultivars of *Agrostis capillaris* L. to cadmium, copper, lead, nickel and zinc. *New Phytol.* 101: 309–315.

Tilstone, G.H. (1996) *The significance of multiple metal tolerances in* Mimulus guttatus *Fischer ex DC.* Ph.D. dissertation, University of Exeter.

Tilstone, G.H. and Macnair, M.R. (1997) Nickel tolerance and copper-nickel co-tolerance in *Mimulus guttatus* from copper mine and serpentine habitats. *Plant and Soil*; *in press.*

Tilstone, G.H., Macnair, M.R. and Smith, S.E. (1997) Does copper tolerance give cadmium tolerance in *Mimulus guttatus? Heredity*; *in press.*

Turner, A.P. and Dickinson, N.M. (1993a) Survival of *Acer pseudoplatanus* L. (sycamore) seedlings on metalliferous soils. *New Phytol.* 123:509–521.

Turner, A.P. and Dickinson, N.M. (1993b) Copper tolerance of *Acer pseudoplatanus* L. (sycamore) in tissue culture. *New Phytol.* 123:523–530.

Walker, P.L. (1990) *Genotypic and phenotypic aspects of metal tolerance in* Holcus lanatus *L.* Ph.D. dissertation, University of Sheffield.

Walley, K.A. Khan, M.S.I. and Bradshaw, A.D. (1974) The potential for the evolution of heavy metal tolerance in plants. *Heredity* 32:309–319.

Wilkins, D.A. (1978) The measurement of tolerance to edaphic factors by means of rootgrowth. *New Phytol.* 80:623–634.

Wilkinson, D.M. and Dickinson, N.M. (1995) Metal resistance in trees: The role of mycorrhizae. *Oikos* 72:298–300.

Wilson, J.B. (1988) The cost of heavy metal tolerance: An example. *Evolution* 42:408–413.

Environmental Stress, Adaptation and Evolution
ed. by R. Bijlsma and V. Loeschcke
© 1997 Birkhäuser Verlag Basel/Switzerland

# Responses of aquatic organisms to pollutant stress: Theoretical and practical implications

Valerie E. Forbes[1] and Peter Calow[2]

[1] *Department of Life Sciences and Chemistry, Roskilde University, P.O. Box 260, DK-4000 Roskilde, Denmark*
[2] *Department of Animal and Plant Sciences, The University of Sheffield, Sheffield S10 2UQ, UK*

*Summary.* Stress associated with exposure to chemical pollutants is a topic of increasing concern. Yet study of population responses to chemical stress remains largely a descriptive science. Being able to predict population responses to toxic chemicals and understanding the mechanisms by which organisms adapt to them (or not) can contribute to the development of ecological and evolutionary theory and may lead to more effective approaches for minimizing undesirable consequences of pollutant stress. Here we examine the effects of pollutant stress at the population level, focusing primarily on marine and freshwater aquatic systems. We provide examples from the ecotoxicological literature that demonstrate how study of the responses of organisms to chemical pollutants can provide insight into key ecological and evolutionary phenomena, and conversely we show how the application of ecological and evolutionary theory can contribute to solving practical problems related to the evaluation of potential chemical hazards (e.g. deciding which aquatic organisms to use in regulatory test systems and developing ecological risk assessment models for aquatic populations). We argue for greater integration between theory development and its application for predicting and managing chemical stress in natural systems.

## Introduction

In this chapter we focus on environmental stress caused by exposure to chemical pollutants. Predicting how organisms respond to chemical pollutants and understanding the mechanisms by which they adapt to chemical stress can provide insight into key ecological and evolutionary processes. Advances in analytical chemistry have revealed the widespread occurrence of potentially hazardous chemicals even in habitats far from industrial activity and have led to growing concern about the possible effects of anthropogenic chemicals on natural systems and the collection of large amounts of new data, much of which are relevant to understanding the ecological and evolutionary roles of stress. In addition, continuing developments in molecular and biochemical methods are improving our understanding of the mechanisms of stress responses and are facilitating the investigation of linkages among stress responses (to both chemicals and other abiotic variables) occurring at different hierarchical levels (e.g. the costs of stress protein induction reflected in life-history responses; see other contributions in this volume, e.g. Loeschcke et al., Feder and Krebs, and Bennett and Lenski).

We begin by making the general connection between stress and fitness. Then we present some recent findings from the ecotoxicological literature that illustrate how responses to chemical stress facilitate an understanding of some important ecological and evolutionary processes. Conversely we shall also touch upon the extent to which an understanding of ecological and evolutionary phenomena can have relevance for a number of practical issues in ecotoxicology.

## The connection between stress and fitness

Stress can be defined in terms of the factors causing it (sometimes referred to as stressors) or in terms of the effects caused by it (i.e. the resulting biological response). For example, Grime (1989) defined stress as "external constraints limiting the rates of resource acquisition, growth or reproduction of organisms"; Sibly and Calow (1989) as "an environmental condition that, when first applied, impairs Darwinian fitness"; and Hoffmann and Parsons (1991) as "an environmental factor causing a change in a biological system which is potentially injurious". In contrast, Bayne (1975) defined stress as "a measurable alteration of a physiological (or behavioral, biochemical or cytological) steady state which is induced by an environmental change, and which renders the individual (or the population, or the community) more vulnerable to further environmental change". Although the word "stress" has been used to represent either an environmental factor causing a biological response or the biological response itself, in all of the above definitions the key feature of stress is in terms of its biological effect.

One way of approaching a definition of stress is to consider stressful any environmental variable that acts as a selection pressure, for under these circumstances, by definition, some fraction of the population must be impaired. This can be made rigorous by reference to measures of fitness and the components of fitness. Fitness is determined by survival (S), reproductive output (n) and time to maturity/between breedings (t). So any factor that impairs S, and/or n and/or t can be defined as a stress, and this affords an opportunity both for quantifying stress effects and distinguishing among them.

Genotypes that are less affected, or not affected at all, in terms of S, n and t by an environmental stress will be favored, and this defines a state of stress tolerance. Alternatively, it is possible that some genotypes that suffer negative effects on one component of fitness (e.g. S) might nevertheless be favored because of enhancement in other components of fitness (e.g. n). It is therefore important to understand the effect of environmental variables on the separate fitness components and their integrated effect on total fitness.

Of course at higher levels of stress all genotypes within a population may suffer impaired fitness that leads to population decline and ultimately

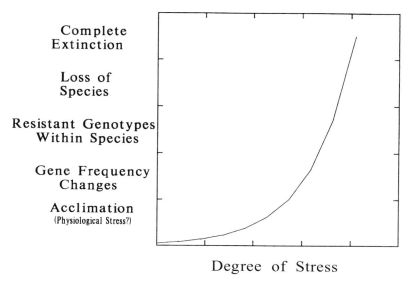

Complete Extinction

Loss of Species

Resistant Genotypes Within Species

Gene Frequency Changes

Acclimation
(Physiological Stress?)

Degree of Stress

Figure 1. Expected gradient of ecological/evolutionary responses to increasing degrees of environmental stress.

extinction. The likelihood of this happening will vary from species to species, some species being more tolerant than others. So there will be differential species loss, but at some level of stress all species will decline in numbers, leading eventually to complete extinction. From this, therefore, we can imagine a gradient of ecological responses to an environmental stress that leads from gene frequency changes to total loss of biodiversity as illustrated in Figure 1. The steepness of this response will clearly vary from one environmental stress to another. We have also included individual acclimation responses in this figure for the sake of completeness, and because these are sometimes referred to in the literature as physiological stress responses (e.g. Depledge, 1994). However, given that these are compensatory and presumably the outcome of prior selection, it is somewhat problematic whether these should strictly be viewed as stress responses.

Here we focus on stress caused by exposure to chemicals. Like other abiotic stresses, it is well known that exposure to chemicals causes changes in biological systems. But can all these changes be deemed to be stress responses? Whether or not they are might depend upon the types of chemicals (novel vs. natural) and/or the types of responses. Synthetic and industrial chemicals are often considered to be more likely causes of stress than natural chemicals. However, the primary difference between synthetic and naturally produced chemicals is the time during which evolutionary responses have had a chance to act prior to our observations; the responses themselves are not unique. Exposure to synthetic chemicals not experienced before might be expected to lead to abnormal responses, just as exposure to natural chemicals outside the normal concentration range might be expected to lead to abnormal responses. We expect that because many commercial chemicals are novel, as far as biological systems are concerned, they are likely to lead to steep response curves (in Fig. 1) and provide an opportunity for using them as convenient probes of the ecological and evolutionary processes just described. Regardless of whether the stress in question is a novel chemical, a natural chemical or some other abiotic variable, identifying the situation as stressful requires that we can define what is a "normal" response.

At the physiological and biochemical levels, identifying the normal response range remains one of the major challenges in the development of biomarkers of pollutant effects (Forbes and Forbes, 1994). In recent years much effort has been devoted toward developing sensitive measures of physiological or biochemical change that can be used as early-warning indicators of pollutant impact. Not only is there presently a lack of understanding of the normal range for many of these biomarkers and of their responses to biotic and abiotic variables other than pollutants, but it is often unclear whether the biomarker response indicates stress (i.e. a reduction in fitness) or acclimatory adjustment to new conditions. Defining the normal range of responses at the ecosystem level has also proved problematic. The concept of an ecosystem distress syndrome (analogous to Selye's general adaptation syndrome for individuals) has been proposed and repeatable features of stressed ecosystems identified (Odum, 1985). However, the question of what constitutes the normal range for many key ecosystem properties has not been fully resolved (Calow, 1995).

## Chemical stress impacts ecological and evolutionary phenomena: Some examples from aquatic systems

### Evolution of tolerance through natural selection

The potential for chemical pollutants to cause ecological and evolutionary change has been recognized for some time. The demonstration of genetically based resistance to a chemical pollutant indicates that the substance has caused stress on the resistant population. The evolution of insecticide-resistant pests was noticed as early as 1908 (Wood, 1981). Early reports of metal tolerance in plants from contaminated habitats appeared during the 1930s (Moriarty, 1983), and since then, the potential importance of heavy metals as agents of selection, particularly in plants, has been firmly established (see Macnair, this volume). A variety of aquatic species has also evolved resistance to heavy metals (Klerks and Weis, 1987).

In order for selection for resistance to chemical pollutants to occur, the concentration of pollutant in the environment must be high enough to impair some fraction of the population, and there must be genetic variation in sensitivity among individuals. Differences in the amount of genetic variation among species help to explain why polluted habitats frequently have a lower diversity of species compared with unpolluted habitats. However, changes in species composition in response to pollution can be due either to inherent differences in tolerance or to differences in the abilities of species to adapt (e.g. because of differences in genetic variability, generation time etc.). Klerks and Levinton (1989) predicted that the extent to which chemical pollution results in increased resistance versus changes in taxonomic composition are likely to depend on the rate and concentration at which pollutants enter a system. Whereas a rapid input of high levels of pollutant will be more likely to eliminate some species, a gradual exposure to lower chemical concentrations will be more likely to be associated with increased resistance and a lesser impact on taxonomic composition. Although many studies have demonstrated an increase in resistance in aquatic populations occupying chemically polluted habitats, very few studies have distinguished between genetic adaptation and physiological acclimation, and of those that have, the evidence for adaptation is less well documented (Klerks and Weis, 1987). Also, as these authors point out, most studies of metazoans compare resistance between field populations from polluted versus clean sites, which biases analyses to those species which have been able to persist under polluted conditions and thereby probably overestimates the occurrence of resistance.

It has been suggested that the presence of genetically based resistance to a chemical may be used as an index of its biological effect (Blanck et al., 1988; Klerks and Levinton, 1989). However, there are two limitations to such an approach that should be kept in mind. The first is that selection for resistance to one chemical may be associated with resistance to a range of chemicals (e.g. if the mode of action of the chemicals is similar or if the adaptation is a very general one, such as reduced permeability of the skin), so that attributing cause to a chemical for which resistance is found may be invalid (Blanck et al., 1988; but see Macnair, this volume). There is some evidence that selection for one type of stress may make a population less resistant to other types of stress (Bush and Weis, 1983). Second, it is obviously invalid to conclude that a lack of resistance indicates a lack of stress. Although a lack of resistance may indicate that the pollutant level is not high enough or in the right form to cause impairment (though this can in principle be tested by exposing appropriate control populations), it may also indicate a lack of genetic variation for resistance in the population in question.

Klerks (1987) investigated the evolution of heavy metal resistance in invertebrates from Foundry Cove, New York, USA, which had been exposed to extremely high levels of cadmium, nickel and cobalt for a period of about 30 years. Although he found no differences in macrofaunal abundance in response to sediment cadmium concentration, he found significant reductions in species diversity at the most polluted sites. He investigated the degree of resistance in two of the most common species inhabiting the cove, the oligochaete *Limnodrilus hoffmeisteri* and the chironomid *Tanypus neopunctipennis*. Interestingly, he was able to demonstrate a genetically based increase in resistance to heavy metals in the oligochaete, but not the chironomid. Furthermore, the increased resistance in the oligochaete was associated with a higher rate of metal uptake from polluted sediment.

In summary, there is a growing body of evidence that indicates that chemical pollutants can be important selective forces in natural aquatic systems. Heavy metals have been the most widely studied chemicals to date, and aquatic organisms have shown acclimatory responses (i.e. metallothionein induction (Benson and Birge, 1985), selection for resistance and reduced taxonomic diversity (Klerks, 1987; Klerks and Levinton, 1989; Klerks and Weis, 1987). Which of these responses is observed in any particular aquatic system (see. Fig. 1) will depend on the rate and concentration at which a pollutant enters the system, the time over which the pollutant has been present relative to the time at which the system is observed, and the inherent sensitivity and degree of genetic variation for resistance in the resident populations.

## Evolution of tolerance through life-history adjustments

In addition to genetically based increases in physiological tolerance to pollutant stress (so-called direct adaptations; Maltby et al., 1987), modifications to life-history traits in response to pollutant stress (so-called indirect adaptations) may also occur. Life-history theory examines the adaptive significance of variability in life-history traits and predicts the type of life-history patterns that should be selected for under particular environmental conditions. The theory predicts that factors reducing adult survival will select for earlier maturation and increased reproductive effort, whereas reduced juvenile survival will select for later maturation and decreased reproductive effort (Sibly and Calow, 1989). Under conditions in which juvenile growth is favored, selection for many small offspring should occur, whereas when conditions for juvenile growth are poor, selection for fewer, larger offspring should be observed (Sibly and Calow, 1985). Larger offspring should take less time to reach reproductive maturity and may reach a larger reproductive size, both of which may have positive effects on fitness.

The extent to which environmental stress will modify life-history patterns depends on several factors, such as the amount of genetic variation in the population, trade-offs among life-history characteristics and matings with non-tolerant individuals (Falconer, 1981). In many species juveniles are more sensitive (in terms of both survival and growth) to pollutant stress than are adults, and life-history theory would therefore predict reduced reproductive output and allocation to fewer, larger offspring in pollutant-stressed compared with clean habitats. Maltby (1991) compared life-history patterns between populations of a freshwater isopod, *Asellus aquaticus*, living upstream and downstream from an old coal mine effluent. She found that long-term chemical stress resulted in less investment in reproduction and the production of fewer, larger offspring and that these changes were genetically based – a pattern consistent with predictions from life-history theory.

In contrast, Postma (1995) compared life-history patterns in chiron-omids, *Chironomus riparius*, collected from four field sites differing in metal contamination and reared in the laboratory under clean conditions. There was no difference in the number of offspring produced among popu-lations (although exposed populations showed a nonsignificant trend toward lower egg production). Larvae from one (but not the other) polluted site had an increased development time when grown under clean conditions compared with larvae from unpolluted sites. In laboratory populations selected for cadmium tolerance, larval growth rates of cadmium-tolerant *C. riparius*, grown in clean conditions, were lower than those of non-tolerant midges. Postma's results are not consistent with the above theo-retical predictions but are illuminating in that they highlight the potential influence of other factors (e.g. costs of tolerance that act to reduce overall fitness; gene flow between tolerant and non-tolerant populations) on the evolution of life-history patterns.

## *Roles of genetics, environment and their interaction in controlling responses to chemical stress*

It is the aim of quantitative genetics to assess the relative importance of genetics, environment, and their interaction in the expression of pheno-typic traits. The genetic basis (or heritability) of a trait will determine its potential for evolutionary change. Various models for the maintenance of sex require the existence of genotype × environment interactions (Bell, 1987; Hughes, 1992). Genotypic × environment interaction may lead to habitat specialization and genetic divergence among populations or may promote the evolution of phenotypic plasticity (Rawson and Hilbish, 1991). The responsiveness of a trait to environmental influences (its degree of phenotypic plasticity) may also play a key role in evolution (e.g. in the origin of novelty, speciation and macroevolution (West-Eberhard, 1989) and may itself be subject to natural selection (Scheiner, 1993)).

Although attention to genotype × environment interactions historically has been minimal in ecotoxicology, a number of recent studies highlights their potential importance in controlling the responses of organisms to chemical stress. Baird et al. (1991) designed a series of experiments to investigate genetic differences among a suite of clones of *Daphnia magna*, one of the most widely used toxicity test species, in tolerance to a range of chemicals. They found that acute chemical tolerance limits (i.e. $EC_{50}s$) varied among clones by several orders of magnitude and that the rank order of tolerances differed for different chemicals. In other words, the relative performance of the different genotypes was strongly dependent on the chemical environment to which they were exposed. These results emphasize the potential importance of genotype $\times$ environment interactions in tolerance to chemical stress and demonstrate that, even within a group of closely related chemicals (e.g. various heavy metals), there may be no single genotype that has consistently superior survival. Another interesting result of this work was that the within-genotype variability in response to a chemical gradient varied among genotypes and chemicals, suggesting that some genotypes were less uniform in their phenotypic response (i.e. had wider tolerance distributions) than others and that the degree of phenotypic homogeneity was chemical-specific (i.e. that there was genotype $\times$ environment interaction for the width of the tolerance distribution).

Møller (1995) examined growth rates in two genotypes of the gastropod *Potamopyrgus antipodarum* and in seven strains of the brine shrimp *Artemia* in response to heavy metal exposure. The rank order of growth in the two *P. antipodarum* genotypes was reversed in cadmium-exposed versus clean conditions, providing evidence for a genotype $\times$ environment interaction. Partitioning the variance in growth of the *Artemia* strains showed that the genotype $\times$ environment interaction component was larger than the environmental component for all but one of the seven strains.

Quantitative genetics models generally partition total phenotypic variance into genetic effects, environmental effects, genotype $\times$ environment interaction effects and a residual component reflecting random variability among individuals (of the same genotype in the same environment). The latter source of variability has been considered to be due either to minor and uncontrollable differences in the external environment or to chance internal events during development (i.e. developmental noise). Several recent studies suggest that random variability among genetically identical individuals may be an important contributor to total phenotypic variance. Random variation among genetically identical *Artemia* was by far the largest contributor to total phenotypic variance in 10 life-history traits examined and accounted for between 47 and 100% (average = 69%) of the total phenotypic variation (Møller, 1995). Forbes et al. (1995) found that chemical stress, at a level sufficient to reduce mean phenotypic performance, increased the within-genotype phenotypic variance in asexual gastropods. Such increasing error variance in phenotypic traits implies that environmental stress increases stochastic influences during development or increases the sensitivity of genotypes to microenvironmental hetero-geneity. The former is supported by studies of fluctuating asymmetry, which has also been shown to increase in response to environmental stress (e.g. Brakefield, this volume; Møller, this volume).

## Relationship between reproductive mode and tolerance to stressful environments

Evolutionary biologists have long sought a satisfactory explanation for the maintenance of sex and have generally conceded that the advantage of sex lies in its perpetuation of genetic variability. The primary advantage of sex is presumably that the greater genetic diversity arising from recombination will favor sexual lineages over their asexual relatives in spatially hetero-geneous environments (tangled bank hypothesis; Bell, 1987) or in response to biotically unpredictable environments (Red Queen hypothesis; Van Valen, 1973). Despite the view that asexual lineages are evolutionary dead ends, the wider geographic distribution of parthenogens and their frequent dominance in extreme or disturbed habitats has led to the hypothesis that such organisms possess widely adapted or, so-called general-purpose genotypes. Empirical tests of the general-purpose genotype hypothesis are limited, and the available results provide equivocal support for predictions of the model (Bierzychudek, 1989; Michaels and Bazzaz, 1989, Møller, 1995; Niklasson and Parker, 1994; Parker and Niklasson, 1995; Weider, 1993). In determining whether parthenogens are better able to exploit environmental extremes than are their sexual relatives, it is relevant to con-sider their responses both to environmental factors that are likely to have been of past selective importance in shaping species distributions and to novel environmental stresses. The general-purpose genotype hypothesis predicts that past selection among clones will eliminate all but the most broadly tolerant to historically relevant environmental factors. Whether sex should be favored under conditions of novel environmental stress, such as that associated with exposure to synthetic chemicals, is also relevant for understanding the maintenance of sex and, as we discuss below, may have important practical implications for dealing with anthropogenic sources of environmental stress.

Very few studies have compared pollutant tolerance between parthe-nogens and their sexual relatives; however, recent comparisons of the per-formance of coexisting sexual and parthenogenetic hydrobiid gastropods to heavy metal exposure have investigated this issue (Forbes et al., 1995; Møller et al., 1996). This system is of particular interest because the parthenogenetic species *Potamopyrgus antipodarum* invaded Europe from New Zealand (where obligate parthenogenic and sexual individuals are found, sometimes sympatrically (Dybdahl and Lively, 1995)) via the spread of remarkably few genotypes (Jacobsen et al., 1996). In contrast to most parthenogenetic species, European (but not New Zealand) populations of *P. antipodarum* are monoclonal, and in the geographically widespread European populations examined to date a total of only three genotypes has been found (Hauser et al., 1992; Jacobsen et al., 1996). In Denmark, the family Hydrobiidae consists of *P. antipodarum* along with three sexually reproducing *Hydrobia* species, all of which exist sympatrically in various combinations. Comparison of acute tolerance to lethal concentrations of cadmium between a monoclonal population of *P. antipodarum* and its sympatric sexual counterpart, *Hydrobia ventrosa*, could detect no difference in tolerance (i.e. $LC_{50}$) between species, although the former had a somewhat narrower tolerance distribution (i.e. smaller confidence limits around the $LC_{50}$). In addition, *P. antipodarum* exhibited faster growth rates than did *H. ventrosa* in both clean conditions and during exposure to sublethal levels of cadmium stress, although neither species could clearly be said to be more tolerant to sublethal cadmium exposure. Environmental sensitivity in growth rate across a sublethal cadmium gradient was analyzed by several measures (coefficient of variation across treatments (Weider, 1993), environmental sensitivity index (Niklasson and Parker, 1994), and joint regression analysis (Falconer, 1981)). The results differed depending upon which method of analysis was chosen, and there were no clear differences in environmental sensitivity between sexual and asexual lineages. Measurements of within-treatment growth rate variability in single clones of *P. antipodarum* compared with genetically mixed *Hydrobia* populations have provided conflicting results (Forbes et al., 1995; Møller et al., 1996); however exposure to cadmium stress appears to increase phenotypic variability in both single clones and sexual populations.

Comparison of sexual and parthenogenetic lineages of the brine shrimp *Artemia* in response to copper exposure found that growth rate in the parthenogenetic clones was not reduced in response to copper stress, whereas growth of Old World sexual populations (believed to be the direct ancestors of the parthenogens) and New World sexual populations varied in response to copper exposure (Møller, 1995). In terms of mean performance across environments, the parthenogenetic populations ranked 2, 4, 6 and 7 (a rank of 1 = highest performance), with the Old World sexual population ranking 5 and the New World sexual populations ranking 1 and 3. The within-treatment coefficient of variation of growth rate (a measure of the phenotypic variability among individuals in a single environment) did not differ between the parthenogenetic clones and the New World sexual populations, but was significantly higher for Old World sexual populations. Møller (1995) also measured a variety of life-history traits in five *Artemia* populations in response to a copper gradient. Since no effect of the copper treatments could be detected for any of the traits in any of the populations, the relative tolerance of parthenogenetic versus sexual lineages could not be compared. However, the highest rank for mean performance (across environments) was found in a parthenogenetic population for three of the traits, and in one of the New World sexual populations for seven of the traits; the means for the remaining three traits did not vary among populations. Comparison of the average within-treatment variability between sexual populations and single parthenogenetic clones indicated that whereas for three traits (reproductive period, total number of offspring and number of nauplii) the clones were less phenotypically variable than were their sexual counterparts, for the remaining seven traits there were no differences in phenotypic variability between sexual populations and single clones (Fig. 2).

With regard to whether sexual reproduction confers an advantage in the face of novel environmental stress, recent evidence that sexual reproduction is more sensitive to pollutant stress than asexual reproduction is of particular interest. Snell and Carmona (1995) measured the effects of four toxic chemicals on asexual and sexual reproduction in the cyclically parthenogenetic rotifer *Brachionus calyciflorus*. They found that in all cases sexual reproduction was more strongly reduced than was asexual reproduction, and at the lowest toxicant concentrations at which significant effects were observed, sexual reproduction was inhibited from 2 to 68 times more than asexual reproduction. The results are consistent with other studies demonstrating the greater sensitivity of sexual reproduction in rotifers to environmental stresses of many types (e.g. food, salinity, temperature; reviewed in Snell and Carmona, 1995). These findings are consistent with the predominance of asexuals at environmental extremes and suggest one mechanism by which environmental stress may act to reduce genetic diversity.

## Implications of ecological and evolutionary theory for some practical problems

Our approach here has been to show how we can use chemical stress to probe ecological and evolutionary processes. Conversely, it is possible to use our understanding of ecological and evolutionary processes to help refine practical approaches for predicting responses of ecological systems to chemicals for ecological risk assessment.

For example, there has been debate on the relative utility of sexual versus asexual genotypes in ecotoxicological tests (Calow, 1992; Forbes and Depledge, 1992, 1993; Forbes and Forbes, 1993; Baird, 1992, 1993). Similarly, selection of end points and their incorporation into predictive models will benefit from a more explicit recognition of ecological theory.

### Choice of organisms for test systems

Because the degree of phenotypic (i.e. tolerance) variability greatly influences the statistical likelihood of detecting effects of toxicant stress on mean performance and in recognition of the importance of repeatability

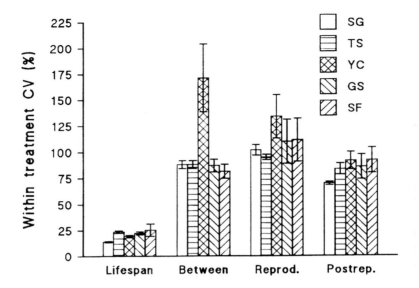

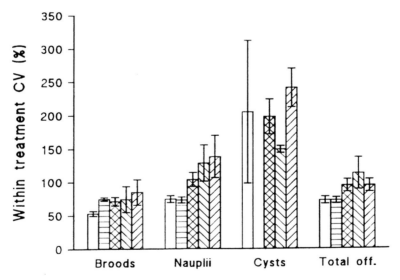

Figure 2. Average within-treatment variability (measured as the coefficient of variation, corrected for sample size) for eight life-history traits measured in five populations of the brine shrimp *Artemia*. "Lifespan" is based only on reproducing females; "Between" is the time between broods; "Reprod" is the length of the reproductive period; "Postrep" is the length of the post-reproductive period; "Broods" is the number of broods per female; "Nauplii" is the total number of nauplii produced per female; "Cysts" is the total number of cysts produced per female; and "Total off." is the total number of offspring produced per female. SG and TS are parthenogenetic Old World populations, GS and SF are New World sexual populations and YC is an Old World sexual population. From Møller (1995).

(Calow, 1992), a tradition has emerged in toxicology and ecotoxicology to employ genetically homogeneous populations (i.e. inbred strains and parthenogenetic clones) to evaluate the effects of chemical stress. However, caution has to be exercised with this approach for two reasons. First, it assumes that genotype × environment influences are negligible. The studies of Baird and colleagues, discussed above, clearly refute this assumption. Also, Forbes et al. (1995) showed that average phenotypic variability, in both clean and heavy metal-exposed conditions, was as high within single clones of a parthenogenetic gastropod as within genetically mixed populations of its sexual relatives. Likewise, the variability in various life-history traits in single clones of *Artemia* was not generally lower than in genetically mixed populations (Møller, 1995). Second, parthenogens may be a poor choice for evaluating potential hazards associated with chemical stress if their fitness is generally insensitive to environmental gradients, including those involving synthetic chemicals, and their use could lead to underestimates of potential chemical hazards for natural species assemblages. The relative insensitivity of particular parthenogenetic lineages to a given chemical may depend upon the degree to which similar chemicals have been prominent selective forces during their evolutionary history (and for an increasing number of ecological systems this will be so), and whether existing biochemical or physiological adaptations are effective for combating novel chemical stresses. In addition, if in cyclically parthenogenetic species sexual reproduction is more sensitive than asexual reproduction to pollutant stress (Snell and Carmona, 1995), toxicity tests that focus only on the asexual phase of the life cycle (which current tests do) will overestimate environmentally safe toxicant concentrations. Further empirical comparisons of the performance of parthenogenetic lineages with their sexual relatives in the face of synthetic chemical stress are therefore essential.

In terms of predicting the effects of chemical stress on natural populations, it appears that significant genotype × environment (= chemical) interactions may limit the precision with which such predictions can be made and may cause practical problems (e.g. which clone to use in standardized test systems) as well as conceptual challenges (e.g. development of reliable concentration-response models). The application of quantitative genetics models can provide a valuable framework for addressing these problems and provides a more scientifically rational approach in that, from both a theoretical and practical perspective, the goal should be to quantify sources of phenotypic variability rather than to eliminate them.

*Life-history-based risk assessment models*

Ecotoxicologists have long been concerned with trying to identify the most sensitive and ecologically meaningful measures of pollutant effects on natural populations. It has been widely assumed in both ecotoxicological studies and regulatory practice that measuring the effect of toxicants on the most sensitive life-cycle trait or the most sensitive stage of the life cycle will provide the most conservative estimate of a toxicant's effect. Likewise, it is a widespread practice among ecologists (ourselves included) to measure various fitness correlates, such as growth, scope for growth or reproductive output, under the implicit assumption that such traits are more or less linearly correlated with fitness. Recent analysis of the effect of cadmium on life-cycle traits of the nematode *Plectus acuminatus* and their incorporation into a model of intrinsic population increase (i.e. fitness) highlight the fallacy of this assumption (Kammenga et al., 1996). In this study, the length of the reproductive period was the most sensitive trait and was reduced by 45% in response to cadmium exposure. This reduction had no measurable effect on fitness. In contrast, prolongation of the juvenile period by 7.5%, the least sensitive trait, reduced fitness by 5%. This study demonstrates the importance of integrating the effects of toxicants on single life-history traits using life-history theory and provides a quantitative approach for doing so.

Examining the effects of toxic chemicals from a life-history perspective can be used to generate testable predictions about the relative sensitivity of different species/community types and can be used to develop more ecologically relevant risk assessment models (Calow et al., 1997). Very briefly, it is likely that the relationship between various life-history traits and population growth rate will differ among species having different life-history strategies (e.g. semelparous v. iteroparous species). The effects of chemicals on population growth rates can be predicted by integrating effects on individual life-history traits into a model of population growth rate. In this way the effects of different chemicals on various life-history traits can be compared, as can the sensitivity of species having different life-history strategies. In principle, the consequences of chemical exposure for individual populations can be integrated into a prediction for the community/ecosystem as a whole by considering the relative frequency of different life-history types in the system of interest.

## Conclusions

Some authors have attempted to define stress either in terms of particular kinds of environmental variables or in terms of a range of biological responses. In our view all of these have foundered because they have failed to make a clear distinction between normality and abnormality. The approach we have developed here does this in terms of Darwinian fitness and its components. In these terms variables that can be described as causing stress are any that lead to selection. Stress responses, whether they are caused by extreme temperatures or exposure to anthropogenic chemicals, involve fundamentally the same processes as any selection pressure. Thus if there are differences, these will be in terms of rates of change rather than the nature of the underlying processes.

In consequence, the responses of biological systems to novel chemical stresses that are now being studied widely from a practical point of view can provide much useful information for addressing fundamental ecological and evolutionary issues. Conversely, ecological and evolutionary understanding is crucial to the effective development of relevant ecotoxicological tests and ecological risk assessment procedures.

# References

Baird, D.J. (1992) Predicting population response to pollutants: In praise of clones. A comment on Forbes & Depledge. *Funct. Ecol.* 6:616–617.

Baird, D.J. (1993) Can toxicity testing contribute to ecotoxicology? *Funct. Ecol.* 7:510–511.

Baird, D.J., Barber, I., Bradley, M., Soares, A.M.V.M. and Calow, P. (1991) A comparative study of genotype sensitivity to acute toxic stress using clones of *Daphnia magna* Straus. *Ecotoxicol. Environ. Safety* 21:257–265.

Bayne, B.L. (1975) Aspects of physiological condition in *Mytilus edulis* L. with respect to the effects of oxygen tension and salinity. *Proceedings of the Ninth European Marine Biology Symposium*, pp. 213–238.

Bell, G. (1987) Two theories of sex and variation. *In*: S.C. Stearns (ed.): *The Evolution of Sex and Its Consequences*, Birkhäuser Verlag, Basel, pp. 117–134.

Benson, W.H. and Birge, W.J. (1985) Heavy metal tolerance and metallothionein induction in fathead minnows: Results from field and laboratory investigations. *Environ. Toxicol. Chem.* 4:209–217.

Bierzychudek, P. (1989) Environmental sensitivity of sexual and apomictic *Antennaria*: Do apomicts have general-purpose genotypes? *Evolution* 43:1456–1466.

Blanck, H., Wängberg, S.-Å. and Molander, S. (1988) Pollution-induced community tolerance: A new ecotoxicological tool. *In*: J. Cairns Jr. and J.R. Pratt (eds): *Functional Testing of Aquatic Biota for Estimating Hazards of Chemicals. ASTM STP 988.* American Society for Testing and Materials, Philadelphia, pp. 219–230.

Bush, C.P. and Weis, J.S. (1983) Effects of salinity on fertilization success in two populations of *Fundulus heteroclitus*. *Biol. Bull.* 164:406–417.

Calow, P. (1992) The three Rs of ecotoxicology. *Funct. Ecol.* 6:617–619.

Calow, P. (1995) Ecosystem health: A critical analysis of concepts. *In*: D.J. Rapport, C.L. Gaudet and P. Calow (eds): *Evaluating and Monitoring the Health of Large-Scale Ecosystems*. NATO ASI Series, Springer, Berlin, pp. 33–41.

Calow, P., Sibly, R. and Forbes, V. (1997) Risk assessment on the basis of simplified life-history scenarios. *Environ. Toxicol. and Chem.; in press*.

Depledge, M.H. (1994) The rational basis for the use of biomarkers as ecotoxicological tools. *In*: M.C. Fossi and C. Leonzio (eds): *Nondestructive Biomarkers in Vertebrates*. Lewis Publishers, Boca Raton, pp. 271–295.

Dybdahl, M.F. and Lively, C.M. (1995) Diverse, endemic and polyphyletic clones in mixed populations of a freshwater snail (*Potamopyrgus antipodarum*). *J. Evol. Biol.* 8:385–398.

Falconer, D.S. (1981) *Introduction to Quantitative Genetics*. Longman, New York.

Forbes, V.E. and Depledge, M.H. (1992) Predicting population response to pollutants: The significance of sex. *Funct. Ecol.* 6:376–381.

Forbes, V.E. and Depledge, M.H. (1993) Testing vs research in ecotoxicology: A response to Baird and Calow. *Funct. Ecol.* 7:509–512.

Forbes, V.E. and Forbes, T.L. (1993) Ecotoxicology and the power of clones. *Funct. Ecol.* 7:511–512.

Forbes, V.E. and Forbes, T.L. (1994) *Ecotoxicology in Theory and Practice*. Chapman & Hall, London.

Forbes, V.E., Møller, V. and Depledge, M.H. (1995) Intrapopulation variability in sublethal response to heavy metal stress in sexual and asexual gastropod populations. *Functional Ecology* 9:477–484.

Grime, J.P. (1989) The stress debate: Symptom of impending synthesis? *Biol. J. Linn. Soc.* 37:3–17.

Hauser, L., Carvalho, G.R., Hughes, R.N. and Carter, R.E. (1992) Clonal structure of the introduced freshwater snail *Potamopyrgus antipodarum* (Prosobranchia: Hydrobiidae), as revealed by DNA fingerprinting. *Proc. Roy. Soc. London B* 249:19–25.

Hoffmann, A.A. and Parsons, P.A. (1991) *Evolutionary Genetics and Environmental Stress*. Oxford University Press, Oxford.

Hughes, D.J. (1992) Genotype-environment interactions and relative clonal fitness in a marine bryozoan. *J. Anim. Ecol.* 61:291–306.

Jacobson, R., Forbes, V.E. and Skovgaard, O. (1996) Genetic population structure of the prosobranch snail *Potamopyrgus antipodarum* (Gray) in Denmark using PCR-RAPD fingerprints. *Proc. Roy. Soc. London B* 263:1065–1070.

Kammenga, J.E., Busschers, M., van Straalen, N.M., Jepson, P.C. and Bakker, J. (1996) Stress-induced fitness reduction is not determined by the most sensitive life-cycle trait. *Func. Ecol.* 10:106–111.

Klerks, P.L. (1987) *Adaptation to metals in benthic macrofauna*, Ph.D. dissertation, State University of New York, Stony Brook, NY.

Klerks, P.L. and Levinton, J.S. (1989) Effects of heavy metals in a polluted aquatic ecosystem. *In*: S.A. Levin, J.R. Kelley and M.A. Harvell (eds): *Ecotoxicology: Problems and Approaches*, Springer-Verlag, New York, pp. 41–67.

Klerks, P.L. and Weis, J.S. (1987) Genetic adaptation to heavy metals in aquatic organisms: A review. *Environ. Poll.* 45:173–205.

Maltby, L. (1991) Pollution as a probe of life-history adaptation in *Asellus aquaticus* (Isopoda). *Oikos* 61:11–18.

Maltby, L., Calow, P., Cosgrove, M. and Pindar, L. (1987) Adaptation in aquatic invertebrates: Speculation and preliminary observations. *Annales Societé Royal Zoologique Belgique* 117 (Suppl. 1):105–115.

Michaels, H.J. and Bazzaz, F.A. (1989) Individual and population responses of sexual and apomictic plants to environmental gradients. *Amer. Nat.* 134:190–207.

Moriarty, F. (1983) *Ecotoxicology: The Study of Pollutants in Ecosystems*. Academic Press, London.

Møller, V. (1995) *Genotypic and environmental sources of variation in population response to pollutant stress: Implications of reproductive mode*, Ph.D. dissertation, Odense University, Denmark.

Møller, V., Forbes, V.E. and Depledge, M.H. (1996) Population responses to acute and chronic cadmium exposure in sexual and asexual estuarine gastropods. *Ecotoxicology* 5:1–14.

Niklasson, M. and Parker, E.D. Jr. (1994) Fitness variation in an invading parthenogenetic cockroach. *Oikos* 71:47–54.

Odum, E. (1985) Trends expected in stressed ecosystems. *BioScience* 35:419–422.

Parker, E.D. Jr. and Niklasson, M. (1995) Desiccation resistance in invading parthenogenetic cockroaches: A search for the general purpose genotype. *J. Evol. Biol.* 8:331–337.

Postma, J.F. (1995) *Adaptation to metals in the midge Chironomus riparius*, Ph.D. dissertation, University of Amsterdam.

Rawson, P.D. and Hilbish, T.J. (1991) Genotype-environment interaction for juvenile growth in the hard clam *Mercenaria mercenaria* (L.). *Evolution* 45:1924–1935.

Scheiner, S.M. (1993) Genetics and evolution of phenotypic plasticity. *Ann. Rev. Ecol. Syst.* 24:35–68.

Sibly, R. and Calow, P. (1985) The classification of habitats by selection pressures: A synthesis of life-cycle and r/K theory. *In*: R.M. Sibly and R.H. Smith (eds): *Behavioural Ecology: Ecological Consequences of Adaptive Behaviour*. Blackwell, Oxford, pp. 75–90.

Sibly, R. and Calow, P. (1989) A life-cycle theory of responses to stress. *Biol. J. Linn. Soc.* 37:101–116.

Snell, T.W. and Carmona, M.J. (1995) Comparative toxicant sensitivity of sexual and asexual reproduction in the rotifer *Brachionus calyciflorus*. *Environ. Toxicol. Chem.* 14:415–420.

Van Valen, L. (1973) A new evolutionary law. *Evolutionary Theory* 1:1–30.

Weider, L.J. (1993) A test of the "general purpose" genotype hypothesis: Differential tolerance to thermal and salinity stress among *Daphnia* clones. *Evolution* 47:965–969.

West-Eberhard, M.J. (1989) Phenotypic plasticity and the origins of diversity. *Ann. Rev. Ecol. Syst.* 20:249–278.

Wood, R.J. (1981) Insecticide resistance: Genes and mechanisms. *In*: J.A. Bishop and L.H. Cooke (eds): *Genetic Consequences of Man-Made Change*. Academic Press, London, pp. 53–96.

Environmental Stress, Adaptation and Evolution
ed. by R. Bijlsma and V. Loeschcke

# Conifers from the cold

Outi A. Savolainen and Päivi K. Hurme

*Department of Biology, University of Oulu, PO Box 333, FIN-90571 Oulu, Finland*

*Summary.* Northern boreal environments have short growing seasons and cold winters. In the northern populations of Scots pine, pollen and seed production are low, seed maturation has low probability, early survival is low and growth is slow. Northern Scots pine populations are genetically differentiated from southern Finnish ones in many of the traits that confer adaptation to harsh conditions, such as cold tolerance and timing of bud set in the fall. As opposed to this pattern, marker gene frequencies through this area are uniform. Abundant pollen flow equalises allelic frequencies, but strong natural selection can lead to differentiation in some parts of the genome. There is also much genetic variation within the populations for important quantitative traits. Molecular genetic methods will provide more detailed information on the nature of the genetic basis of this variation. While the general patterns of variation in Scots pine are similar to those of many other conifers, there are differences in the physiological mechanisms of adaptation.

## Introduction

Life in the high northern (and southern) latitudes requires plants and animals to adapt in many ways. Growth and reproduction must be timed to coincide with favourable environmental conditions. The growing season is short – the number of days with average temperature over 5°C in the boreal zone is in many areas less than 150 – which means that the time available for growth and reproduction is short. Further, the organisms must not just grow in summer but also survive the winter, either by escaping, as do migratory birds and spring annual plants, or by various mechanisms that allow survival in the cold. The minimum temperatures in the northern boreal zone may be lower than −40°C, and can go as low as −67°C. Havranek and Tranquillini (1995) have suggested that survival during winter and recovery from winter dormancy may be the most important characteristics of conifer tree ecophysiology. The different tissues of the plants must be able to withstand the cold and commence life processes again the following spring. Full winter dormancy includes forming resting buds, suspending growth processes, reducing metabolic activity, enhancing frost and desiccation resistance, and undergoing changes in cellular and cytoplasmic structure (Havranek and Tranquillini, 1995). Mortality in the winter often occurs through injuries that show up in late winter, due to frost drought and oscillating freezing and thawing.

Adaptation to cold stress is being studied intensively by scientists in different fields (e.g. Thomashow, 1990; Hughes and Dunn, 1996). Here we will concentrate on the genetic and evolutionary aspects of the problem. We try to combine results from quantitative and population genetics in understanding the evolution of adaptation to harsh conditions. The most important traits are likely to be related to the timing of life history events, such as timing of germination, date of bud burst and bud set, date of male and female flowering and timing of seed maturation. The timing of cold acclimation and level of frost resistance will also be important (e.g. Sakai and Larcher, 1987; Havranek and Tranquillini, 1995). How different are southern and northern populations genetically in general, and with respect to these adaptive traits? How much variation can be found within populations? What evolutionary factors account for observed patterns? This paper will deal with northern conifers, with a special emphasis on Scots pine. From a population biology perspective, the climatic conditions impose similar selection pressures on all species, and thus parallels can be found. Where appropriate, we will make reference to findings in other plants and animals. Understanding the evolution and genetic basis should also allow for better tree breeding and for predictions of responses to changing environmental conditions.

## Ecology and life history of northern conifers

Boreal conifers are long-lived, outcrossed, wind-pollinated species. We use Scots pine to exemplify conifers because its life history is rather well known, thanks to the efforts of forest biologists (e.g. Sarvas, 1962, 1972; Eiche, 1966; Koski, 1970; Hänninen, 1990). This species has the widest distribution of all 110 species of the genus *Pinus* (Mirov, 1967), ranging from Turkey or Spain in the south (about 38°N) to northern Finland (68°N), and western Scotland (6°W) to eastern Siberia (135°E). Our studies have been mainly on Finnish populations, so we continue by describing the ecological theatre where this evolutionary play has been and is still being acted out.

All of Finland was still covered by glaciers 10,000 years ago. Scots pine colonised the northernmost areas about 7000–8000 years ago, and actually did reach a little futher to the north in an earlier warm period (Hyvärinen, 1987). The most extensive range was between 7000 and 5000 B. P. Later, colder climates caused a retreat of the species from higher latitudes and altitudes. A comparison of growing conditions in the south (about 60°N) and north (about 68°N) shows that the environmental cline is steep: for instance, the January mean temperature is − 13.2°C in the far north and − 6.9°C in the south. These mean temperatures are supposed to be correlated with minimum temperatures, critical for plant survival (see Sykes and Prentice, 1995). Temperature sums and length of growing season (number of days with mean temperature over 5°C) are displayed in Figure 1. Other relevant climatic data are shown in Table 1 or were listed and referen-

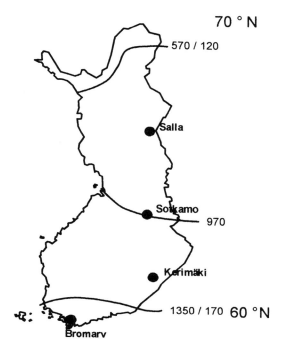

Figure 1. Origins of Finnish populations of Scots pine (Salla, Sotkamo, Kerimäki and Bromarv). Temperature sums (degree days) and lengths of the growing season (number of days with mean temperature above 5°C) are also depicted.

Table 1. Genetical and ecological differences in Scots pine between latitudes 60° and 68°N

| Climatological character or trait | ~60°N | ~68°N | Ref. |
|---|---|---|---|
| January mean temperature | − 6.9°C | − 13.9°C | |
| July mean temperature | 16.4°C | 13.5°C | |
| Length of thermal growing season (days)* | 174 | 121 | |
| Temperature sum (degree days) | 1350 | 570 | |
| Bud set (sowed 1 June 1994)† | 24 September | 2 September | 1 |
| Frost hardiness on 26 Sep (sowed 1 June 1994)† | − 13°C | − 26°C | 1 |
| Beginning of pollen shedding | 30 May | 16 June | 2 |
| Mean female flowering time †† | 14 June | 11 June | 3 |
| Probability of seed maturation | 98% | 5% | 4 |
| Seed set (seeds/m²) | 120 | 50 | 5 |
| Height at 140 years | 23 m | 19 m | 6 |
| Survival from 2–3 years to height of 2.5 m | 75% | 25% | 7 |
| Mortality from seedling to 20 years | 31% | 97% | 8 |

* Number of days with mean temperature above 5°C.
† Traits measured in common garden experiments.
†† Common garden experiment in southern Finland; northern trees will flower after a lesser temperature sum.
References: Hurme et al., 1997 (1); Luomajoki, 1993 (2); Chung, 1981 (3); Henttonen et al., 1986 (4); Koski and Tallqvist, 1978 (5); Ilvessalo, 1937 (6); Persson, 1994 (7); Eriksson et al., 1980 (8).

ces given by Savolainen (1996). Within Finland the environmental variation is simple, as latitude explains most of the climatic variation. In other areas (e.g. parts of Sweden and western North America) the patterns are more complex, because longitude, altitude and aspect all have an influence.

The life history traits of Scots pine in the north and south differ in many aspects, as described in Table 1. The annual seed production in the south is on average more than twice as high as in the north (Koski and Tallqvist, 1978; see Fig. 2 in Savolainen, 1996). In the south, seeds practically always mature; in the north full maturation of pine seeds is a rare event; and in the northernmost areas at timberline there is a 5% chance of 50% maturation (Henttonen et al., 1986). Similarly, there is a large difference in pollen production (Koski and Tallqvist, 1978), with much lower average and higher between-year variance in the north than in the south. Anthesis (pollen release) and receptivity of female strobili occur much later in the north than in the south (Sarvas, 1972; Luomajoki, 1993). The annual seed crop is approximately $100/m^2$ (1 million/ha). These seeds establish a stand which will have on average 33,000 ten-year-old trees per hectare in an even-aged *Calluna*-type forest (Lönnroth, 1925). Thus, even if a new forest were due to 1 year's seed crop, only 1 in 30 would survive from seed to sapling. In most studies, survival estimates are obtained from experimental plantations such as those of Eiche (1966), in which trees were 2–3 years old when planted and were protected against some herbivore damage by fencing and against pathogens by spraying. Thus, these experiments measure only a part of the natural mortality. The average survival from young seedling (2–3 years) to a mean height of 2.5 m was 75% in the south (latitude 62°N), but only 25% in the north (68°N) (Persson, 1994, Fig. 2). Growth in northern harsh environmental conditions is slow. In similar environments, 140-year-old trees are over 20-m tall in the south; in the north they are shorter by several metres (Ilvessalo, 1937).

Scots pine has a vast distribution, and we are here comparing just southern and northern Finland or Sweden, which, for our topic, is a relevant comparison. As is evident from the data, the stress imposed by the harsh conditions increases dramatically from latitude 60°N to 68°N. The growing season for reproduction is short in the north: in more southern latitudes (58°N to 63°N) anthesis can occur when a constant 15% of the annual heat sum is reached; but starting at latitude 63°N, the proportion of heat sum needed for anthesis increases steeply, and in the northernmost areas about 50% of the heat sum is required (Sarvas, 1972; Luomajoki, 1993). As mentioned, seed maturation probability decreases from about 98 to 5% (Henttonen et al., 1986). Eriksson et al. (1980) estimated that the mortality of local pine populations from seedling to age 20 years, at a similar altitude (360 m), was 31% at latitude 61°N, whereas at latitude 68°N it was 97%. These figures suggest that in fact northern pine populations are not very well adapted to the northern climate. The northern distribution limit of Scots pine is not a physical boundary, but the conditions gradually become too harsh. Woodward (1995) has suggested that winter cold tolerance is an important factor in determining the distribution limits of conifers. Further south the most important environmental effects may be from other kinds of stress, such as drought in Mediterranean conditions (e.g. Notivol and Alia, 1996). Our choice of comparison was naturally motivated by the availability of results, but we think it is also biologically well justified: there really is a threshold in cold stress in this area.

## Distribution of genetic variation in Scots pine

The above comparisons of northern and southern populations were in natural environments. It is clear that any trees will have lower survival and grow poorly in harsh conditions. However, we are interested in the extent of genetic differences between populations: How much have the northern populations become genetically differentiated in the relatively low number of generations since colonisation?

Genetic differences between populations are evident in reciprocal transfer experiments (e.g. Eiche, 1966; Eriksson et al., 1990; Beuker, 1994; Persson, 1994). Northern populations transferred to the south have increased survival (about 10% for each degree of latitude), and correspondingly, southern provenances transferred to the north have poor survival (Eriksson et al., 1980). Northern populations transferred to the south still have worse growth than the local southern populations (Eriksson et al., 1980; Beuker, 1994). These results on transfer experiments show that the populations are genetically differentiated. While there are some indications that the environmental conditions of seed development can influence later growth and survival, for Scots pine these effects are of short duration (Ruotsalainen et al., 1995).

### Neutral quantitative traits in markers

Many studies have compared northern and southern populations of Scots pine. Several studies describe the patterns of variation at marker loci, allozymes (Gullberg et al., 1985; Szmidt and Muona, 1985; Muona and Harju, 1989) and various DNA markers (Karhu et al., 1996). Variation of ribosomal DNA phenotypes has also been examined (Karvonen and Savolainen, 1993). These results show that within Finland and Sweden all populations have about equal amounts of variability as measured by expected heterozygosity or similar statistics. Further, there are hardly any differences between the populations in allelic frequencies; the proportion of variation between populations is of the order 0.02 (Karhu et al., 1996). This pattern of variation is not exclusive to marker genes. Koski (1970), in his study of pollen flow, examined variables that describe cone morphology. The ratio

of the apical part of the apophysis of fertile cone scales to the width of apophysis served as an example of a trait where the proportion of variation between populations was 3%, as measured in a common garden experiment.
This uniformity of variation over populations holds true for marker variability of most other conifers (Hamrick et al., 1992).

## Deleterious recessives and genetic load

Inbreeding depression and genetic load are not directly related to cold
stress. However, the general vigour of plants influences survival. For
instance, only undiseased plants will be capable of proper cold acclimation
(Sakai and Larcher, 1987). Also, inbreeding depression and genetic load
may be informative with respect to the structure of populations. Northern
populations of Scots pine have significantly less genetic load than do
southern ones, as measured by the estimates of numbers of embryonic
lethals. The measurements were all made in common garden exeriments in
southern Finland (Kärkkäinen et al., 1996). The reasons for difference are
presently not well understood. It is possible that there is more inbreeding in
the north than in the south, due to increased partial selfing. This could
result in purging of lethals. On the other hand, the intensity of selection
against inbreds may also differ between the environments (Dudash, 1990;
Bijlsma et al., this volume).

## Timing of reproduction and growth, and development of frost hardiness

Chung (1981) compared the timing of anthesis and female strobili receptivity of northern and southern Scots pine clones in a common garden experiment. Northern clones become receptive earlier than did southern ones
(see Tab. 1), and they release pollen before southern ones do. The clinal
variation over latitudes is regular. This demonstrates the genetic differentiation of the populations.

Cessation of vegetative growth was studied earlier, e.g. by Mikola (1982),
who found that in common garden experiments southern populations set
bud much later than northern ones. However, no estimation was made of
within-population genetic variation. One goal of our studies has been to
complement earlier work with this information (Hurme et al., 1997).

We chose four populations from different latitudes in Finland (Fig. 1).
When seeds were sown on 1 June, bud set of seedlings took place between
the beginning of August and the end of October. Seedlings from the
northern populations set buds earlier than those from the southern populations when grown in a common environment (see Fig. 2(A)). Populations
differed significantly, and 36% of the total variation was between the populations. On average, every increment of 1 degree of latitude caused bud set

to take place about 3 days earlier. Variation of bud set within populations was extensive as well, and there was a large additive genetic component (P. K. Hurme, T. Repo, O. Savolainen, unpublished data). Our results are in agreement with Mikola's (1982), but clinal variation in his study was twice as steep as in ours, as measured by the difference in the bud set date of northern and southern populations. This could perhaps be due to the earlier start of his experiment (sowing the seeds on 1 May).

Between-population differences of frost hardiness are well known in Scots pine (Hagner, 1970; Toivonen et al., 1991; Aho, 1994). We also found a latitudinal cline in frost hardiness (Hurme et al., 1997). The distribution of between versus within population variability could not be estimated, because single seedling estimates of frost hardiness were not available (Fig. 2(B)). Frost hardiness developed most rapidly in the northernmost populations. Correlations between bud set and frost hardiness were high (0.69–0.97) at the population level. After most of the buds had formed (about 90% on 30 September), frost hardening was accelerated. At the population level, bud set was followed by the development of frost hardiness.

Within-population variation in phenological traits and frost tolerance is known to occur in Scots pine, but has not been well characterised. The phenotypic data of Mikola (1982) suggest extensive within-population variation. Our recent experiments both for bud set and cold tolerance confirm this (P. K. Hurme, T. Repo, O. Savolainen, unpublished data). Other studies have demonstrated that for Scots pine frost hardiness, about 30% of total phenotypic variability was due to additive effects in central Swedish populations (Norell et al., 1986). Aho (1994) showed that there were differences in cold tolerance due to male parents within populations, but did not quantify the effects. Nilsson and Walfridsson (1995) found large variability in cold tolerance between clones, also in Swedish populations.

We have described in some detail, as a case study, patterns of variation in Scots pine. Next we compare findings in Scots pine with those of other conifer species.

## Differences between populations in timing of growth and frost tolerance

Traits related to climatic adaptation have been studied in many species since the pioneering work of de Vilmorin (1863) and, for example, Langlet (1936) (see review of Langlet, 1971). The main emphasis has been on examining differences between populations, to the neglect of characterising in more detail variation within populations. The comparison of species or populations for phenological or frost tolerance traits is not easy. The exact traits measured, age at measurement, pretreatments and methods of measurement vary. Correlations between frost tolerance of different

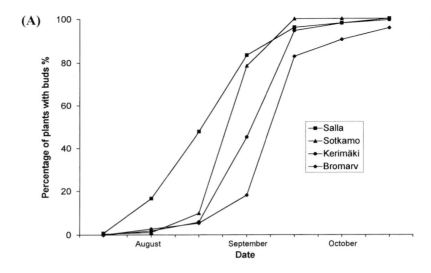

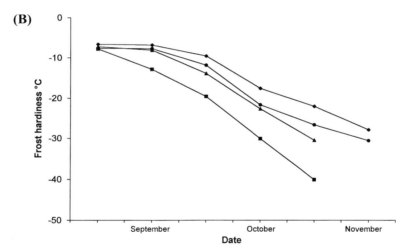

Figure 2. (A) Cumulative distribution of terminal bud forming in first-year seedlings of Scots pine populations from different latitudes in Finland in 1994. (B) Development of frost hardiness in first-year seedlings of Scots pine populations with electrolyte leakage method on needles.

tissues or organs can be low (Sakai and Larcher, 1987; Aitken et al., 1996). Further, common garden environmental conditions may not be a relevant environment for all genotypes. Because of these difficulties with the data, we have not tried to base our comparisons on detailed statistics but describe the patterns at a more general level. Some examples of the common garden studies are listed in Table 2. To describe the extent of differences between populations, or the extent of clinal variation, two statistics have been used.

Table 2. Examples of variation in growth traits and frost hardiness in several conifer species. Number of populations (Pop), latitudinal range of populations (Lat) and age of the populations (Age) are also depicted

| Species and traits studied (Reference) | Measures | Pop | Lat (°N) | Age (years) | Geographical cline mainly |
|---|---|---|---|---|---|
| *Picea abies* (1) altitude × growth start altitude × growth cessation | $R^2$ (%) 5 73 | 36 | 48–60 | 6–7 | latitudinal and elevational |
| *Pseudotsuga menziesii* (2) date of bud set frost damage total height | variance between breeding zones (%) > 50 | 14 | 42–44 | 1 | elevational |
| *Pinus ponderosa* (3) spring frost injury winter frost injury growth start growth cessation | variance between populations (%) 49 21 31 63 | 62 | 36–41 | 2, 4 | elevational |
| *Pinus contorta* (4) height survival | variance between populations (%) 37 22 | 53 | 49–63 | 3, 6, 10, 20 | latitudinal, longitudinal and elevational |
| *Picea mariana* (5) seedling height date of needle flushing | variance between populations (%) 10–25 | 75 | 45–55 | 1–2 | elevational |
| *Pinus monticola* (6) bud set bud burst | variance between populations (%) 0–0.1 | 12 | ~48 | 3–6 | |
| *Larix occidentalis* (7) altitude × bud set altitude × frost hardiness altitude × bud burst altitude × height | $R^2$ (%) 47 21 39 53 | 82 | 45–49 | 2 | elevational |
| *Larix occidentalis* (8) geog. variables × survival no frost-free days × survival survival | $R^2$ (%) 30 21 variance between populations (%) 13 | 143 | 43–51 | 2 | latitudinal and elevational |

References: Skroppa and Magnussen, 1993 (1); Loopstra and Adams, 1989 (2); Rehfeldt, 1990 (3); Xie and Ying, 1995 (4); Parker et al., 1994 (5); Rehfeldt, 1989 (6); Rehfeldt, 1982 (7); Rehfeldt, 1995 (8).

First, for some populations the variation has been partitioned into a between (genetic)- and within (genetic and environmental)-populations component. For other cases, the investigators have regressed the phenological or frost tolerance results on a single environmental variable such as latitude, altitude, length of frost-free period and so on. The $R^2$ values are proportions of variation explained by the multiple regressions. An exhaustive summary is neither feasible nor necessary.

Most conifer populations are highly differentiated with respect to phenological variables such as bud flushing and bud set as well as for the timing and degree of development of cold tolerance. These are the traits most often studied, but presumably many others also differ between populations. Fryer and Ledig (1972), for instance, found that there was a steep elevational cline in the photosynthetic optimum temperature of *Abies balsamea*, with the temperature decreasing nearly 1°C for each 100-m increase in elevation (between 700 and 1500 m).

Norway spruce, Douglas fir (*Pseudotsuga menziesii*) and lodgepole pine (*Pinus contorta*) have all been extensively studied. In North America, in addition to latitude, altitude and longitude (describing continentality) also can account for much of the variation, in some studies for more than 80% of the total variation in multiple regressions (e.g. Rehfeldt, 1989). As is shown in Table 2, the $R^2$ values vary between studies, from 5 to over 80%. The relative within-population variation differs between studies and traits. Skroppa and Magnussen (1993) found that for starting growth, altitude accounted for only 5% of the variation, whereas for growth cessation the estimate was 71%. There are also differences in the steepnees of the slope of the regression between species, and even areas. Rehfeldt (1989) found in *P. menziesii* that a higher elevations the cline in height growth flattened. Whereas at lower elevations a 240-m difference in altitude resulted in genetically differentiated populations, at higher altitudes 350 m was required to have genetically distinct populations. Xie and Ying (1995) showed that, in a common garden experiment, measurements on older trees of the same cohort resulted in larger relative differences between populations. The reasons for such findings will be discussed below. In *Larix occidentalis* clines in growth traits were quite flat; only 34% of the variation could be accounted for by environmental variables (Rehfeldt, 1995).

Most conifers show variation in phenology-related traits. Red pine (*Pinus resinosa*) grows over quite an extensive range in North America but is known to have hardly any genetic variation at enzyme loci or RAPD (random amplified polymorphic DNA) markers (see Mosseler et al., 1992). However, Fowler (1964) reported variation between different populations in frost tolerance and timing of start of growth. *P. monticola* may be a rare exception. Populations representing much of the ecological range were not differentiated with respect to growth traits like bud set and bud burst, even if there was ample within-population variability. Thus, the species presumably has much phenotypic plasticity (Rehfeldt, 1979). One specific provenance of *Larix sibirica* (Raivola) is known for growing very well over a wide range of environments, thus also perhaps representing exceptional phenotypic plasticity (Heikinheimo, 1956).

## Variation within populations in phenological traits and frost tolerance

While there are extensive studies on between-population variation, detailed study of within-population variation has received less attention. Tree breeders traditionally estimate heritabilities to give an indication of potential response to selection. Some examples of these studies are considered below.

Bud burst and bud set seem to have high heritabilities in Douglas fir (Li and Adams, 1993), especially in pole-sized trees (as opposed to seedlings), but cambial phenology had much lower additive genetic variation (Li and Adams, 1994). Fall cold hardiness of Douglas fir had low heritability, even if there was significant between-family variation (Aitken et al., 1996). In a later study, Aitken and Adams (1996) showed that the heritabilities for winter cold hardiness were lower than those in the fall. Genetic correlations between cold hardiness of different tissues (needle, stem and bud) were low. Furthermore, at least in Scots pine, the phenotypic data and the initial genetic data suggest that the variability is not reduced in populations at the species margin (e.g. Hurme et al., 1997). However, in other species, e.g. *Abies sachalinensis*, reduced variability has been found at the highest altitudes (Sakai and Larcher, 1987, p. 149).

We discussed earlier between-population correlations, but in this context it is also of interest to consider within-population genetic correlations between cold tolerance, duration of growth period and height growth. The within-population genetic correlations can provide information on the genetic basis of the traits. Further, they also suggest how selection on one trait will influence others. In Scots pine, Nilsson and Walfridsson (1995) found, in a set of Swedish clones, that shoot elongation rhythm was uncorrelated with cold tolerance throughout the year. However, early cold acclimation in the autumn was related to poor height growth. In Douglas fir, timing of bud burst and bud set were only weakly correlated with cold injury, and bud phenology was a poor predictor of fall cold hardiness (Aitken et al., 1996). Further, genetic correlations seem to be dependent on the environment and age of trees. Correlations between bud burst and bud set were high in pole-sized trees (0.96) but quite low in transplant experiments of seedlings (Li and Adams, 1993).

These few results highlight some difficulties in comparative studies of frost tolerance and bud phenology. First, the traits to be studied need to be defined carefully, because environmental conditions, age, and organ or tissue measured can all have an influence on the results. Plant breeders often estimate heritabilities: a ratio of additive genetic to total phenotypic variation, which calculates the prospect of short-term selection gain. For examining the absolute amount of additive genetic variation (relative to the mean), which gives some idea of the possible long-term response to selection, it would be more informative to estimate additive genetic coefficients of variation (AGCV = square root $(V_A)$/mean) (Houle, 1992; Cornelius, 1994). These data are rarely available, and scales of measurement can cause problems. The statistics require an absolute scale of measurement, which for the phenology or cold tolerance traits may be difficult to obtain, as relative scales (such as Julian dates) are often used. Thus, most informative comparisons may be just within a species or population, and on exactly the same trait. Heritabilities are easier to estimate, but they need not always exactly reflect the AGCV. Cornelius (1994), for instance, listed median heritabilities of height and wood specific gravity (0.25 and 0.48), and their AGCVs (0.08 and 0.05), which suggests that, despite the higher heritability, additive genetic variation in wood specific gravity would not allow selection to proceed as far as for height.

## Migration-selection balance

The pattern of variation of high differentiation for growth rhythm and frost tolerance traits and little differentiation for marker genes can be interpreted in the context of Scots pine (and general conifer) ecology. It is well known that gene flow through pollen migration is quite strong. This was demonstrated by the work of Koski (1970) in direct measures of pollen flow, either to areas were no Scots pine grew, or with radioactively marked pollen. More than 50% of pollen came from further away than 50 m. Later studies with paternity analysis, using genetic markers, have confirmed that the migrating pollen participates in fertilisation (e.g. Harju and Muona, 1989). Within a seed orchard (seed production populations) of a few hectares, 20–40% of the seeds were fertilised by pollen from outside the orchard. Further, at the time of pollination prevailing winds are from the south, making south-north pollen flow especially efficient. In northern Finland in some years at least, local pollen is hardly shed at all during the time female strobili are receptive (Kärkkäinen, 1991). Efficient gene flow will account for the uniform distribution of random molecular markers, which probably are not much influenced by selection. On the other hand, the genes influencing growth rhythm traits or frost tolerance experience similarly efficient gene flow. Natural selection will eliminate, through frost damage, seedlings that continue growing too late. Those that cease growth too early may soon be outcompeted by taller rivals, as suggested by genetic correlations between growth and development of frost tolerance (Nilsson and Walfridsson, 1995), which will give rise to stabilizing selection. This kind of migration selection balance should yield a difference in the pattern of variation between neutral markers and selected traits, which we have seen. A second prediction is that seeds and seedlings should be less selectively differentiated than adults for the selected traits. For this purpose, comparison of the same traits in a cohort at different ages, before and after the effects of selection, should be possible. We do not have suitable data for Scots pine. However, Xie and Ying (1995) found that older populations were more clearly differentiated with respect to survival and growth than were young ones.

Similar patterns of variation are found among populations of northern *Drosophila* species. Variation of diapause in *Drosophila*, measured as critical day length (where 50% of the females enter reproductive arrest), is highly differentiated between populations (Muona and Lumme, 1981, and references therein). There is hardly any differentiation with respect to most marker genes in northern *Drosophila* populations (Lankinen, 1976). As in pine trees, there is much migration between *Drosophila* populations.

There is much within-population variation in phenological and frost tolerance traits in pine trees. Even the northernmost pine populations have high variability in all traits examined. Tigerstedt (1994) suggested that the variation is a result of variable selection and is a direct adaptation. It seems to us that there may in fact be strong directional selection of more "northern" traits. The extensive pollen migration from the south may maintain variation, even in the face of rather strong directional selection. In the northernmost populations, the extensive migration from the south may hinder adaptation (at least in the short run). If this is the case, one should be able to show that migration from the south reduces the viability of offspring. A comparison of local north × north progenies with northern open-pollinated (partly southern pollen) seeds should help to answer the question. Rousi (1983) provides such data, but the north × north crosses were between so-called plustrees. These have been phenotypically selected on the basis of their superior growth. In this data set, the survival of the local population and the controlled crosses between northern trees were equal, with no visible influence of the southern pollen. However, as the plustrees were selected on the basis of their supposedly better growth, they may have longer growth periods and thus lower survival than average trees. Nilsson (1995) has suggested that the fertilising pollen may actually come from the north, based on progeny tests of seed orchard clones. This does not seem likely, given what we know about the biology of the species and wind directions, and the issue remains to be examined.

However, this general interpretation of balance between migration and strong selection should hold for other conifers as well, and has been found to hold for lodgepole pine by comparison of markers and quantitative traits (Yang et al., 1996). Variation in rates of migration and selection intensities should account for differences in the slope of the clines between species.

## Mechanisms of adaptation in different species

Sakai (1983) made a comparative study of the cold resistance of a large number of conifer species. Within several important conifer genera, very hardy species have evolved, e.g. *Pinus, Picea, Abies* and *Larix*. In these genera, some species can withstand temperatures below − 70°C. The most interesting comparison for us is between Scots pine and Norway spruce.

It is well known that many conifers and other trees have photoperiodic ecotypes (e.g. Vaartaja, 1959; Sakai and Larcher, 1987). Norway spruce has a clear photoperiodic reaction for bud set. There is a critical night length: shorter nights result in bud set, longer ones do not (e.g. Ekberg et al., 1979). There are latitudinal, longitudinal and altitudinal clines, corresponding to environmental conditions. As opposed to this pattern, Scots pine seedlings will eventually set bud whatever the night length. Hänninen et al. (1990) have suggested that using the critical night length as opposed, to for example, temperature sums, should result in the most efficient use of the growing season, but this seems not to have evolved in Scots pine. A photoperiodic optimum can, however, be determined for growth of Scots pine seedlings (see Ekberg et al., 1979). It has also been suggested that temperature sum has a substantial impact on growth cessation and but set of Scots pine (Koski and Sievänen, 1985). Scots pine and Norway spruce differ in the schedule of events during cold acclimation. Scots pine buds develop rest and hardiness in parallel, whereas Norway spruce buds lose rest during development of hardiness. The species thus react differently to increased temperatures during acclimation. Norway spruce will stay dormant in the early stages of acclimation, whereas Scots pine may restart growth (Dormling, 1993).

Pines have a different mechanism for cold tolerance than do other conifer species. Other genera avoid freezing damage by so-called extra organ freezing, where water is removed during slow cooling, but pines have extra-cellular freezing (Sakai, 1983). Similar freezing conditions cause more damage to Scots pine than to Norway spruce, at least in some stages of cold acclimation (Dormling, 1993).

Another difference between Norway spruce and Scots pine is their behaviour in transfer experiments. Populations of Norway spruce can be transfered quite long distances southwards or northwards. For example, populations from eastern Poland grow well in southern Scandinavia (Skroppa and Magnussen, 1993). Scots pine, on the other hand, does not tolerate long transfers to the north (Eriksson et al., 1980; Persson, 1994). It is not clear how and why the critical night-length cline has evolved in Norway spruce, if the damage due to lack of dormancy is not severe.

Scots pine and Norway spruce also have different histories of spreading to Fennoscandia after glaciation. Scots pine followed the withdrawing ice from southeastern Finland, whereas Norway spruce colonised from the east later, about 5000 years ago (Alho, 1990). Norway spruce had wide areas to colonise, but was able to do so only after the temperature had decreased after a warmer peak. Thus, the time span of adaptation has been different.

## Concluding remarks

Cold acclimation and frost tolerance are physiologically complex phenomena, which gives rise to some difficulties in studies of their genetics. They are polygenic, even though proper studies of the genetic basis are not yet really available. Further, their evolution is constrained, in addition to selection, by migration and the history of the populations. It is also evident that random molecular markers do not give much information on the patterns of variation of these traits. We try to point out below some areas that need further study, and where population genetics data should and could be used.

### *Molecular genetics methods offer possibilities for improved genetic studies*

We can see two important developments in molecular genetics methods that will influence evolutionary studies of phenological traits and cold tolerance. First, it has become possible to study the genetic basis of quantitative variation using QTL (quantitative trait locus) mapping. Genetic maps are now available on many trees (Devey et al., 1996, and references therein). These maps, combined with the availability of suitable crosses, allow estimation of the number of loci influencing the differences in a trait, mapping of the loci and study of their mode of genetic action. Using these approaches, wood quality trait loci have been mapped in *P. taeda* (Groover et al., 1994), and some phenology trait loci in a *Populus* interspecific cross (Bradshaw and Stettler, 1995). Such studies will also increase our understanding of the genetics of timing and frost tolerance-related traits. They could also provide efficient marker genes (closely linked to the actual loci) for further study of these traits in natural or breeding populations.

Second, understanding of the molecular nature of the acclimation process is increasing (Thomashow, 1990; Leborgne et al., 1995). The processes, the genes involved and their regulation are all being studied. Genes involved in freezing tolerance have been isolated and characterised, e.g. in *Arabidopsis thaliana* (Welin et al., 1995), and in many crop species (e.g. Houde et al., 1995). The physiological changes during acclimation in gymnosperms have been studied in, for example, *P. banksiana* (Zhao et al., 1995), *Picea glauca* (Binnie et al., 1994) and *P. monticola* (Ekramoddoullah and Taylor, 1996). Genes related to frost tolerance are being characterised in *P. lambertiana, P. monticola* and *P. strobus* (Ekramoddoullah et al., 1995). These studies provide candidate loci which can also be used in population genetics studies. The importance of a gene can be assessed by studying its patterns of variation in natural populations. For instance, in the tobacco budworm random molecular markers are not genetically differentiated, but a sodium channel locus related to pyrethroid insecticide resistance is highly geographically differentiated (Taylor et al., 1995). Frost resistance and phenology should show the same kinds of steep clines for allelic frequencies of candidate genes as are being observed for the traits themselves. It will be possible to combine candidate locus studies with examination of phenotypic variation as has been done in *Drosophila* (McKay and Langley, 1991).

## Using information on genetic variation

The information on patterns of genetic variation can be put to several uses. First, it is of importance for conservation genetics. It is evident that when the species is differentiated with respect to genetic variation for important traits for survival, genetic variation should be conserved both in natural populations and for plant-breeding purposes.

Second, this variation has implications for considering the consequences of climate change. Past changes have sometimes caused very rapid changes in vegetation. When *P. mariana* spread to the north in North America, the vegetation at one site changed from tundra to dominant *P. mariana* in just 150 years (MacDonald et al., 1993) due to interactions between growth form and migration. Present predictions of effects of climate change are often based on the assumption that trees stay as they are (as e.g. in Sykes and Prentice, 1995). However, both short-term physiological changes and genetic changes also occur. We know that there is much genetic variability even within populations. As conditions change, the selection regime will be altered. As temperatures increase, photoperiods will not change. Thus, mere migration of southern trees will not suffice to produce a new adapted population. Selection, migration and recombination will generate new genotypic constitutions of populations, which may allow them to adapt to the new combinations of day length and temperatures that do not exist now. There are good reasons to add a genetic component to prediction models.

*Acknowledgments*
We thank the Finnish Forest Research Institute for a longtime collaboration. Tapani Repo has proved indispensable for the frost tolerance work. This research has been supported by the Research Council for Environment and Natural Resources. We are grateful to Drs. Inger Ekberg and Tapani Repo for their comments on the manuscript.

# References

Aho, M.-L. (1994) Autumn frost hardening of one-year-old *Pinus sylvestris* (L.) seedlings: Effect of origin and parent trees. *Scand. J. For. Res.* 9:17–24.

Aitken, S. and Adams, W.T. (1996) Genetics of fall and winter cold hardiness of coastal Douglas-fir in Oregon. *Can. J. For. Res* 26:1828–1837.

Aitken, S. Adams, W.T., Schermann, N. and Fuchigami, L.H. (1996) Family variation for fall cold hardiness in two Washington populations of coastal Douglas-fir [*Pseudotsuga menziesii* var. *menziesii* (Mirb.) Franco]. *For. Ecol. Manag.* 80:187–195.

Alho, P. (1990) Suomen metsittyminen jääkauden jälkeen. The history of forest development in Finland after the last Ice Age. *Silva Fenn.* 24:9–18.

Beuker, E. (1994) Long-term effects of temperature on the wood production of *Pinus sylvestris* (L). and *Picea abies* (L.) Karst in old provenance experiments. *Scand. J. For. Res.* 9:34–45.

Binnie, S.C., Grossnickle, S.C. and Roberts, D.R. (1994) Fall acclimation patterns of interior spruce seedlings and their relationship to changes in vegetative storage proteins. *Tree Physiol.* 14:1107–1120.

Bradshaw, H.D. Jr. and Stettler, R.F. (1995) Molecular genetics of growth and development in Populus. IV. Mapping QTLs with large effects on growth and phenology traits in a forest tree. *Genetics* 139:963–973.

Chung, M.S. (1981) Flowering characteristics of *Pinus sylvestris* L. with special emphasis on the reproductive adaptation to local temperature factor. *Acta For. Fenn.* 169:1–69.

Cornelius, J. (1994) Heritabilities and additive genetic coefficients of variation in forest trees. *Can. J. For. Res.* 24:372–379.

Devey, M.E., Bell, J.C., Smith, D.N., Neale, D.B. and Moran, G.F. (1996) A genetic linkage map for *Pinus radiata* based on RFLP, RAPD and microsatellite markers. *Theor. Appl. Genet.* 92:673–679.

de Vilmorin, L.H. (1863) Exposé historique et descriptif de l'Ecole forestière des Barres, Commune de Nogent-sun-Vernisson (Loiret). Soc. Agric. Fr. Mém. 1862, Pt. 1., pp. 297–353.

Dormling, I. (1993) Bud dormancy, frost hardiness and frost drought in seedlings of *Pinus sylvestris* and *Picea abies*. *In*: P.H. Li and L. Christersson (eds): *Advances in Plant Cold Hardiness*. CRC Press, Boca Raton, Florida, pp. 285–298.

Dudash, M.R. (1990) Relative fitness of selfed and outcrossed progeny in a self-compatible, protandrous species, *Sabatia angularis* L. (*Gentianaceae*): A comparison in three environments. *Evolution* 44:1129–1139.

Eiche, V. (1966) Cold damage and plant mortality in experimental provenance plantations with Scots pine in northern Sweden. *Stud. For. Suec.* 36:1–218.

Ekberg, I., Eriksson, G. and Dormling, I. (1979) Photoperiodic reactions in conifer species. *Hol. Ecol.* 2:255–263.

Ekramoddoullah, A.K.M. and Taylor, D.W. (1996) Seasonal variation of western white pine (*Pinus monticola* D. Don) foliage proteins. *Plant Cell. Physiol.* 37:189–199.

Ekramoddoullah, A.K.M., Taylor, D. and Hawkins, B.J. (1995) Characterization of a fall protein of sugar pine and detection of its homologue associated with frost hardiness of western white pine needles. *Can. J. For. Res.* 25:1137–1147.

Eriksson, G., Andersson, S., Eiche, V., Ifver, J. and Persson, A. (1980) Severity index and transfer effects on survival and volume production of *Pinus sylvestris* in northern Sweden. *Stud. For. Suec.* 156:1–31.

Fowler, D.P. (1964) Effects of inbreeding in red pine. *Pinus resinosa* Ait. *Silvae Genet.* 13:170–177.

Fryer, J.H. and Ledig, F.T. (1972) Microevolution of the photosynthetic temperature optimum in relation to the elevational complex gradient. *Can. J. Bot.* 50:1231–1233.

Groover, A., Devey, M., Fiddler, T., Lee, J., Megraw, R., Mitchel-Olds, T., Sherman, B., Vujcic, S., Williams, C. and Neale, D. (1994) Identification of quantitative trait loci influencing wood specific gravity in an outbred pedigree of loblolly pine. *Genetics* 138:1293–1300.

Gullberg, U., Yazdani, R., Rudin, D. and Ryman, N. (1985) Allozyme variation in Scots pine (*Pinus sylvestris* L.) in Sweden. *Silvae Genet.* 34:193–200.

Hagner, M. (1970) A genecological investigation of the annual rhythm of *Pinus contorta* Dougl. and a comparison with *Pinus sylvestris* L. *Stud. For. Suec.* 81:1–23.

Hamrick, J.L., Godt, M.J. and Sherman-Broyles, S.L. (1992) Factors influencing levels of genetic diversity in woody plant species. *New Forests* 6:95–124.

Hänninen, H. (1990) Modelling bud dormancy release in trees from cool and temperate regions. *Acta For. Fenn.* 213:1–47.

Hänninen, H., Häkkinen, R., Hari, P. and Koski, V. (1990) Timing of growth cessation in relation to climatic adaptation of northern woody plants. *Tree Physiol.* 6:29–39.

Harju, A. and Muona, O. (1989) Background pollination in *Pinus sylvestries* seed orchards. *Scand. J. For. Res.* 4:513–520.

Havranek, W.M. and Tranquillini, W. (1995) Physiological processes during winter dormancy and their ecological significance. *In:* W.K. Smith and T.M. Hinckley (eds): *Ecophysiology of Conifers.* Academic Press, New York, pp. 95–124.

Heikinheimo, O. (1956) Tuloksia ulkomaisten puulajien viljelystä Suomessa. *Commun. Inst. For. Fenn.* 46:1–129.

Henttonen, H., Kanninen, M., Nygren, M. and Ojansuu, R. (1986) The maturation of *Pinus sylvestries* seeds in relation to temperature climate in northern Finland. *Scand. J. For. Res.* 1:243–249.

Houde, M., Daniel, C., Lachapelle, M., Allard, F., Laliberte, S. and Sarhan, F. (1995) Immuno-localization of freezing-tolerance-associated proteins in the cytoplasm and nucleoplasm of wheat crown tissues. *Plant J.* 8:583–593.

Houle, D. (1992) Comparing evolvability and variability of quantitative traits. *Genetics* 130:195–204.

Hughes, M.A. and Dunn, M.A. (1996) The molecular biology of plant acclimation to low temperature. *J. Exp. Bot.* 47:291–305.

Hurme, P., Repo, T., Savolainen, O. and Pääkkönen, T. (1997) Climatic adaptation of bud set and frost hardiness in Scots pine (*Pinus sylvestris* L.). *Can. J. For. Res.; in press.*

Hyvärinen, H. (1987) History of forests in northern Europe since the last glaciation. *Ann. Acad. Scient. Fenn. Series A III. Geologia-Geographica* 145:7–18.

Ilvessalo, I. (1937) Perä-Pohjolan luonnon normaalien metsiköiden kasvu ja kehitys. [Growth of natural normal stands in central North-Suomi (Finland)]. *Commun. Inst. For. Fenn.* 24:1–168.

Karhu, A., Hurme, P., Karjalainen, M., Karvonen, P., Kärkkäinen, K., Neale, D. and Savolainen, O. (1996) Do molecular markers reflect patterns of differentiation in adaptive traits of conifers? *Theor. Appl. Genet.* 93:215–221.

Kärkkäinen, K. (1991) *Itsesiitos, kukintamuuntelu ja sukusiitosheikkous pohjoisissa mäntypo-pulaatioissa.* Ph.D. Dissertation, University of Oulu, Department of Genetics.

Kärkkäinen, K., Koski, V. and Savolainen, O. (1996) Geographical variation in the inbreeding depression of Scots pine. *Evolution* 50:111–119.

Karvonen, P. and Savolainen, O. (1993) Variation and inheritance of ribosomal DNA in *Pinus sylvestris* L. (Scots pine). *Heredity* 71:614–622.

Koski, V. (1970) A study of pollen dispersal as a means of gene flow. *Commun. Inst. For. Fenn.* 70:1–78.

Koski, V. and Sievänen, R. (1985) Timing of growth cessation in relation to variations in the growing season. *In:* P.M.A. Tigerstedt, P. Puttonen and V. Koski (eds): *Crop Physiology of Forest Trees.* Helsinki University Press, Helsinki, pp. 167–193.

Koski, V. and Tallqvist, R. (1978) Tuloksia monivuotisista kukinnan ja siemensadon määrän mittauksista metsäpuilla (Results of long-time measurements of the quantity of flowering and seed crop of forest trees). *Folia For.* 364:1–60.

Langlet, O. (1936) Studier över tallens fysiologiska variabilitet och dess samband med klimatet. Ett bidrag till kännedomen om tallens ekotyper (Studien über die physiologische Variabilität der Kiefer und deren Zusammenhang mit dem Klima. Beiträge zur Kenntnis der Ökotypen von *Pinus silvestris* L.). *Medd. Statens Skogsförsöksanst.* 4:1–470.

Langlet, O. (1971) Two hundred years genecology. *Taxon* 20:653–722.

Lankinen, P. (1976) *Drosophila virilis – ryhmän lajien malaattidehydrogenaasilokusten muun-telusta.* Ph.D. dissertation, University of Helsinki.

Leborgne, N., Teulieres, C., Cauvin, B., Travert, S. and Boudet, A.M. (1995) Carbohydrate content of *Eucalyptus gunnii* leaves along on annual cycle in the field and during induced frost-hardening in controlled conditions. *Trees* 10:86–93.

Li, P. and Adams, W.T. (1993) Genetic control of bud phenology in pole-size trees and seedlings of coastal douglas-fir. *Can. J. For. Res.* 23:1043–1051.

Li, P. and Adams, W.T. (1994) Genetic variation in cambial phenology of coastal Douglas-fir. *Can. J. For. Res.* 24:1864–1870.

Lönnroth, E. (1925) Untersuchungen über die innere Struktur und Entwicklung gleichaltiger naturnormaler Kiefernbestände basiert auf Material aus der Südhälfte Finnland. *Acta For. Fenn.* 30(1):1–269.

Loopstra, C.A. and Adams, W.T. (1989) Patterns of variation in first-year seedling traits within and among Douglas-fir breeding zones in Southwest Oregon. *Silvae Genet.* 38:5–6.

Luomajoki, A. (1993) Climatic adaptation of Scots pine (*Pinus sylvestris* L.) in Finland based on male flowering phenology. *Acta For. Fenn.* 237:1–27.

MacDonald, G. Edwards, T.W.D., Moser, K.A., Pienitz, R. and Smol, J.P. (1993) Rapid response of treeline vegetation to past climate warming. *Nature* 361:243–246.

McKay, T.C. and Langley, C.H. (1991) Molecular and phenotypic variation in the achaete-scute region of *Drosophila melanogaster*. Nature 348:64–66.

Mikola, J. (1982) Bud-set phenology as an indicator of climatic adaptation of Scots pine in Finland. *Silva Fenn.* 16:178–184.

Mirov, N.T. (1967) *The Genus Pinus*. The Ronald Press Company, New York.

Mosseler, A., Egger, K.N. and Hugher, G.A. (1992) Low levels of genetic diversity in red pine confirmed by random amplified polymorphic DNA markers. *Can. J. For. Res.* 22:1332–1337.

Muona, O. and Harju, A. (1989) Effective population sizes, genetic variability and mating system in natural stands and seed orchards of *Pinus sylvestris*. *Silvae Genet.* 38:221–228.

Muona, O. and Lumme, J. (1981) Geographical variation in the reproductive cycle and photoperiodic diapause of *Drosophila phalerata* and *D. transversa (Drosophilidae:* Diptera). *Evolution* 35:158–167.

Nilsson, J.-E. (1995) Genetic variation in the natural pollen cloud of *Pinus sylvestris*: A study based on progeny testing. *Scand. J. For. Res.* 10:140–148.

Nilsson, J.-E. and Walfridsson, E.A. (1995) Phenological variation among plus-tree clones of *Pinus sylvestries* (L). in northern Sweden. *Silvae Genet.* 44:20–28.

Norell, L., Eriksson, G., Ekberg, I. and Dormling, I. (1986) Inheritance of autumn frost hardiness in *Pinus sylvestris* L. seedlings. *Theor. Appl. Genet.* 72:440–448.

Notivol, E. and Alia, R. (1996) Variation and adaptation of *Pinus sylvestris* L. provenances in Spain. Abstracts. Diversity and adaptation in forest ecosystems in a changing world. Vancouver, Canada, 5–9 August, 1996.

Parker, W.H., van Niejenhuis, A. and Charrette, P. (1994) Adaptive variation in *Picea mariana* from northwestern Ontario determined by short-term common environment tests. *Can. J. For. Res.* 24:1653–1661.

Persson, B. (1994) Effects of provenance transfer on survival in nine experimental series with *Pinus sylvestries* (L.) in northern Sweden. *Scand. J. For. Res.* 9:275–287.

Rehfeldt, G.E. (1979) Ecotypic differentiation in populations of *Pinus monticola* in North Idaho: Myth or reality? *Am. Nat.* 114:627–636.

Rehfeldt, G.E. (1982) Differentation of *Larix occidentalis* populations from the northern Rocky Mountains. *Silvae Genet.* 31:13–19.

Rehfeldt, G.E. (1989) Ecological adaptation in Douglas-fir (*Pseudotsuga menziesii* var. *glauca*): A synthesis. *For. Ecol. Manag.* 28:203–215.

Rehfeldt, G.E. (1990) Genetic differentiation among populations of *Pinus ponderosa* from the upper Colorado river basin. *Bot. Gaz.* 151:125–137.

Rehfeldt, G.E. (1995) Genetic variation, climate models and the ecological genetics of *Larix occidentalis*. *For. Ecol. Manag.* 78:21–37.

Rousi, M. (1983) The thriving of the seed orchard progenies of northern Finland at Kittilä. *Folia For.* 547:1–14.

Ruotsalainen, S., Nikkanen, T. and Haapanen, M. (1995) Effect of seed-maturing conditions on the growth and hardines of one-year-old *Pinus sylvestries* seedling. *Forest Genetics* 2:189–198.

Sakai, A. (1983) Comparative study on freezing resistance of conifers with specical reference to cold adaptation and its evolutive aspects. *Can. J. Bot.* 61:2323–2332.

Sakai, A. and Larcher, W. (1987) *Frost Survival of Plants: Responses and Adaptation to Freezing Conditions*. Ecological Studies 62. Springer-Verlag, Berlin.

Sarvas, R. (1962) The development of the tree species composition of the forests of southern Finland during the past two thousand years. *Commun. Inst. For. Fenn.* 55:1–14.

Sarvas, R. (1972) Investigations on the annual cycle of development of forest trees: Active period. *Commun. Inst. For. Fenn.* 76:1–110.

Savolainen, O. (1996) Pines beyond the polar circle: Adaptation to stress conditions. *Euphytica* 92:139–145.

Skroppa, T. and Magnussen, S. (1993) Provenance variation in shoot growth components of Norway spruce. *Silvae Genet.* 42:111–120.

Sykes, M.T. and Prentice, I.C. (1995) Boreal forests futures: Modelling the controls on tree species range limits and transient responses to climate change. *Water, Air and Soil Pollution* 82:415–428.

Szmidt, A.E. and Muona, O. (1985) Genetic effects of Scots pine *Pinus sylvestries* (L.) domestication. *In*: H.-R. Gregorius (ed.): *Population Genetics in Forestry*. Lecture Notes in Biomathematics 60. Springer-Verlag, Berlin, pp. 241–252.

Taylor, M.F.J., Shen, Y. and Kreitman, M.E. (1995) A population genetic test of selection at the molecular level. *Science* 270:1497–1499.

Thomashow, M.F. (1990) Molecular genetics of cold acclimation in higher plants. *Adv. Genet.* 28:99–131.

Tigerstedt, P.M.A. (1994) Adaptation, variation and selection in marginal areas. *Euphytica* 77:171–174.

Toivonen, A., Rikala, R., Repo, T. and Smolander, H. (1991) Autumn colouration of first year *Pinus sylvestries* seedlings during frost hardening. *Scand. J. For. Res.* 6:31–39.

Vaartaja, O. (1959) Evidence of photoperiodic ecotypes in trees. *Ecol. Monogr.* 29:91–111.

Welin, B.V., Olson, Å. and Palva, E.T. (1995) Structure and organization of two closely related low-temperature-induced *dhn/lea/rab*-like genes in *Arabidopsis thaliana* L. Heynh. *Plant Mol. Biol.* 29:391–395.

Woodward, F.I. (1995) Ecophysiological controls of conifer distributions. *In*: W.K. Smith and T.M. Hinckley (eds): *Ecophysiology of Coniferous Forests*. Academic Press. pp. 79–94.

Xie, C.Y. and Ying, C.C. (1995) Genetic architecture and adaptive landscape of interior lodgepole pine (*Pinus contorta* ssp. latifolia) in Canada. *Can. J. For. Res.* 25:2010–2021.

Yang, R.-C., Yeh, F.C. and Yanchuk, A.D. (1996) A comparison of isozyme and quantitative genetic variation in *Pinus contorta* ssp. *latifolia* by $F_{ST}$. *Genetics* 142:1045–1052.

Zhao, S., Colombo, S.J. and Blumwald, E. (1995) The induction for freezing tolerance in jack pine seedlings: The role of root plasma membrane $H^+$-ATPase and redox activities. *Physiol. Plant.* 93:55–60.

# Genetic variation and environmental stress

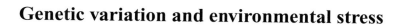

Environmental Stress, Adaptation and Evolution
ed. by R. Bijlsma and V. Loeschcke
© 1997 Birkhäuser Verlag Basel/Switzerland

# Phenotypic plasticity and fluctuating asymmetry as responses to environmental stress in the butterfly *Bicyclus anynana*

Paul M. Brakefield

*Institute of Evolutionary and Ecological Sciences, Leiden University, P.O. Box 9516, NL-2300 RA Leiden, The Netherlands*

*Summary.* Butterflies of the genus *Bicyclus* inhabiting wet-dry seasonal environments in Africa express striking seasonal polyphenism. This paper describes this example of phenotypic plasticity in the context of an evolutionary response to alternative seasons, one of which is favourable for growth and reproduction while the other is a stress environment, limiting in terms of larval growth and adult survival. The seasonal forms reflect alternative adult phenotypes which involve both morphological (wing pattern) and life history traits. The genetic and physiological coupling of these traits involves mediation by a common hormonal system. Finally, I show that the eyespot patterns on the wings of these butterflies also offer potential for studying the mechanisms of fluctuating asymmetry and its interactions with environmental stress.

## Introduction

Phenotypic plasticity occurs when variability in an environmental stimulus leads individuals of the same genotype to develop into alternative phenotypes (see Stearns, 1989, 1992). Because phenotypic plasticity can be an adaptation to variable environments, it is becoming increasingly recognised that understanding its regulation and evolution may offer general insights into the genetic and developmental bases of morphological evolution (e.g. Via, 1993; Via et al., 1995; Gotthard and Nylin, 1995; Brakefield et al., 1996; Pigliucci, 1996). However, phenotypic plasticity has been considered less frequently in the context of the evolution of responses to stress environments (see Scharloo, 1989; Harvell, 1990; Spitze, 1992).

Many of the 80 or so species of *Bicyclus* butterflies inhabit highly seasonal environments in Africa south of the Sahara. These species are characterised by a wet season adult form which develops and flies in a favourable climate and a dry season form which develops in progressively less favourable conditions before persisting as adults through a long, dry season until reproducing. The most obvious difference in wing pattern between these alternative phenotypes is the presence of large, conspicuous ventral eyespots in the wet season form and their near absence in the dry season form. In this chapter, I will review our understanding of seasonal polyphenism in the tropical butterfly, *B. anynana*, as gained from an integrated approach. This will indicate how the phenomenon can be viewed as an example of adaptive phenotypic plasticity in response to variability in environmental stress.

In addition, I will show how the eyespots of *B. anynana* provide potential for understanding mechanisms of fluctuating asymmetry in morphological traits in response to stress. Fluctuating asymmetry (FA) is a term describing the unsigned difference between the phenotypic values of characters on the left and right sides of individual organisms (Møller, this volume). When fitness depends on morphology, individuals which can develop the phenotype reliably or show greater developmental stability should be more fit (Palmer, 1996). The departure of individuals from bilateral symmetry, as measured by FA, has frequently been suggested as an appropriate index of genetic or environmental health (e. g. Leary and Allendorf, 1989) and of the effects of stress (Parsons, 1990, 1992). Experimental work shows that females of some organisms prefer to mate with more symmetric males (Watson and Thornhill, 1994) and that some pollinating insects are more likely to visit symmetric flowers (Møller and Eriksson, 1995). While the proximate cause of increased FA may frequently be environmental stress, the ability to execute developmental programmes correctly and uniformly in the face of such stress must have a genetic basis. The measurement of FA is thus an attempt to assess the ability of an individual to stabilise or canalise development to achieve the morphogenetic ideal of perfect symmetry. Studies of eyespots may provide answers to some of the questions about the mechanisms of FA and developmental stability in the face of stress environments.

## Population biology of *Bicyclus* species in seasonal environments

Many species of satyrine butterfly are adapted to the wet-dry seasonal environments of Africa (Fig. 1; Brakefield and Larsen, 1984; Brakefield, 1987). The wet season form of each of five sympatric species of *Bicyclus* in a forest edge biotope in Malawi flies actively during months with high rainfall from late December until May (Brakefield and Reitsma, 1991; Windig et al., 1994; Brakefield and Mazzotta, 1995; Roskam and Brakefield, 1996). The adult butterflies are active. They rest on, and fly amongst the luxuriant growth of herbs and grasses. There are two overlapping generations in the rains, the first arising from eggs laid by dry season form butterflies after the early rains of late November and December. Some males from the first emergence probably survive long enough to mate with females from the second. Butterflies of the second generation of the wet season form lay eggs which produce adults of the dry season form at the transition from wet to dry in May and June. These latter butterflies must survive at least 6 months before they can oviposit on freshly growing grasses at the beginning of the next wet season. Most of the duration of the dry season in Malawi is cool. The dry season form butterflies tend to rest

Figure 1. The alternative seasonal forms of the evening brown *Melanitis leda* feeding on ripe banana. The butterfly to the left is a wet season form with conspicuous eyespots. The other is of the dry season form without eyespots and resembles a dead leaf.

inactively on the carpet of dead leaf litter. Temperatures rise again several weeks before the rains begin. During this latter period the butterflies become more active, mate and mature their eggs (Brakefield and Reitsma, 1991).

Butterflies of the wet season form fly in a hot and humid period with widely available larval food plants and adult food. In similar conditions in the laboratory, females mate soon after eclosion and can begin to oviposit within 2–3 days (Kooi et al., 1997). They lay 300–400 eggs over 3 and 4 weeks (Brakefield and Kesbeke, 1995). Eggs are laid on a variety of grasses, especially *Oplismenus* spp. (Kooi et al., 1996). Pre-adult development occurs in 3–4 weeks (see Brakefield and Mazzotta, 1995). Adults feed on fallen fruit, including from *Ficus* trees (Brakefield and Kesbeke, 1995). It is thus a season of apparently freely available larval and adult resources with a favourable climate for growth and adult activity.

In contrast, adults of the dry season form develop as larvae under comparatively low temperature and humidity. They have no choice but to feed on grasses which become increasingly senescent and desiccated as larval growth progresses. Pre-adult development is much longer at about 6–8 weeks (see Brakefield and Mazzotta, 1995). After adult eclosion in a common environment in the laboratory, females of the dry season form tend to take longer to mate and commence egg laying than those of the wet season form. The dry season form adults must survive a long period of predation risk (especially from ground-browsing lizards) before egg laying can occur. There may also be periods of food shortage, although they feed opportunistically on fallen fruit (Brakefield and Reitsma, 1991). The butterflies show a dramatic development of fat bodies, which are likely to be an important energy reserve contributing to survival (Brakefield and Reitsma, 1991; Kooi et al., 1997). The female reproductive organs are inactive through the cool months.

## Phenotypic plasticity in wing pattern

Species of *Bicyclus* inhabiting wet-dry seasonal environments express striking phenotypic plasticity in wing pattern (Brakefield and Reitsma, 1991; Windig et al., 1994; Roskam and Brakefield, 1996). The wet season form flying in the rains has large, conspicuous eyespots, submarginal chevrons and, usually, a median pale band on the ventral surface of both wings. In contrast, these pattern elements are all absent or greatly reduced in the dry season form (Fig. 2, top row). These seasonal phenotypes are produced in *B. anynana* after larval development at high (ca. 23°C) or low (ca. 18°C) temperatures, respectively.

The plasticity in wing pattern is an adaptive response to the seasonal changes in climate and resting background (Brakefield and Larsen, 1984; Brakefield and Reitsma, 1991). The dry season butterflies rest on dead, brown leaf litter and are highly cryptic; any conspicuous markings on the exposed ventral wing surfaces can attract browsing lizards. In contrast, eyespots are favoured in the highly active butterflies within the green herbage layer of the wet season because they can deflect bird or lizard attacks away from the vulnerable body. Survival analyses of marked cohorts of each form in each season have supported this hypothesis (N. Reitsma unpublished data).

The theory of norms of reaction (see van Noordwijk, 1989; Stearns, 1989, 1992) provides a useful framework for examining the evolution of phenotypic plasticity. A reaction norm maps the phenotypic response, strictly speaking of a single genotype, on to different environments. We use four different rearing environments (17, 20, 23 and 27–28°C) to plot the phenotypic response of single families, stocks or selected lines of *B. anynana*. While intermediate phenotypes in most species are infrequent or rare in the field, they are produced readily when larvae are reared at intermediate temperatures in the laboratory (Roskam and Brakefield, 1996). Thus the seasonal polyphenism with discrete phenotypes associated with each season in nature probably occurs because such conditions of intermediate temperatures do not persist for long enough during periods of larval development in the field (see Brakefield and Mazzotta, 1995).

Figure 2. The seasonal polyphenism in *B. anynana* together with butterflies from artificially selected lines. All butterflies are representative females displaying their right ventral wings (as they would at rest). Specimens to the left were reared at 17°C, and to the right, at 27°C. The top row shows the dry season form (left) and wet season form as reared from the unselected laboratory stock. The middle row illustrates butterflies of the LOW selected line, and the bottom row, those of the HIGH selected line. (From Brakefield et al., 1996.)

## Genetic variation for plasticity in wing pattern

We have used artificial selection experiments to examine genetic variation for eyespot size and plasticity in a laboratory stock of *B. anynana* (Holloway et al., 1993; Holloway and Brakefield, 1995; Brakefield et al., 1996). Direct responses to selection are rapid and continuous over many generations, indicating high levels of additive variance and the influence of numerous genes of small phenotypic effect. Estimates of realised heritability are high at around 50–60%. Furthermore, rearing of selected lines over our standard temperature gradient demonstrates highly positive genetic covariances across environments; the alleles producing phenotypic effects at one rearing temperature tend to produce similar effects at other temperatures.

Upward and downward selection on ventral eyespot size, beginning at intermediate temperatures, produces selected lines with highly divergent "bundles" of the norms of reaction for individual families (Fig. 3). This selection experiment has now yielded a LOW-line which only produces the dry season form across all temperatures, while only wet season forms are reared from the HIGH-line (Fig. 2). The high level of genotype × environment interactions in our unselected stock was eroded during the course of selection. Work using a molecular marker (*Distal-less*, see below) has shown that the LOW- and HIGH-lines diverge developmentally in the early pupal stage during the final stages of determination of the eyespot pattern. Analysis of crosses between the lines show that they differ at a number of gene loci (Brakefield et al., 1996).

## Plasticity in life history traits

Any factor which influences development time, including temperature, thermoperiod (Brakefield and Mazzotta, 1995; Brakefield and Kesbeke, 1997) and food quality (Kooi et al., 1996), has an effect on adult wing pattern. Within a cohort of larvae reared in a single temperature regime, butterflies with relatively short development times tend to have larger ventral eyespots than those which develop more slowly. This relationship is also reflected at a genetic level in correlated responses of life history traits to selection on eyespot size. Thus the LOW-line tends to show relatively slow pre-adult development, while HIGH-line individuals develop more rapidly with higher growth rates. A corresponding effect of this genetic covariance between wing pattern and development time is also observed when selection is applied directly to the life history trait. FAST-lines produce butterflies with larger eyespots than SLOW-lines (Brakefield and Kesbeke, in preparation).

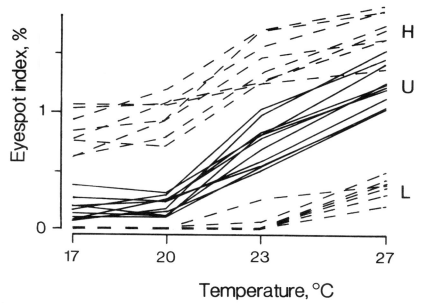

Figure 3. Sets of norms of reaction for groups of families form the LOW-line (L), the HIGH-line (H) and the unselected stock (U, solid lines). Reaction norms were measured after 20 (L) or 16 (H) generations of artificial selection on ventral eyespot size. The area of the black centre of the largest hindwing spot relative to wing area was measured by image analysis. Each reaction norm is based on four subsamples of females (n usually > 4) from a single family reared at each of the four standard temperatures. (After Brakefield et al., 1996.)

## Hormonal mediation of phenotypic plasticity

The ecdysteroid hormones are involved in mediating the phenotypic plasticity in both wing pattern and development time. Microinjections of 20-hydroxyecdysone into early pupae of the LOW-line fated to produce the dry season form can shift eyespot development in the direction of butterflies of the wet season form with larger eyespots and a pale medial band (Fig. 4). Furthermore, adult eclosion occurs earlier in the injected pupae compared with controls treated with Grace's solution (Koch et al., 1996). We have monitored the dynamics of ecdysteroid titres around the time of pupation. There is a more rapid increase in ecdysteroids following pupation in pupae fated to produce adults of the wet season form as well as in pupae from lines selected for rapid development (P. M. Brakefield, F. Kesbeke and P. B. Koch, unpublished data).

Figure 4.   The effects of ecdysteroid hormones on seasonal polyphenism in *B. anynana*. Ventral wing surfaces of butterflies of the LOW selected line. The top row shows untreated controls. The middle row illustrates the effect of injection of 0.5 µg of 20-hydroxyecdysone into 0–6-h-old pupae; the butterflies exhibit wet season characteristics. The bottom row shows the effect of control injections with Grace's medium. (From Koch et al., 1996.)

## Developmental mechanisms and symmetry in eyespots

The phenotypic plasticity of eyespot size is a response to alternative seasons, one of which can be described as a stress environment. In addition, the extent of symmetry of the eyespots on the left and right wings of individual butterflies may also be sensitive to the effects of stress during growth and development. While the proximate cause of increased variance between sides (FA) may frequently be environmental stress, the ability to execute developmental programmes correctly and uniformly in the face of such stress must have a genetic basis. The measurement of FA may enable assessment of the ability of an individual to canalise development to the morphogenetic ideal of perfect symmetry. Because we have a good understanding of the developmental pathway of butterfly eyespots, they can provide the potential for understanding the mechanisms of FA in morphological traits and its relation to environmental stress.

Butterfly wings show spectacular diversity. Their colour patterns result from a two-dimensional matrix of scale cells, each of which contains a single colour pigment. An eyespot is made up of concentric rings of scale cells with different colour pigments. *B. anynana* has a series of eyespots, each consisting of a central white pupil, an inner black disc and an outer gold ring. The L-R pairs of eyespots represent serial developmental homologues, with each eyespot based on a common developmental mechanism (Monteiro et al., 1994).

The eyespot developmental pathway can be considered to involve four phases (Nijhout, 1991; Brakefield et al., 1996). Before pupation, in the final instar larva, a pre-pattern is established in the growing wing discs which includes the location of potential eyespot "organisers" known as foci. This pre-pattern shortly before pupation becomes a series of specified foci corresponding to the centres of each adult eyespot. After pupation, the groups of cells at each focus induce a signal to surrounding cells, apparently by diffusion of a chemical morphogen. The surrounding cells – the scale cells to be – then interpret this signal to gain positional information and become fated to express specific colours. However, the synthesis of the colour pigments does not occur until some days later just before adult eclosion. The signal produced by each eyespot focus in *Bicyclus* can be thought of as a cone-shaped information gradient with the annuli of cells at increasing distance from the focus experiencing and responding in a threshold manner to the progressively lower morphogen concentrations. Strong support for this pathway comes from the ability to produce ectopic eyespots by grafting the focal cells to different positions on a developing wing (see Nijhout, 1991; French and Brakefield, 1995) and from studies of mutants, selected lines and different species using the molecular probe for the protein product of the developmental gene *Distal-less* (Brakefield et al., 1996).

The diameter of the eyespots of *B. anynana* can be measured on both left and right wings with very low measurement error. The experimental designs of Palmer (1994) can be applied to partition out measurement error to yield the most rigorous, between-sides variance, measure of fluctuating asymmetry called FA10. Brakefield and Breuker (1996) found that butterflies reared in a favourable environment from two groups of selected parents, one with high and one with low eyespot FA, did not differ in FA10. There was, therefore, no evidence for heritable variation for FA in an

Figure 5. A mutant female *B. anynana* exhibiting an extreme example of asymmetry across left and right wings in the pattern of eyespots in the posterior part of the forewing (see arrows).

outcrossed stock under favourable conditions. However, an additional study of a mutant stock showed the potential for genetic influences on FA. Figure 5 shows a butterfly which is heterozygous for the *Cyclops* mutant and homozygous for the *Spotty* mutant (see Brakefield et al., 1996). It exhibits a particularly striking example of a deviation from bilateral symmetry for the modified pattern of forewing eyespots (cf. Fig. 2). Measurements of *Spotty* heterozygotes show that an additional "mutant" eyespot specified by the *Spotty* allele expresses higher FA10 than flanking "wild-type" eyespots, perhaps because they have no history of visual selection and canalisation (Brakefield and Breuker, 1996).

## Eyespot FA and the effects of stress environments

In a further experiment, different stresses were applied at specific times during development (P.M. Brakefield, unpublished data). Larvae were from an outcrossed stock and reared outside the stress "window" in favourable conditions (healthy maize plants, 27°C and high relative humidity (RH). The stresses used were:

1) starvation: larvae starved for 48 h immediately after their final larval moult. About 75% of the larvae died during or immediately after the

stress. The surviving individuals subsequently showed no mortality as pupae (cf. controls);

2) cold shock I: larvae (as pre-pupae) given a cold shock of 3°C for 24 h beginning 1–6 h before pupation (some mortality with pupation always occurring after return to 27°C);

3) cold shock II: pupae given the same cold shock but starting 1 h after pupation (no mortality).

In addition, a cohort of butterflies (4) was reared in the favourable conditions from a line which had been selected over about 20 generations for fast pre-adult development at 27°C. Comparisons with untreated controls (5) using replicate measurements of female butterflies showed effects on adult eyespot FA only in the third group, which received a cold shock in the period of pattern determination (3; Fig. 6). For these latter butterflies, the large posterior eyespot on the dorsal surface of the forewing showed significantly increased FA. However, the small anterior eyespot showed no such effect. Wing dimensions showed more FA in both groups treated with a cold shock but not after starvation.

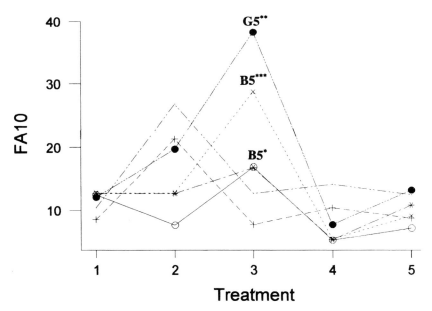

Figure 6. Effects of different stresses on the variance of eyespot size over left and right dorsal forewings of female *B. anynana* (FA10). The treatments (1–5) are detailed in the text. Statistical comparisons with the controls (5) are made by *F*-test with only the indicated test values being significant (*, $p<0.05$; **, $p<0.01$; ***, $p<0.001$). B5 = black centre of fifth eyespot (two different axes); G5 = gold outer ring of fifth eyespot (anterior-posterior axis). Not labelled: G5 (proximal-distal axis); black centre of second eyespot (two axes). Sample sizes were in treatment sequence: 32, 15, 40, 59 and 53.

These results show that eyespot symmetry can be sensitive to stress. They also illustrate how differences in the nature and timing of environmental stress can have divergent effects on FA. The outer gold ring of the posterior eyespot tends to show higher FA than the inner black region. This may follow from the expected bell-shaped curve of the information gradient: changes in concentration may be steeper around the boundary from black to gold than for the outer border of the eyespot (gold to brown). The effects of a stress may differ across eyespots because of subtle differences in the timing or sensitivity of the developmental mechanisms.

## Concluding remarks and future research

*Bicyclus* butterflies inhabiting wet-dry seasonal environments in Africa exhibit striking seasonal polyphenism. The phenotypes occurring in the two seasons reflect alternative responses, one to a favourable environment and the other to a stress environment. The phenotypic plasticity has evolved partly in response to seasonality of predation and resting background, and partly to favour growth and survival through a dry, stress environment in combination with rapid metamorphosis and reproduction in a favourable wet season. The plasticity is mediated by hormonal mechanisms. This leads to the observed genetic coupling of fundamental life history traits and the plasticity in wing pattern. In effect the species of *Bicyclus* have evolved the ability to express divergent life history forms in alternate seasons, each form involving two suites of coordinated traits.

We can explore the consequences of genetic covariances among traits for responses to (artificial) selection. This can be combined with an examination of the underlying physiological and developmental mechanisms. Future research will focus on the flexibility of the genetic covariances and physiological mechanisms underlying the phenotypic plasticity in wing pattern and life history traits. It may be very difficult, or very unlikely in evolutionary terms, to be able to uncouple the wing pattern and life history traits or to reverse the direction of genetic covariance. These relationships may constrain the ease with which *Bicyclus* can adapt to changing climate. Where there is a matching of the functional demands of the environment such that small eyespots are favoured by selection for crypsis in harness with delayed metamorphosis and physiological traits to maximise survival (e.g. fat bodies and ovarian dormancy), plasticity may evolve extremely rapidly. On the other hand, the reverse of large eyespots coupled with slow development may be most unlikely to evolve. The former situation is the type of selection regime which we believe occurs in Malawi and elsewhere in wet-dry regions of East or Central Africa. However, it may not occur throughout the seasonal areas of Africa (J.C. Roskam, personal communication).

The eyespot patterns of *Bicyclus* also provide the potential for studying the effects of environmental or genetical stresses during ontogeny on the extent of bilateral symmetry within individual butterflies. We will use additional "window" experiments with a wider variety of stresses to further explore the interactions between developmental mechanisms of eyespot formation, FA and stress. In addition, we will examine whether the seasonal forms of *B. anynana* and other species differ in developmental stability. It will be interesting to determine whether the underlying mechanisms of phenotypic plasticity and of fluctuating asymmetry lead to dry season forms which consistently exhibit higher levels of FA.

## References

Brakefield, P.M. (1987) Tropical dry and wet season polyphenism in the butterfly *Melanitis leda* (Satyridae): Phenotypic plasticity and climatic correlates. *Biol. J. Linn. Soc.* 3:175–191.

Brakefield, P.M. and Breuker, C.J. (1996) The genetical basis of fluctuating asymmetry for developmentally integrated traits in a butterfly eyespot pattern. *Proc. R. Soc. Lond. B.* 263: 1557–1563.

Brakefield, P.M. and Kesbeke, F. (1995) Raised adult lifespan and female fecundity in tropical fruit-feeding *Bicyclus* butterflies. *Proc. Exper. Appl. Entomol.* 6:93–98.

Brakefield, P.M. and Kesbeke, F. (1997) Genotype × environment interactions for insect growth in constant and fluctuating temperature regimes. *Proc. R. Soc. Lond. B.*; *in press.*

Brakefield, P.M. and Larsen, T.B. (1984) The evolutionary significance of dry and wet season forms in some tropical butterflies. *Biol. J. Linn. Soc.* 22:1–12.

Brakefield, P.M. and Mazzotta, V. (1995) Matching field and laboratory environments: Effects of neglecting daily temperature variation on insect reaction norms. *J. Evol. Biol.* 8:559–573.

Brakefield, P.M. and Reitsma, N. (1991) Phenotypic plasticity, seasonal climate and the population biology of *Bicyclus* butterflies. *Ecol. Entomol.* 16:291–303.

Brakefield, P.M., Gates, J., Keys, D., Kesbeke, F., Wijngaarden, P.J., Monteiro, A., French, V. and Carroll, S.B. (1996) Development, plasticity and evolution of butterfly eyespot patterns. *Nature* 384:236–242.

French, V. and Brakefield, P.M. (1995) Eyespot development on butterfly wings. II. The focal signal. *Dev. Biol.* 168:112–123.

Gotthard, K. and Nylin, S. (1995) Adaptive plasticity and plasticity as an adaptation: A selective review of plasticity in animal morphology and life history. *Oikos* 74:3–17.

Harvell, C.D. (1990) The ecology and evolution of inducible defences. *Q. Rev. Biol.* 65: 323–339.

Holloway, G. and Brakefield, P.M. (1995) Artificial selection of reaction norms of wing pattern elements in *Bicyclus anynana. Heredity* 74:91–99.

Holloway, G.J., Brakefield, P.M. and Kofman, S. (1993) The genetics of wing pattern elements in the polyphenic butterfly, *Bicyclus anynana. Heredity* 70:179–186.

Koch, P.B., Brakefield, P.M. and Kesbeke, F. (1996) Ecdysteroids control eyespot size and wing color pattern in the polyphenic butterfly *Bicyclus anynana. J. Insect Physiol.* 42:223–230.

Kooi, R.E., Brakefield, P.M. and Rossie, E.M.-Th. (1996) Effects of food plant on phenotypic plasticity in the tropical butterfly *Bicyclus anynana. Ent. Exper. Appl.* 80:149–151.

Kooi, R.E., Bergshoeff, C., Rossie, E.M.-Th. and Brakefield, P.M. (1997) *Bicyclus anynana* (Lepidoptera: Satyrinae): Comparison of fat content and egg-laying in relation to dry and wet season temperatures. *Proc. Exper. Appl. Entomol.* 8:17–22.

Leary, R.F. and Allendorf, F.W. (1989) Fluctuating asymmetry as an indication of stress: Implications for conservation biology. *Trends Ecol. Evol.* 4:214–217.

Møller, A.P. and Eriksson, M. (1995) Pollinator preference for symmetrical flowers and sexual selection in plants. *Oikos* 73:15–22.

Monteiro, A.F., Brakefield, P.M. and French, V. (1994) The evolutionary genetics and developmental basis of wing pattern variation in the butterfly *Bicyclus anynana. Evolution* 48:1147–1157.

Nijhout, H.F. (1991) *The Development and Evolution of Butterfly Wing Patterns.* Smithsonian Institute Press, Washington.

Palmer, A.R. (1994) Fluctuating asymmetry analyses: A primer. *In*: T. Markow (ed.): *Developmental Instability: Its Origins and Evolutionary Implications*. Kluwer Academic Publishers, The Hague, pp. 355–364.

Palmer, A.R. (1996) Waltzing with asymmetry. *BioScience* 46:518–532.

Parsons, P.A. (1990) Fluctuating asymmetry: An epigenetic measure of stress. *Biol. Rev.* 65: 131–145.

Parsons, P.A. (1992) Fluctuating asymmetry: A biological monitor of environmental and genomic stress. *Heredity* 68:361–364.

Pigliucci, M. (1996) How organisms respond to environmental changes: From phenotypes to molecules (and vica versa). *Trends Ecol. Evol.* 11:168–173.

Roskam, J.C. and Brakefield, P.M. (1996) Comparison of temperature-induced polyphenism in African *Bicyclus* butterflies from a seasonal savannah-rainforest ecotone. *Evolution* 50: 2360–2372.

Scharloo, W. (1989) Developmental and physiological aspects of reaction norms. *BioScience* 39:465–471.

Spitze, K. (1992) Predator-mediated plasticity of prey life history and morphology: *Chaoborus americanus* predation on *Daphnia pulex*. *Am. Nat.* 139:229–247.

Stearns, S.C. (1989) The evolutionary significance of reaction norms. *BioScience* 39:436–446.

Stearns, S.C. (1992) *The Evolution of Life Histories*. Oxford University Press, Oxford.

van Noordwijk, A.J. (1989) Reaction norms in genetical ecology. *BioScience* 39:453–459.

Via, S. (1993) Adaptive phenotypic plasticity: Target or by-product of selection in a variable environment? *Am. Nat.* 142:352–366.

Via, S., Gomulkiewicz, R., de Jong, G., Scheiner, S.M., Schlichting, C.D. and van Tienderen, P.H. (1995) Adaptive phenotypic plasticity: Consensus and controversy. *Trends Ecol. Evol.* 10:212–217.

Watson, P.J. and Thornhill, R. (1994) Fluctuating asymmetry and sexual selection. *Trends Ecol. Evol.* 9:21–25.

Windig, J.J., Brakefield, P.M., Reitsma, N. and Wilson, J.G.M. (1994) Seasonal polyphenism in the wild: Survey of wing patterns of five species of *Bicyclus* butterflies in Malawi. *Ecol. Entomol.* 19:285–298.

Environmental Stress, Adaptation and Evolution
ed. by R. Bijlsma and V. Loeschcke
© 1997 Birkhäuser Verlag Basel/Switzerland

# Environmental stress and the expression of genetic variation

Nicole L. Jenkins, Carla M. Sgrò and Ary A. Hoffmann

*School of Genetics and Human Variation, La Trobe University, Bundoora, Victoria 3083, Australia*

*Summary.* We have started to test the effects of environmental extremes on the expression of genetic variation for traits likely to be under selection in natural populations. We have shown that field heritability may be high for stress response traits in contrast to morphological traits, which tend to show lower levels of heritable variation in nature compared with the laboratory. Selection for increased stress resistance can lead to a number of other evolutionary changes, and these may underlie trade-offs between favourable and stressful environments. Temperature extremes can have a marked influence on the heritability of life history traits. Heritabilities for fecundity can be high when parental flies are reared at low temperatures and under field conditions. The expression of genetic variation for development time is somewhat more complex when temperature extremes are considered. Populations at species margins may be ideal for studying the effects of environmental stress on evolution.

## Introduction

Animals often experience environmental changes that lead to large reductions in their fitness. The continued monitoring of natural populations of animals for longer periods of time is revealing that most populations experience at least sporadically stressful conditions. Although populations may appear relatively stable over shorter periods spanning a few years, most change drastically in size when studied over long intervals (Pimm, 1991). Stressful conditions in nature may regularly arise from the direct effects of sudden climatic changes, such as involving the processes of drought or flooding, or the effects of extreme hot or cold temperatures. These environmental changes can also have many indirect effects, influencing competitive interactions, predator-prey interactions and susceptibility to diseases.

There is little doubt that stressful conditions have had a large impact on the evolution of organisms. Many animals possess characteristics that enable them to deal with stressful conditions. Some species migrate long distances to evade seasonally extreme conditions, others evade extreme stress by entering an inactive phase, as in the case of hibernation in mammals and diapause in insects. Animals have also evolved many mechanisms for coping with sudden environmental shifts; for instance, rapid changes at the cellular level involving the synthesis of enzymes and heat shock proteins occur in response to most stresses (Lindquist and Craig, 1988; Feder and Krebs, this volume).

Stressful conditions have traditionally been considered to act only as selective agents, favouring some genetic variants and selecting against others within a population. Natural selection is associated with biotic and abiotic factors in the environment acting upon phenotypic variation present in populations. The usual situation depicted in textbooks is that of genotypes giving rise to phenotypes, which are in turn subjected to selection (e.g. Futuyma, 1986; Ridley, 1993). In the traditional view of evolution by natural selection, variation arises by chance. A particular inherited characteristic might increase fitness in a new environment because individuals with this characteristic leave more progeny. As a consequence, it will increase in frequency in a population as adaptation occurs to the new environment. In this scenario, the environment does not influence the extent to which a characteristic is inherited, or the likelihood that a new characteristic is produced.

An alternative view is that the stressful conditions themselves can somehow trigger evolutionary changes. Extreme conditions can influence both the expression of phenotypic variability and the extent to which it is genetically determined (Jablonka and Lamb, 1995; Parsons, 1996). Waddington (1957) proposed that extreme conditions can lead to the disruption of normal development, leading in turn to the appearance of characteristics not evident when organisms develop under normal conditions. For instance, in *Drosophila melanogaster* stress can produce abnormal morphological characteristics when individuals are exposed to stresses early in their development; these abnormalities can become expressed under non-stressful conditions once selection has been undertaken for several generations (Waddington, 1953, 1956; Milkman, 1960). The production of such novelties could provide a source of variation upon which natural selection can act. However, the extreme conditions used to generate these mutant phenotypes are likely to be absent or rate under field conditions, and the morphological peculiarities produced, such as ultrabithorax in *Drosophila*, are unlikely to survive in nature.

Nevertheless, these findings raise the question of whether environmental conditions play a more direct role in determining the evolution of commonly studied traits, such as life history character, physiological traits and size-related traits. Perhaps major evolutionary shifts in these characters require novel phenotypic shifts that only occur when environments are stressful. The expression of many life history traits appears to be buffered against minor changes in environmental conditions, particularly when life history characters are closely related to fitness (Stearns and Kawecki, 1994; Neyfakh and Hartl, 1993). Stressful conditions in nature might enhance the expression of genetic variation in such traits, allowing them to evolve more rapidly when the development of organisms is less well buffered.

In this chapter, we briefly describe experimental *Drosophila* data that bear on these questions. First, we consider the field heritability of traits likely to be under selection. If environmental conditions in nature are stressful, and if these alter the expression of genetic variation in traits, then heritabilities in nature will not necessarily be smaller than in the laboratory, even though this is commonly assumed. Second, we summarize recent data from our laboratory on responses to selection for increased stress resistance. In particular, we consider whether increased resistance generally leads to a correlated set of changes in other traits. Finally, we discuss the effects of environmental extremes on the expression of genetic variation in life history traits. We ask if temperature extremes experienced in one generation impact on the expression of genetic variation in subsequent generations.

If these results are to be related to natural populations, we need to start studying populations that are likely to experience repeatedly stressful conditions. One area where these populations might exist is at the margins of species distributions, often thought to be defined by some form of environmental stress (Hoffmann and Blows, 1994). In the final section of this paper, we outline our approach for placing the previous work in the context of stressful conditions experienced by natural populations. By studying evolutionary processes at species borders, we hope to test the impact of environmental stress on evolution by natural selection.

## Field heritabilities are not always lower than laboratory estimates

Given the difficulty of following individuals of most species, including *Drosophila*, across generations in the field, an alternative is to estimate field heritability by comparing field-caught parents and their offspring raised in the laboratory (Prout, 1958; Riska et al., 1989). If there are no genotype-environment interactions, and the additive genetic variance is the same in the laboratory and the field, the measured heritability is equivalent to the actual heritability in nature (Lande, in Coyne and Beechman, 1987). Riska et al. (1989) extended Lande's approach to show how field heritability could be obtained even when additive genetic variances differed in the two environments.

There are two reasons why heritabilities might be lower in the field than in the laboratory. First, by inflating the overall phenotypic variance, increased environmental variance in the field will result in lower heritability than that measured under controlled laboratory conditions (Falconer and Mackay, 1996). Second, fitness-related traits may exhibit low heritabilities as a consequence of directional selection (Fisher, 1930; Mousseau and Roff, 1987; Falconer and Mackay, 1996). This effect of selection is more likely to be expressed under the conditions where selection took place (i.e. field conditions) rather than in a novel laboratory environment.

In *Drosophila*, field heritability estimates of wing length and abdominal bristle number (Coyne and Beecham, 1987), thorax length (Prout and Barker, 1989; Ruiz et al., 1991), wing length (Hoffmann, 1991) and courtship song characters (Aspi and Hoikkala, 1993) have suggested that heritability does tend to be lower in nature compared with the laboratory. This was also found for wing length in the milkweed bug, *Oncopeltus fasciatus* (Groeters and Dingle, 1996). Maternal effects have not been reported in any of these studies.

In contrast to such results, Jenkins and Hoffmann (1994) obtained a high estimate of field heritability for a measure of heat resistance in *D. simulans*. Heritability estimates tended to be larger in the field than in the laboratory, and large maternal genetic effects were evident under field conditions but not in the laboratory. The knockdown measure of heat resistance used in these experiments is not necessarily correlated to mortality due to heat stress (Hoffmann et al., 1997). Nevertheless, it may have more relevance as a measure of natural heat resistance than mortality, as incapacitation may lead to many potential fitness disadvantages, such as an increased risk of predation.

Field heritability has also been estimated by using full-sib families of the field cricket, *Gryllus pennsylvanicus*, raised in the laboratory or in field cages (Simons and Roff, 1994). In this study heritability in a number of predominantly morphological traits such as femur length was lower in the more variable field environment. This reduction was due to both increased environmental variance and a reduction in genetic variance. In contrast, the heritability of development time was not significantly different between the two environments. This trait also exhibited the lowest genetic correlation across environments, suggesting that it may be influenced by different genes in the different environments.

We have recently examined the heritability of early fecundity in *D. melanogaster* collected from the field (Sgrò and Hoffmann, 1998). Flies were obtained from a pile of rotting fruit in an orchard near Melbourne in three seasons: summer, autumn and spring. This allowed us to examine whether levels of heritable variation changed with different parental temperatures. Field parents were collected as virgins directly from the fruit pile soon after eclosion. The temperature of the heap was measured at places where larval activity was found: temperatures were relatively constant, and around 30°C in summer, compared with 17°C in autumn. In the spring collection, the fruit pile had largely decomposed, so greater temperature fluctuations were experienced by the larvae during this season. Fecundity of field flies was measured over 10 days at 14°C for flies collected in autumn and spring and at 28°C for flies collected in summer. Offspring from each collection were tested at both of these temperatures as well as at 25°C.

The regression analyses (Tab. 1) indicate that heritability estimates for fecundity in field-collected adults varied between collections. Estimates were positive in summer, and particularly large (84%) when progeny were reared at 25°C. However, heritability estimates were negative for parents collected in spring and tested at 14°C. The negative regressions were

Table 1. Heritability estimates for dam-daughter comparisons for fecundity in flies collected from the field and their offspring raised and tested under three lab temperatures. Probabilities are for two tailed t-tests testing if regression coefficients for dam-daughter comparisons are significantly different from zero. Heritabilities were computed from standardized coefficients if phenotypic variances between environments influenced estimates

| Origin of field parents | Laboratory testing temperature (°C) | N | $h^2 \pm SE$ | p |
|---|---|---|---|---|
| Summer (tested at 28°C) | 28 | 105 | $0.41 \pm 0.19$ | 0.016 |
| | 25 | 109 | $0.84 \pm 0.20$ | <0.001 |
| | 14 | 107 | $0.08 \pm 0.04$ | 0.016 |
| Autumn (tested at 14°C) | 28 | 128 | $-0.67 \pm 0.54$ | 0.219 |
| | 25 | 121 | $-0.43 \pm 0.47$ | 0.473 |
| | 14 | 135 | $0.38 \pm 0.10$ | <0.001 |
| Spring (tested at 14°C) | 28 | 196 | $-0.16 \pm 0.14$ | 0.262 |
| | 25 | 198 | $-0.34 \pm 0.14$ | 0.017 |
| | 14 | 203 | $-0.32 \pm 0.14$ | 0.012 |
| Lab estimates* | 28 | 147 | $0.47 \pm 0.16$ | <0.001 |
| | 25 | 172 | $0.12 \pm 0.14$ | 0.655 |
| | 14 | 160 | $0.49 \pm 0.15$ | <0.001 |

* Daughter-granddaughter comparison from the spring collection where both were tested at lab temperature.

significant for two of the offspring temperatures (25 and 14°C) indicating that parents with high fecundities produced offspring with relatively low fecundities. In the laboratory, the narrow-sense heritability estimates were intermediate when flies were raised at extremes, but low when they were raised under optimal conditions. These findings indicate that heritabilities for fecundity in field-collected flies can vary markedly between collections and even change in sign. They also show that heritabilities are not necessarily lower in the field and that they depend on environmental conditions experienced in both parent and offspring generations.

Together the results for heat resistance and fecundity in *Drosophila* and development time in crickets suggest that heritabilities under field conditions are not necessarily lower in the field than in the laboratory. Findings based on morphological traits do not necessarily apply to the heritability of fitness-related traits in natural populations. Additional work is required on the field heritability of such traits to determine the generality of this finding. In particular, the influence of parental conditions on genetic variance needs further consideration.

## Selection for stress resistance can alter many other traits

When stressful periods occur, traits involved in stress responses are expected to be under selection. As these traits change, they may also alter other traits correlated with them that are not apparently under direct selection. Consequently, stressful periods may have indirect effects on a range of traits. Laboratory selection experiments may be a useful method of determining the potential for complex evolutionary responses under stress. In addition, any underlying physiological basis to the selection response in the laboratory may allow extrapolation of these studies to potential responses and correlations in field populations.

The most extensive set of experiments in our laboratory has been undertaken for desiccation stress in *D. melanogaster* and *D. simulans* (Hoffmann and Parsons, 1989, 1993a, b). In these experiments, lines were selected for increased resistance to desiccation, and scored for a number of potentially correlated traits. The response to selection was generally rapid, reflecting a high realized heritability for this trait. For instance, in *D. melanogaster* the realized heritability was estimated to be approximately 50–65%, although the response was lower in *D. simulans*, at around 20–30% (Hoffmann and Parsons, 1993b).

Selected *D. melanogaster* lines also exhibited increased resistance, relative to the control lines, to starvation, heat, radiation, ethanol and acetic acid stress (Tab. 2). A decrease in metabolic rate, early fecundity and early behavioural activity, as well as an increase in male longevity was also observed in the selected lines. It was postulated that the generalized increase in stress resistance and the changes in the life history traits may be the result of the lower metabolic rate in the lines selected for increased desiccation resistance.

Correlated responses have also been described for other *Drosophila* species. In *D. simulans*, increased desiccation resistance is associated with increased resistance to starvation and toxic ethanol (Tab. 2). As in the

Table 2. Summary of correlated responses in lines of different *Drosophila* species selected for increased desiccation resistance. Compiled from data in Hoffmann and Parsons (1989, 1993) and from Blows and Hoffmann (1993)

| Trait | *D. melanogaster* | *D. simulans* | *D. serrata* |
|---|---|---|---|
| basal metabolic rate | ↓ | – | ↓* |
| starvation resistance | ↑ | ↑ | ↑ |
| heat resistance | ↑ | – | – |
| intense γ-radiation resistance | ↑ | – | nc |
| toxic ethanol exposure | ↑ | ↑ | – |
| water loss rate | ↓ | ↓ | – |
| body size | nc | nc | nc |
| lipid content | nc | – | nc |
| early fecundity | ↓ | – | – |
| early behavioural activity | ↓ | ↓ | – |
| male longevity | ↑ | – | – |

↑ = increase, ↓ = decrease, nc = no change, – = not done.
* In *D. serrata*, there was a correlated response for metabolic rate only under dry conditions.

case of *D. melanogaster*, the selection response may be associated with metabolic rate because selected lines have decreased behavioural activity. Rates of water loss are also lower in both species. Selection for desiccation resistance in another species, *D. serrata*, also resulted in increased resistance to starvation, although there was no increase in radiation resistance (Blows and Hoffmann, 1993). In *D. serrata*, a lower metabolic rate was only evident under desiccating conditions, suggesting that life history trade-offs evident for *D. melanogaster* may not occur in this species.

Indirect effects of selection have been observed for other stresses. Watson and Hoffmann (1996) selected both *D. melanogaster* and *D. simulans* for increased cold resistance. Flies were selected with and without prior hardening to increase cold resistance. Cross-generational effects were detected in all lines, but lines were eventually produced with diverse responses to cold stress. Comparisons of selected and control lines indicated marked effects of selection on early fecundity in both *D. melanogaster* and *D. simulans*. Selected lines in both species showed reduced fecundities which were demonstrated to be specifically associated with the selection response. This is shown in Figure 1, which presents the mean fecundities of $F_1$s generated from crosses between the independent replicate selection lines or replicate control lines. By scoring fecundities in these crosses (rather than the lines themselves), we can control for any inbreeding that might have taken place while the lines were maintained. The results indicate that crosses between control lines resulted in females with relatively higher fecundities than those made between the selection lines, regardless of whether hardening was a component of the selection regime.

Other workers using *Drosophila* have also detected an association between stress resistance and life history traits. For instance, Chippindale et al. (1996) reported a strong selection response, indicative of significant genetic variation, following 60 generations of selection for increased starvation resistance in *D. melanogaster*. There was a correlated increase in development time and a decrease in viability in the selected lines relative to their controls. These correlations were suggested to have arisen as a consequence of increased resource sequestering in early life stages.

However, not all responses to stressful conditions are associated with correlated changes in life history traits. For instance, selection for increased knockdown resistance to heat in *D. melanogaster* led to lines which were two- to fivefold more resistant to this stress (Hoffmann et al., 1997). This selection response did not alter life history traits, and the response was highly specific to the measure of heat resistance that was used. The selection response was associated with allelic changes in two heat shock genes (McColl et al., 1996), suggesting that such genes may have specific effects on stress resistance and do not necessarily have life history costs.

In summary, our experiments and those of other *Drosophila* workers show that increased stress resistance in *Drosophila* may often be associated with changes in other traits, including those likely to influence fitness

### D. melanogaster

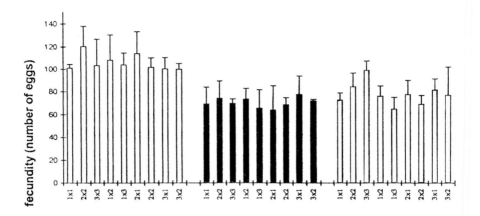

### D. simulans

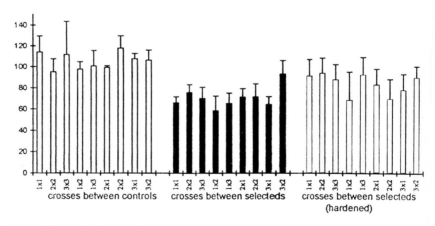

Figure 1. Effects of selection for increased cold resistance on early fecundity in *D. melano-gaster*. The graph shows the results of crosses among replicate selection or control lines (indicated as 1, 2 and 3). Crosses were undertaken to control for potential inbreeding effects on correlated responses. Error bars are standard deviations. Lines had been selected for increased cold resistance following preliminary cold hardening or without prior hardening. (After Watson and Hoffmann, 1996.)

under favourable conditons. Changes in energy reserved and/or metabolic rate may underlie many of these associations.

These findings are important for several reasons. First, they show that responses to stress can have a number of secondary effects unrelated to the stress. This means that organisms under continuously stressful conditions may be expected to show life history changes (particularly decreased early fecundity) that reduce their fitness under more favourable conditions. This could form the basis of a trade-off between fitness in different environments. Second, the correlated responses suggest that trade-offs in life history traits may be complicated by survival under stressful conditions. For instance, much of the discussion on life history trade-offs concerns that between early reproduction and late reproduction or longevity. Because longevity is correlated with survival under stressful conditions, the life history trade-off could involve stress resistance rather than longevity of late reproduction, particularly for organisms exposed to repeated stressful conditions (Parsons, 1995).

Many trait values seen in natural populations may reflect the effects of infrequent stressful conditions. Once environmental variability is present, fitness estimates need to be averaged across environments, and the mean fitness of a genotype is its geometric mean fitness rather than its arithmetic mean fitness (Cohen, 1966; Gillespie, 1973). As a consequence, averaging fitness measures will overestimate the overall fitness across environments, and the fitness of a genotype or phenotype in unfavourable conditions can have a particularly large influence on its overall fitness. An individual that fails to survive an infrequent stress will have an overall fitness of zero regardless of how well it performs under normal conditions.

## The expression of genetic variation can increase under extremes

Because organisms normally experience environmental conditions that are continually changing, parents and their progeny will often experience different conditions. This may consequently influence the rate of evolution of traits in natural environments. Increases in the environmental variance will lead to a decrease in the similarity of parents and their progeny, and therefore decrease the rate at which a trait responds to selection, while a reduction in the mean of a trait will decrease the extent to which a trait is modified by selection (see Houle, 1992). The additive genetic variance in a trait may depend on environmental conditions, influencing its heritability and the extent to which it changes under selection. Finally, similarity between parents and progeny may be altered if the genes that contribute to variation in a trait in one environment do not contribute in another environment.

Most studies on the effects of environmental conditions on heritable components are usually concerned with genotype-environment interactions and measure the same or related genotypes in different environments (e.g. Scheiner and Lyman, 1991; Gebhardt and Stearns, 1992; Thomas and Barker, 1993; Simons and Roff, 1994; Via and Conner, 1995). The approach commonly followed in these experiments is to expose different strains or different members of the same family to a range of conditions which do not generally encompass extremes. The effects of parental conditions on the expression of genetic variation in progeny are not considered in such studies.

To examine cross-generation effects on the expression of genetic variation, we have considered the effects of temperature extremes on genetic and environmental variances for two life history traits (development time and fecundity) and one morphological trait (wing length) in D. *melanogaster* (Sgrò and Hoffmann, 1997). The effects of fluctuating conditions were considered directly by comparing parental and offspring generations exposed to the same or different conditons, rather than by considering only genotypes tested in the same generation across different conditions. This allows us to answer the following questions:

(1) Does the expression of genetic variation in the traits change with environmental extremes?
(2) Is the expression of genetic variation in progeny influenced by parental conditions?
(3) Is there a strong positive genetic correlation between environments?

Parents were reared and tested for development time and fecundity (over 10 days), at either 14 or 28°C. These temperatures represent extremes for rearing D. *melanogaster* in the laboratory. The offspring of both groups of parents were raised and tested at three temperatures (14, 25 and 28°C). This generated combinations of parental and progeny temperatures (14–14, 14–25, 14–28 etc.). Sample sizes were around 100 families for each temperature combination. Simple parent-offspring regressions were used to determine heritabilities when parents and offspring are measured in the same environment.

Interpretations are more complex when parents and progeny are reared in different environments. In this case, the midparent regression for offspring in environment 1 and parents in environment 2 provides an estimate (Riska et al., 1989) of

$$\beta_{(O1,\ P2)} = \frac{\gamma \sigma_{A1} \sigma_{A2}}{\sigma_{P2}^2},$$

where $\gamma$ is the additive genetic correlation between the trait in the two environments, $\sigma_{A1}^2$ and $\sigma_{A2}^2$ are the additive genetic variances in environments 1 and 2, and $\sigma_{P2}^2$ is the phenotypic variance in environment 2. The regression across environments therefore cannot separate additive genetic effects from correlations between environments. In addition, regression across environments can lead to misleading heritability estimates when the variances in the parent and progeny generations differ (Ward, 1994). For instance, if heritabilities are the same in the parental and offspring environments, but the phenotypic and additive variances are greater in the offspring environment, the regression coefficient will be inflated by the larger estimate for $\sigma_{A1}^2$. To overcome this problem, standardized regression coefficients were computed to determine whether heritabilities differ between environments and/or genetic correlations between environments are less than 1 (Sokal and Rohlf, 1981). These coefficients can also be used to test whether parental conditions influence the expression of genetic variation.

For fecundity, the regression analyses indicated large and highly significant within- and across-environment coefficients when parents were exposed to 14°C regardless of the offspring temperature (Tab. 3). In contrast, regression coefficients were smaller and tended to be non-significant when parents were reared at 28°C. We also estimated the evolvability measure ($I_A$) recommended by Houle (1992), given as $I_A = V_A/\bar{x}^2$. This measures the extent to which a trait will be altered by directional selection. Evolvability estimates were relatively higher when parents were raised at 14°C (Tab. 3), suggesting that the proportional increase in fecundity under selection would be greater at 14°C than at 28°C.

Whilst standardization of fecundity scores reduced differences between the coefficients (Tab. 3), those for the 28–25 and 28–14 comparisons were significantly lower than coefficients for comparisons involving 14°C parents. This suggests that the parental environment, rather than simply the fluctuating extremes themselves, influenced the expression of genetic variation in the progeny environment. As coefficients were similar for all comparisons involving 14°C parents, genetic correlations between environments are probably close to 1.0 when parents are raised at 14°C.

For development time, regression coefficients were negative and significant in the 14–14 comparison (Tab. 3), suggesting that parents with slow development times produced daughters and sons with relatively fast developmental times. When both parents and progeny were reared at the extreme high temperature of 28°C, regressions were significant and positive, producing heritability estimates of 0.46–0.64. Estimates of evolvability were lower for development time than for fecundity, and negative in the case of the 14–14 comparison, indicating slower rates of evolutionary change. Regressions were particularly large in some comparisons across temperatures, presumably because of different phenotypic variances at 14, 25 and 28°C. To overcome this problem, we can consider the standardized regression coefficients. These were similar for the 28–28 and 28–14 comparisons, but lower for the 28–25 comparison. When parents were reared at 14°C, standardized coefficients were significantly larger when offspring were reared at the opposite temperature extreme, suggesting differences in the additive genetic variances and/or genetic correlations in the across-environment comparisons.

Table 3. Dam-daughter comparisons for fecundity and development time. Regression coefficients (b) and standardized regression coefficients ($b_s$) are given

| Temperature (°C) | | Fecundity | | | Development time | | |
|---|---|---|---|---|---|---|---|
| Parents | Offspring | $b \pm SE$ | $b_s \pm SE$ | Evolvability | $b \pm SE$ | $b_s \pm SE$ | Evolvability |
| 14 | 28 | $0.75 \pm 0.24$** | $0.29 \pm 0.09$** | 17.13 | $0.49 \pm 0.05$*** | $0.72 \pm 0.07$*** | 0.07 |
| 14 | 25 | $1.07 \pm 0.19$*** | $0.46 \pm 0.09$*** | 19.31 | $0.04 \pm 0.08$ | $0.04 \pm 0.09$ | 0.003 |
| 14 | 14 | $0.19 \pm 0.05$*** | $0.33 \pm 0.08$*** | 12.70 | $-0.43 \pm 0.18$* | $-0.22 \pm 0.09$* | $-0.04$ |
| 28 | 28 | $0.11 \pm 0.05$* | $0.23 \pm 0.10$* | 2.69 | $0.42 \pm 0.12$*** | $0.32 \pm 0.10$*** | 0.03 |
| 28 | 25 | $-0.07 \pm 0.05$ | $-0.14 \pm 0.10$ | $-1.59$ | $0.01 \pm 0.01$ | $0.04 \pm 0.10$ | 0.01 |
| 28 | 14 | $0.02 \pm 0.02$ | $0.07 \pm 0.10$ | 0.62 | $4.57 \pm 0.69$*** | $0.56 \pm 0.09$*** | 0.02 |

* $p < 0.05$, ** $p < 0.01$, *** $p < 0.001$.

In contrast to these complex results for the life history traits, genetic variation for wing length was generally not affected by the temperature extremes. Genetic correlations across the extremes estimated from the parent-offspring comparison were close to 1 for this trait. Overall, the expression of genetic variation in life history traits therefore seemed to be influenced substantially by environmental temperature, whereas the expression of morphological traits was not.

These results indicate that responses of life history traits to selection may depend on environmental conditions in different ways. On the one hand, fecundity will respond to selection more rapidly when parents experience 14°C, and estimates of the genetic correlation across environments indicate that any change in fecundity obtained by selection will be expressed across all temperatures. The reason for parental effects on heritable variation is unknown. On the other hand, any genetic changes in development time will not necessarily apply across environments. Particularly striking is the unexpected negative covariance when both generations were exposed to 14°C. Our results suggest a complex interaction between gene expression and environmental conditions in the case of this life history trait. Perhaps parental exposure to 14°C acts as an "environmental switch" which causes progeny that are predisposed to develop rapidly to curtail their development. The evolvability estimates indicate fluctuating temperature extremes could lead to rapid responses to selection on development time.

Overall, the data suggest that heritabilities for life history traits will be high under some temperature extremes and low under others, whereas those for morphological traits will be relatively more constant. This contrasts with the general notion that morphological traits have higher heritabilities than life history traits (Roff and Mousseau, 1987; Mousseau and Roff, 1987), and instead emphasizes the effect of the environment (both of the parents and progeny) on the expression of heritable variation.

## Species borders provide opportunities to study stressful effects in natural populations

So far, we have discussed that there may be significant genetic variation in natural populations for both stress resistance and life history traits. Under controlled laboratory conditions, selection for increased resistance can lead to a host of evolutionary responses. The expression of genetic variation in life history traits may also be enhanced under stressful conditions. In order to determine the applicability of such results to the evolution of natural populations, we are interested in natural populations that are likely to often encounter stress, such as those at the limit of a species distribution.

In the absence of obvious geographic limits such as mountains, the border populations at the limit of a species distribution are generally referred to as marginal. Ecologically marginal populations have been defined as those that experience environmental stress, have a fluctuating population size and a high probability of extinction (Soulé, 1973). Factors affecting the distribution and abundance of species are often unknown. The environmental conditions just beyond the species border often do not appear to be markedly different. Why don't populations at the edge of the species range evolve and expand into these areas?

Understanding why an organism occupies a particular range could allow prediction of responses to environmental change. When a species occupies a distinct geographic range without an obvious physical reason, distributions are often assumed to be limited by some form of stress, such as climatic, competition or predation stress, or a combination of different stresses. An ecological approach to studying borders is used to determine relevant environmental factors and traits that constrain the species distribution. A complementary evolutionary approach is used to determine what prevents those traits from evolving at the genetic level.

Most ecological studies of species have tended to focus on local populations rather than deal specifically with species borders. A comparison of species with both similar and differing distributions, combined with comparisons of central and marginal populations within species, may yield information about the stresses limiting species (Hoffmann and Blows, 1994). Computer simulation programs, such as CLIMEX and BIOCLIM, have been designed to help predict likely climatic influences on the abundance and distribution of organisms. We have combined such ecological approaches with genetic approaches to assess important factors in limiting the distribution of D. serrata, a species that appears to be limited to the east coast in Australia and is also found in Papua New Guinea.

Extensive field collections along the east coast of Australia, by many researchers over a number of years, indicate that there may be large seasonal fluctuations in the southern distribution and abundance of D. serrata. Post-winter collections suggest a southern border at Seal Rocks, approximately 300 km north of Sydney, whereas late summer and autumn collections yield D. serrata in Wollongong, approximately 70 km south of Sydney. Failure to collect this species south of Seal Rocks after winter may be due to the southern population becoming seasonally extinct or simply occurring at such low numbers as to be undetectable by normal collection methods.

CLIMEX analysis using the field collection data combined with laboratory data on the performance of D. serrata under extreme conditions indicates that the southern border may be related to cold stress, whilst the western border is likely to be limited by desiccation stress. This is consistent with findings for other animals with distributions similar to D. serrata (e.g. Law, 1994). Comparisons of D. serrata with other drosophilid species with differing distributions including D. melanogaster, D. simulans, D. immigrans and D. birchii for cold and desiccation resistance (Hoffmann, 1992; N.L. Jenkins, unpublished data) are also consistent with the importance of these stresses in limiting the distribution of *D. serrata*. We are currently concentrating on the potential influence of cold stress on the southern border.

The expectation of traits important in determining species borders is that means will differ between central and marginal populations, and that genetic variation may also differ between these populations. In *D. serrata* high levels of cold resistance are found post-winter at marginal sites, relative to more northern populations, but not in autumn (N.L. Jenkins, unpublished data).

In evolutionary terms, there have been a number of hypotheses proposed to explain species borders. It is thought that lack of expansion into adjacent areas may occur due to low overall levels of genetic variation in marginal populations, low heritability of relevant traits due to directional selection or to high environmental variability in marginal environments. Other hypotheses proposed include favoured genotypes occurring too rarely, and the possibility of genetic trade-offs between fitness in favourable and stressful environments (with the latter occurring less frequently) or trade-offs between fitness traits, such as fecundity and stress resistance, constraining evolution of these traits in marginal conditions (Hoffmann and Parsons, 1991; Hoffmann and Blows, 1994).

Data from selection for desiccation resistance in *D. serrata* suggested low levels of genetic variability in marginal populations compared with that in central populations (Blows and Hoffmann, 1993). A limitation of this study in terms of its application to natural populations is that it was based on selection of populations after many laboratory generations. Also, whilst the levels of genetic variation were lowest in the marginal population, the initial population means were not dissimilar. This result may be indicative of a general trend for stress response traits. Subsequent fieldwork and computer simulations have indicated that cold is implicated more strongly for the southern border; consequently estimates of genetic variation for cold resistance were obtained for border and more central populations.

Response to cold appears to be a threshold trait in *D. serrata*, whereby individuals either survive indefinitely or die within 48 h after a cold shock, regardless of the length or severity of the cold shock. This precludes the use of simple parent-offspring regressions to determine the heritability of cold resistance; however, the length of time taken to die varies, with this also reflecting the fitness of flies subsequent to the cold stress. Flies can therefore be given a rank according to whether they have survived the cold shock and, if not, how long they survived. Field females and males were collected from the southern border population (Forster), and sites 100 km (Taree), 300 km (Coffs Harbour) and 380 km (Grafton) north of the border. Progeny were obtained from field females prior to a cold stress of 60 min at $-2°C$, and the first and second laboratory generations also tested for their cold tolerance. Males were crossed to virgin females from a labora-

Table 4. Spearman's rank correlation for cold resistance in field flies and their laboratory-reared progeny

| Comparison | Females | | Males | |
| --- | --- | --- | --- | --- |
| | $n$ | Correlation | $n$ | Correlation |
| Forster | 29 | 0.34* | 55 | 0.11 |
| Taree | 35 | 0.20 | 54 | 0.05 |
| Coffs Harbour | 31 | 0.26 | 56 | −0.10 |
| Grafton | 34 | 0.29* | 67 | 0.07 |

* $p < 0.05$.

tory stock to generate laboratory generations. Table 4 depicts the Spearman's rank correlation for field female and male flies and their progeny. Due to differing responses between the sexes, comparisons were not made between individuals of different sexes. From these results it can be seen that, contrary to expectations, there is genetic variation within the border population (Forster) for the cold resistance of females. Moreover, there is no evidence that levels of genetic variability are relatively lower in this population compared with three other populations north of the border. However, the data for males are not significant despite being based on a relatively larger number of individuals. Perhaps a low level of heritable variation in male D. serrata precludes further range expansion in this species.

Overall levels of genetic variation do not appear to be limiting the expansion of D. serrata. Using flies reared under laboratory conditions, we have found that the border population exhibits significant differences between isofemale lines for cold resistance, development time and viability at different temperatures. There is also no evidence for consistent trade-offs between cold resistance and performance in optimal or stressful environments, as measured by development time and viability. However, there is a significant association between survival after cold shock and low productivity in field females, suggesting that there may be a trade-off between fecundity and cold resistance, such as that found by Watson and Hoffmann (1996) for D. melanogaster and D. simulans (see above). Further investigation is required to determine if this trade-off is limiting the distribution of D. serrata.

## Future directions

From the preceding discussion it is clear that studies of genetic variation in optimal environments, which comprise the bulk of previous work, are not necessarily indicative of the situation in natural populations. When organisms are pushed by extreme conditions, the patterns of genetic variation may often be more complex than those observed under constant laboratory conditions. Environmental extremes can result in higher levels of genetic variation, and the parental environment may also have a significant effect. Long-term monitoring of natural populations is needed to assess the impact of seasonal stress on genetic variation in nature. The effect of stressful conditions fluctuating within generations also remains to be investigated. Finally, genetic studies in marginal areas are needed to assess the influence of extreme conditions on rates of evolutionary change.

*Acknowledgments*
We thank Stuart Barker, Shane McEvey and Mark Blows for field collection data, and Kuke Bijlsma, Volker Loeschcke and Gawain McColl for comments on the manuscript. Our research is supported by grants from the Australian Research Council.

# References

Aspi, J. and Hoikkala, A. (1993) Laboratory and natural heritabilities of male courtship song characters in *Drosophila montana* and *D. littoralis*. *Heredity* 70:400–406.

Blows, M.H. and Hoffmann, A.A. (1993) The genetics of central and marginal populations of *Drosophila serrata*. I. Genetic variation for stress resistance and species borders. *Evolution* 47(4):1261–1270.

Chippindale, A.K., Chu, T.J.F. and Rose, M.R. (1996) Complex trade-offs and the evolution of starvation resistance in *Drosophila melanogaster*. *Evolution* 50(2):753–766.

Cohen, D. (1966) Optimizing reproduction in a randomly varying environment. *Journal of Theoretical Biology* 12:119–129.

Coyne, J.A. and Beecham, E. (1987) Heritability of two morphological characters within and among natural populations of *Drosophila melanogaster*. *Genetics* 117:727–737.

Falconer, D.S. and Mackay, T.F.C. (1996) *Introduction to Quantitative Genetics*, Fourth Edition Longman, London.

Fisher, R.A. (1930) *The Genetical Theory of Natural Selection*. Clarendon Press, Oxford.

Futuyma, D.J. (1986) *Evolutionary Biology*, Second Edition, Sinauer, Sunderland.

Gebhardt, M.D. and Stearns, S.C. (1992) Phenotypic plasticity for life-history traits in *Drosophila melanogaster*. III. Effect of the environment on genetic parameters. *Genet. Res.* 60:87–101.

Gillespie, J. (1973) Polymorphism in random environments. *Theor. Pop. Biol.* 4:193–195.

Groeters, F.R. and Dingle, H. (1996) Heritability of wing length in nature for the milkweed bug, *Oncopeltus fasciatus*. *Evolution* 50(1):442–447.

Hoffmann, A.A. (1991) Heritable variation for territorial success in field-collected *Drosophila melanogaster*. *Am. Nat.* 138:668–679.

Hoffmann, A.A. and Blows, M.W. (1994) Species borders: Ecological and evolutionary perspectives. *Trends Ecol. and Evol.* 9:223–227.

Hoffmann, A.A. and Parsons, P.A. (1989) Selection for increased desiccation resistance in *Drosophila melanogaster*: Additive genetic control and correlated responses for other stresses. *Genetics* 122:837–845.

Hoffmann, A.A.and Parsons, P.A. (1991) *Evolutionary Genetics and Environmental Stress*. Oxford University Press, Oxford.

Hoffmann, A.A. and Parsons, P.A. (1993a) Selection for adult desiccation resistance in *Drosophila melanogaster*: Fitness components, larval resistance and stress correlations. *Biol. J. Linn. Soc.* 48:43–54.

Hoffmann, A.A. and Parsons, P.A. (1993b) Direct and correlated responses to selection for desiccation resistance: A comparison of *Drosophila melanogaster* and *D. simulans*. *J. Evol. Biol.* 6:643–657.

Hoffmann, A.A., Dagher, H., Hercus, M. and Berrigan, D. (1997) Comparing different measures of heat resistance in selected lines of *Drosophila melanogaster*. *J. Insect Physiol.* 43(4):393–405.

Houle, D. (1992) Comparing evolvability and variability of quantitative traits. *Genetics* 130:185–204.

Jablonka, E. and Lamb, M. (1995) *Epigenetic Inheritance and Evolution.* Oxford University Press, Oxford.

Jenkins, N.L. and Hoffmann, A.A. (1994) Genetic and maternal variation for heat resistance in *Drosophila* from the field. *Genetics* 137:783–789.

Law, B.S. (1994) Climatic limitation of the southern distribution of the common blossom bat *Syconycteris australis* in New South Wales. *Australian Journal of Ecology* 19:366–374.

Lindquist, S. and Craig, E.A. (1988) The heat-shock proteins. *Ann. Rev. Genet.* 22:631–677.

McColl, G., Hoffmann, A.A., McKechnie, S.W. (1996) Response of two heat shock genes to selection for knockdown heat resistance in *Drosophila melanogaster. Genetics* 143:1615–1627.

Milkman, R.D. (1960) The genetic basis of natural variation. II. Analysis of polygenic systems in *Drosophila melanogaster. Genetics* 45:377–391.

Mousseau, T.A. and Roff, D.A. (1987) Natural selection and the heritability of fitness components. *Heredity* 59:181–197.

Neyfakh, A.A. and Hartl, D.L. (1993) Genetic control of the rate of embryonic development: Selection for faster development at elevated temperatures. *Evolution* 47:1625–1631.

Parsons, P.A. (1995) Inherited stress resistance and longevity: A stress theory of ageing. *Heredity* 85:216–221.

Parsons, P.A. (1996) Stress, resources, energy balances and evolutionary change. *Evol. Biol.* 29:39–72.

Pimm, S.L. (1991) *The Balance of Nature? Ecological Issues in the Conservation of Species and Communities.* University of Chicago Press, Chicago.

Prout, T. (1958) A possible difference in genetic variances between wild and laboratory populations. *Drosophila Information Service* 32:148–149.

Prout, T. and Barker, J.S.F. (1989) Ecological aspects of the heritability of body size in *Drosophila buzzatii. Genetics* 123:803–813.

Ridley, M. (1993) *Evolution.* Blackwell, Boston.

Riska, B., Prout, T. and Turelli, M. (1989) Laboratory estimates of heritability and genetic correlations in nature. *Genetics* 123:865–871.

Roff, D.A. and Mousseau, T.A. (1987) Quantitative genetics and fitness: Lessons from *Drosophila. Heredity* 58:103–118.

Ruiz, A, Santos, M., Barbadilla, A., Quezada-Diaz, E., Hasson, E. and Fontdevila, A. (1991) Genetic variation for body size in a natural population of *Drosophila buzzatii. Genetics* 128:739–750.

Scheiner, S.M. and Lyman, R.F. (1991) The genetics of phenotypic plasticity. II. Response to selection. *J. Evol. Biol.* 4:23–50.

Simons, A.M. and Roff, D.A. (1994) The effect of environmental variability on the heritabilities of traits of a field cricket. *Evolution* 48:1637–1649.

Sgrò, C.M. and Hoffmann, A.A. (1997) Effect of temperature extremes on genetic variances for life history traits in *Drosophila melanogaster* as determined from parent-offspring comparisons. *J. Evol. Biol.; in press.*

Sgrò, C.M. and Hoffmann, A.A. (1998) Heritable variation for fecundity in field-collected *D. melanogaster* and their offspring reared under different environmental temperatures. *Evolution; in press.*

Sokal, R.R. and Rohlf, F.J. (1981) *Biometry,* Second Edition W.H. Freeman, New York.

Soulé, M. (1973) The epistasis cycle: A theory of marginal populations. *Ann. Rev. Ecol. Syst.* 4:165–187.

Stearns, S.C. and Kawecki, T.J. (1994) Fitness sensitivity and the canalization of life-history traits. *Evolution* 48:1438–1450.

Thomas, R.H. and Barker, J.S.F. (1993) Quantitative genetic analysis of body size and shape of *Drosophila buzzatii. Theor. App. Genet.* 85:598–608.

Via, S. and Conner, J. (1995) Evolution in heterogeneous environments: Genetic variability within and across different grains in *Tribolium castaneum. Heredity* 74:80–90.

Waddington, C.H. (1953) The genetic assimilation of an acquired character. *Evolution* 7:118–126.

Waddington, C.H. (1956) Genetic assimilation of the bithorax phenotype. *Evolution* 10:1–13.

Waddington, C.H. (1957) *The Strategy of the Genes.* Georg Allen & Unwin, London.

Ward, P.J. (1994) Parent offspring regression and extreme environments. *Heredity* 74:80–90.

Watson, M.J.O. and Hoffmann, A.A. (1996) Acclimation, cross-generation effects and the response to selection for increased cold resistance in *Drosophila. Evolution* 50:1182–1192

Environmental Stress, Adaptation and Evolution
ed. by R. Bijlsma and V. Loeschcke

# Worldwide latitudinal clines for the alcohol dehydrogenase polymorphism in *Drosophila melanogaster*: What is the unit of selection?

Wilke van Delden and Albert Kamping

*Department of Genetics, University of Groningen, Kerklaan 30, 9751 NN Haren, The Netherlands*

*Summary.* Geographical clines may reflect the action of natural selection on genetic polymorphisms. In *Drosophila melanogaster* several latitudinal clines occur for many characters like allozymes, inversions and quantitative traits. The identical nature of these clines on the various continents, both on the Northern and Southern Hemispheres strongly suggests adaptation to specific stress factors. The alcohol dehydrogenase (*Adh*) polymorphism shows high frequencies of the *S* allele in tropical regions and declines with latitude. The reasons for this cline are difficult to determine because of the entanglement with other polymorphisms varying with latitude.

In this paper the tentative connections with other polymorphisms like *alpha-Gpdh*, *In(2L)t*, body size and development time are reviewed with respect to the possible environmental stress factors involved. It is concluded, also from recent experiments, that the *(2L)t* inversion plays a dominant role in resistance to high temperature and is partly responsible for the *Adh* cline. Further research is aimed at the specific selective forces acting on *Adh*, focussing on the physiological and life history aspects.

## Introduction

The presence of genetic variation is a prerequisite for adaptation of a population to newly emerging stress factors, either biotic or abiotic. But even in constant and stable environments genetic polymorphisms may be maintained due to some form of balancing selection like overdominance or frequency-dependent selection. Heterozygosity *per se* has been associated with high fitness and the ability of a population or species to resist environmental stress and to occupy larger and more diversified habitats (Mitton and Grant, 1984). The mere presence of genetic variation, or the existence of genetic differences among populations, however, does not prove the action of selection, because the variants may be selectively neutral, and genetic drift may be involved in causing genetic differentiation. Geographical variation is thus no proof of selection. Still, the extent of geographic variation is tremendous, and genetic variants are often found to be associated with environmental stress factors (Mayr, 1963; Endler, 1973). Especially in the case of geographic clines which are repeatedly correlated with environmental variables, selection may be supposed (Hoffmann and Parsons, 1991). Clines may occur over relatively small distances, like those associated with altitude. The classical studies on plant ecotypes of Clausen and coworkers (Clausen et al., 1940) have unambiguously demonstrated the selective nature of such clines. Also, large-scale clines are numerous, and among them latitudinal clines take a dominant position. Many morphological traits vary with latitude and do so in an identical way on several continents. This repeated occurrence has led to the formulation of several ecogeographical rules like Allen and Bergmann's rules (see review in Mayr, 1963). Although the adaptive nature of such latitudinal clines is generally implied, the exact nature both at the genetic and physiological levels is often obscure. This certainly holds for the many latitudinal clines observed in several *Drosophila* species (Lemeunier et al., 1986; Sperlich and Pfriem, 1986; Lemeunier and Aulard, 1992). Such clines have been found for biochemical, physiological, morphological, behavioural and life history traits. In many cases underlying genetic variation has been demonstrated.

In this chapter we will deal with allozyme variation, a form of biochemical variation revealed by protein electrophoresis, and more specifically focus on the alcohol dehydrogenase polymorphism in *D. melanogaster*. It appears that most organisms show high levels of allozyme variation (reviews in Nevo et al., 1984; Hamrick and Godt, 1990). However, the adaptive significance of these allozyme polymorphisms was questioned by authors who proposed that this kind of molecular variation was for the greater part selectively neutral and maintained by mutation and genetic drift (Kimura, 1968; Kimura, 1983). The neutralist view that allozyme variation was merely biochemical noise was strongly opposed by others who claimed that allozyme polymorphisms were modelled by natural selection (see Lewontin, 1974).

In the resulting neutralist-selectionist debate several, often ingenious, attempts have been made to decide about the nature, either neutral or selective, of allozyme polymorphisms. Initially, many efforts involved rather indirect statististical approaches in which, for example, distributions of allele frequencies or heterozygosities derived from natural populations were compared with the expectations derived from the neutral theory. Due to uncertainties with respect to some essential parameters involved in such tests, like effective population size and gene flow, the outcomes were often ambiguous.

A more straightforward approach departs from the biochemical properties of the variants at a particular enzyme locus (Clarke, 1975). When allozyme variants differ in properties like enzyme activity, enzyme stability, pH optimum or substrate specificity, these genotypic differences may lead to fitness differences under appropriate conditions. Such a functional approach allows *a priori* predictions about genotypic fitness differences which then can be tested experimentally. This causal analysis of the genotype-phenotype relation has been successfully applied to several allozyme polymorphisms in a number of organisms, varying from *Drosophila* to fish species, which provided ample evidence for selection under relevant environmental conditions (Powers et al., 1991). The picture emerging from

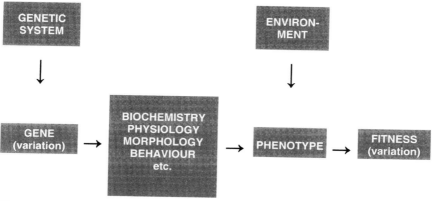

Figure 1. General scheme for the analysis of the relation between genetic variation and stress tolerance. Depending on the available knowledge, investigations may either depart from the DNA or the phenotype/fitness level.

such well-analysed cases is that under a range of conditions selection is weak or absent, but that under particular – generally stressful – conditions, considerable selection pressures may occur. Such analyses demand profound knowledge of the metabolic pathways involved, the associated physiological processes, and the phenotypic and fitness effects. It should be clear that, when properly carried out, these studies will both provide essential information about the question of the presumed neutrality of enzyme variation and also contribute to the understanding of the pathway from DNA to fitness in adaptation processes including reaction to stress (Fig. 1).

The alcohol dehydrogenase (*Adh*) polymorphism in *D. melanogaster* is perhaps the most intensively studied allozyme polymorphism. In the following sections we will review this polymorphism in the light of its role in adaptation to stress factors while focussing on its geographic distribution pattern. We will more explicitly deal with several complications, e.g. those arising from interactions with other loci and with chromosomal inversions.

## The *Adh* polymorphism

Populations of *D. melanogaster* from various parts of the world are nearly always found to be polymorphic for two common electrophoretic alleles $Adh^F$ and $Adh^S$, though some other alleles may occasionally be found at low frequencies (summary in Chambers, 1988). The geographic distribution of *Adh* is characterised by latitudinal clines, which have been found for both Northern and Southern Hemispheres. Invariably the frequency of $Adh^S$ is high at the equator and declines with latitude. This clinal behaviour has been found in North America and Australasia (Parkash and Shamina, 1994;

Bubli et al., 1996), Europe and Africa (David et al., 1989; Bénassi and Veuille, 1995; Bubli et al., 1996) and most recently in Central and South America (Van 't Land et al., 1993; Van 't Land et al., 1995; Van 't Land, 1997). The latter populations, for example, showed high $Adh^S$ frequencies, varying from 0.85–0.96 for Panamanian populations (latitudinal range from 7°45'N to 9°19'N) and even as high as 0.99 in Ecuador (latitude 2°13'S), while $Adh^S$ frequencies varied clinally from 0.34–0.07 over a long North-South transect in Chile (latitudinal range from 18°28'S to 41°30'S). Such manifold repeated clines strongly point to the occurrence of selection (Endler, 1977). With respect to putative selective agents Van 't Land (1997) found positive correlations of $Adh^S$ frequency with average annual temperature, average monthly minimum temperature, average monthly maximum temperature and average of total annual rainfall. Studies of clines in other continents showed comparable correlations (see Van Delden, 1982).

It may be asked if the functional approach outlined above provides an explanation for the $Adh$ cline. $D.$ $melanogaster$ ADH oxidises quite a variety of primary and secondary alcohols to aldehyde and ketones, respectively. ADH-FF (from $Adh^{FF}$ homozygotes) differs from ADH-SS (from $Adh^{SS}$) in a number of biochemical properties (summaries in Van Delden, 1982; Chambers, 1988, 1991; Heinstra, 1993). Characteristic for $D.$ $melanogaster$ ADH is that ADH-FF has greater ADH $in$ $vitro$ enzyme activity than does ADH-SS. And though $in$ $vitro$ measurements may deviate from those $in$ $vivo$ (see Heinstra, 1993), it has been convincingly shown that $Adh^{FF}$ genotypes survive longer on toxic concentrations of ethanol (and other alcohols) than do $Adh^{SS}$ genotypes, while $Adh^{SF}$ heterozygotes (with ADH activities intermediate between those of $AdH^{FF}$ and $Adh^{SS}$) have intermediate survival.

Compared with most $Drosophila$ species, $D.$ $melanogaster$ has in general a high resistance to alcohol stress. One reason for the high stress tolerance is that $D.$ $melanogaster$ ADH has a peculiar dual function: it oxidises ethanol to aldehyde, but it is also responsible for the immediate conversion of this highly toxic intermediate in acetate (at least in larvae (Heinstra et al., 1983)). The sibling species $D.$ $simulans$ lacks this ability and is consequently much more susceptible to ethanol. As many $Drosophila$ species oviposit on decaying fruit, where, due to yeast action, ethanol is produced, the efficiency of ethanol conversion and consequently varying detoxifying properties may determine to which extent a particular habitat can be exploited. In the case of $D.$ $melanogaster$ and $D.$ $simulans$, the latter species does not occur in sites with high ethanol concentrations, such as wineries. The differential survival of $Adh$ genotypes in $D.$ $melanogaster$ is probably the reason for small-scale geographic variation in $Adh$-allele frequencies, as sometimes observed between wineries and surrounding areas (summary in Van Delden, 1982)), where the $Adh^F$ frequency was higher in wineries.

It is not easy to see how the increase in $Adh^F$ frequency with latitude could be attributed to selection for more alcohol-resistant genotypes. The ecology of natural populations of *D. melanogaster* is not sufficiently elaborated to provide evidence of higher ethanol concentrations in relevant sites in temperate compared with tropical regions. In this respect it is interesting to note that in the Mediterranean regions of the West Coast of North America, of Eastern Australia and of Europe a particular trend in *Adh* frequencies is observed. In these areas there is a much steeper and non-linear decline in $Adh^S$ frequency than in non-Mediterranean regions. This "Mediterranean instability" (David and Capy, 1988; David et al., 1989) has recently also been observed along the East Coast of South America (Van 't Land, 1997).

In addition to *in vitro* ADH activity differences among *Adh* genotypes, *in vitro* differences in enzyme stability at high temperature also exist (reviewed in Van Delden, 1982; Chambers, 1988). ADH-SS possesses greater *in vitro* thermal stability than do ADH-FF and ADH-FS. It is tempting to relate the high $Adh^S$ frequency in tropical areas to higher *in vitro* ADH temperature stability. In this respect it has been reported that $Adh^S$ frequencies decline with altitude in mountainous regions (Grossman et al., 1970; Pipkin et al., 1976). In laboratory experiments Johnson and Powell (1974) found that when adult flies were subjected to heat shocks, the $Adh^S$ frequency was increased among the survivors compared with controls in some of the strains tested. However, Van Delden and Kamping (1980) observed that the higher *in vitro* thermal stability of ADH-SS was not correlated with higher survival at high temperature. They found that in all genotypes exposed to 35°C for 24 h, ADH activity did not decrease, and concluded that the *in vivo* results did not follow the *in vitro* results. Concerning genotypic differencies in survival at 35°C, it was in fact found that $Adh^{FS}$ survived more often than both homozygotes. This suggests overdominance under high-temperature conditions, which is in agreement with earlier observations (Van Delden et al., 1978) that polymorphic populations kept at 29.8°C for a period of 3 years showed a slight initial increase in $Adh^S$ frequency in comparison with populations kept at 25 °C. At both temperatures, however, allele frequencies stabilised at what appeared to be equilibrium values. In experiments in which both the *Adh* and $\alpha$-glycerophosphate dehydrogenase ($\alpha Gpdh$) polymorphisms were studied with respect to adult survival at high temperature, Oudman et al. (1992) also found that $Adh^{SF}$ flies survived more often than both homozygotes, while no effect of the $\alpha Gpdh$ genotype was observed.

## Associations with other polymorphisms

A further potential complication arising in the interpretation of the latitudinal *Adh* cline is that other allozyme polymorphisms also show latitudinal clines (see Bubli et al., 1996 for references). Here we will limit ourselves to the $\alpha Gpdh$ locus, which is also located on the second chromo-

some (map position 2–20.5). Several reports describe latitudinal clines for
$\alpha Gpdh$ on various continents (references in Van Delden and Kamping,
1989; Oudman, 1993; Van 't Land, 1997) in which $\alpha Gpdh^F$ frequencies
decrease with latitude. In fact, $Adh^S$ and $\alpha Gpdh^F$ frequencies appear to be
positively correlated over latitude with high frequencies of these alleles in
the tropics. Cavener and Clegg (1981) suggested that both loci are functionally linked through the NAD$^+$/NADH ratio, which may lead to fitness
interactions. Oudman (1993) studied both loci in combination and determined the two-locus genotypic effects for characters like developmental
time, adult weight, protein content, triglyceride content and ADH and
$\alpha$GPDH enzyme activity at three different temperatures. It was found
(Oudman et al., 1991) that several characters were affected by both $Adh$ and
$\alpha Gpdh$ genotypes, and that epistatic interactions occurred. In an additional
detailed study of survival at 35°C of flies which had been cultured as larvae
at either 20, 25 or 29°C, Oudman et al. (1992) found significant effects of
the $Adh$ genotype, though not for the $\alpha Gpdh$ genotype, together with
significant interaction between both loci. It was remarkable that $Adh^{SS}$
$\alpha Gpdh^{FF}$ was among the less heat-resistant genotypes, as this is a highly
common genotype in tropical areas. The contribution of the observed
$Adh$-$\alpha Gpdh$ interaction in heat resistance to the worldwide cline for these
polymorphisms is not easily seen. However, traits other than heat resistance
may also be influenced by $Adh$ and $\alpha Gpdh$. If such traits have selective
optima that differ with latitude, corresponding clines for these allozymes
could occur.

In $D.$ $melanogaster$ there is, apart from allozymes, an extensive list of
other, genetically based, morphological and physiological characters varying with latitude and systematically differing between tropical and temperate regions (review in Lemeunier et al., 1986). Among these traits body
size and pre-adult development time especially seem to be important life-
history traits in $D.$ $melanogaster$. Body size, and associated traits, increase
with latitude (David et al., 1977; James et al., 1995), whereas development
time decreases. The reasons for these apparently adaptive phenomena are
unclear. James and Partridge (1995) suggest that lower temperatures are in
an unexplained way permissive of higher growth efficiency.

With respect to the latitudinal cline for $Adh$, it is remarkable that $Adh$
genotypes, in addition to various biochemical properties, also vary in
development time (references in Oudman et al., 1991) such that $Adh^{FF}$ has
a shorter pre-adult development time than does $Adh^{SS}$, while $Adh^{FS}$ is intermediate. Van Delden and Kamping (1979) concluded that the differences
in development time were due to genotypic differences in the time needed
to reach critical weight for pupation. This gave pronounced differences in
larval survival under crowded conditions, with higher survival of $Adh^{FF}$ and
$Adh^{FS}$ and consequently a decline in $Adh^S$ frequency. Similar differences
were found for $\alpha Gpdh$; in this case the development time of $\alpha Gpdh^{FF}$
generally exceeded that of $\alpha Gpdh^{SS}$ whereas, on average, $\alpha Gpdh^{FS}$ was

intermediate (Oudman et al., 1991 and references herein). Analysis of variance (ANOVA) with temperature, *Adh* and *αGpdh* genotype as main factors provided, in addition to significant main effects, significant interaction terms for *Adh* × *αGpdh* interaction and for interaction of both loci with temperature. The *Adh* × *αGpdh* interaction mainly concerned the relative position of the heterozygotes: in some cases they were equal to one homozygote, in others to the other homozygote. The relation of *Adh* genotype with body weight is less consistent than with development time, though there is a tendency for lower weight in $Adh^{SS}$ compared with $Adh^{FF}$. The overall longer development time found in tropical regions compared with temperate regions is thus paralleled by a relative high frequency of $Adh^{SS}$ $αGpdh^{FF}$, the genotype with the longest development time.

When focussing on the *Adh* polymorphism, the question arises, why do *Adh* genotypes differ in an important life-history trait like development time? In addition to its detoxifying role in the conversion of ethanol in toxic concentrations, ADH can also be seen as an important link in the intermediary metabolism of ethanol degradation and lead to the synthesis of highly useful metabolic products like lipids, glutamate, pyruvate or trehalose (see Heinstra, 1993, for a review). In addition it was found that ADH may also play a role in carbohydrate metabolism and is involved in the pathway leading from glucose to lipid synthesis (Geer et al., 1985; Heinstra, 1993). Biochemical differences among *Adh* genotypes may thus lead to fitness differences, even under less stressful conditions. This is confirmed by the experiments of Van Delden and Kamping (1988), who demonstrated the rather rapid loss of an *Adh*-null allele from populations polymorphic for *Adh*-null and *Adh*-positive alleles on regular food. Bijlsma-Meeles and Bijlsma (1988) have performed the elaborate procedure prescribed by Prout's (1971) method for estimating the relative fitness of *Adh* genotypes under normal laboratory conditions without ethanol stress. They found fitness differences for two adult fitness components: overdominance for female fecundity and lower male virility for $Adh^{SS}$ compared with $Adh^{FF}$ and $Adh^{SF}$. The resulting predictions of allele frequency changes were confirmed by laboratory experiments in which populations with perturbated allele frequencies were followed over time. All frequencies in these populations, as in the ones previously described by Van Delden et al. (1978) converged to equilibrium values, close to the frequency in the original population. Kreitman and Hudson (1991) compared the DNA sequences of 11 *D. melanogaster* lines for the *Adh* region and surrounding regions for segregating sites and found excess silent polymorphism around position 140, the site of the amino acid replacement difference between $Adh^S$ and $Adh^F$. They concluded that this finding supported the hypothesis of a balanced polymorphism at or near this site. In a previous paper Kreitman (1983) had already shown that the frequency of amino acid polymorphisms was much lower than would be expected on the level of polymorphic silent nucleotide sites, suggesting strong selective constraints.

The molecular findings and the outcome of the fitness and perturbation tests in the laboratory point to overdominance as the mechanism for maintaining the *Adh* polymorphism in nature. The relative fitness proportions of the homozygotes could then vary with latitude and provide different equilibrium values. The reason for the overall heterozygote superiority stays unclear.

## Association with *In(2L)t*

In addition to the latitudinal clines for allozymes and quantitative traits, a cline for the cosmopolitain second chromosome inversion (*2L)t* 22D3-E1; 34A8-9 also exists (see references in Van Delden and Kamping, 1989, 1991). *In(2L)t* frequencies may be as high as 0.50 in tropical regions and are less than 0.01 at high latitudes. In a recent screening of populations in Panama, Ecuador and Chile, Van 't Land (1997) confirmed the *In(2L)t* cline but also showed considerable frequency variation among tropical sites only slightly varying in latitude, suggesting additional microscale variation. The $\alpha Gpdh$ locus is included in the chromosomal region containing *In(2L)t*, while *Adh* lies just outside. The inversion is nearly always associated with the $Adh^S$ and $\alpha Gpdh^F$ alleles. Van Delden and Kamping (1989) showed that *In(2L)t* frequencies were modified by temperature both in a tropical greenhouse and in laboratory populations. At 29°C or higher temperatures *In(2L)t* frequencies in these populations, polymorphic for *In(2L)t*, *Adh* and $\alpha Gpdh$, increased to values up to 0.40, whereas at lower temperatures frequencies generally decreased. Not unexpectedly in view of the association with *In(2L)t*, $Adh^S \alpha Gpdh^F$ frequencies covaried with *In(2L)t* and, for example, increased at high temperature, while considerable linkage equilibrium between both allozyme loci was maintained. This raises the question whether $Adh^S$ and $\alpha Gpdh^F$ are merely hitchhiking with *In(2L)t* or still are (partly) selected for their own properties. Comparison of egg-to-adult survival in a strain monomorphic for $Adh^S$ $\alpha Gpdh^F$ but polymorphic for the Standard (*ST*) and for *In(2L)t* karyotype showed significantly higher survival of *ST/In(2L)t* and *In(2L)t/In(2L)t* karyotypes with respect to *ST/ST* at 29.5 and 33°C but not at 20 and 25°C (Tab. 1). This points to an effect on pre-adult survival of *In(2L)t* independent of *Adh* and $\alpha Gpdh$ and suggests that the high *In(2L)t* frequencies in the tropics may be caused by higher resistance to high-temperature stress, most probably due to the inclusion in the inversion of particular gene variants.

Also for adult survival at 35°C, *ST/ST* is inferior to the other karyotypes (Tab. 1). There is a heterozygote superiority both for pre-adult and adult survival at high temperatures. Better adaptation to high-temperature stress for karyotypes involving *In(2L)t* was also observed for quite another trait: restoration of male fertility. Males of the three karyotypes were exposed to 33°C for 2 days, which induced 100% sterility in all karyotypes. Therafter

Table 1. Survival in relation to temperature of various karyotypes

| Karyotype | Egg-to adult survival (in percentages) | | | Adult survival (in hours) | |
|---|---|---|---|---|---|
| | 25°C | 29.5°C | 33°C | 35°C | |
| | | | | ♀♀ | ♂♂ |
| *ST/ST* | 69.7[a] | 64.1[b] | 28.2[c] | 50.2[c] | 37.9[b] |
| *ST/In(2L)t* | 81.1[a] | 82.2[a] | 68.2[a] | 60.1[a] | 41.0[a] |
| *In(2L)t/In(2L)t* | 73.4[a] | 70.2[b] | 51.6[b] | 57.6[b] | 41.4[a] |

Figures within each column which share a common superscript letter are not significantly different at the 5% level.

the individual males were replaced at 25°C for 4 days and subsequently individually tested for restored fertility. The results are presented in Figure 2, which shows significantly higher restoration of male fertility for both *In(2L)t/In(2L)t* and *In(2L)t/ST*. Similar results (not shown) were obtained for restoration of female fertility after exposure to high temperature.

It thus appears that several aspects of adaptation to high-temperature stress, namely egg-to-adult survival, adult survival and restoration of fertility after high-temperature shocks, are associated with *In(2L)t* either

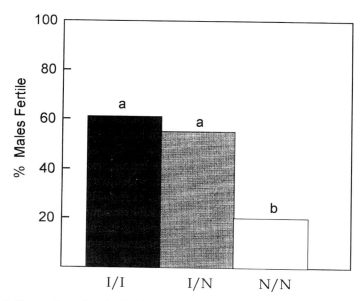

Figure 2. Restoration of male fertility after heat shock. Sixty males of each karyotype (I = *In(2L)t*; N = *St*) were individually exposed to 33°C for 2 days. Thereafter the males were placed at 25°C for 4 days and subsequently individually tested for restored fertility. The number of matings producing offspring was recorded. Significant differences (p < 0.05) are indicated by different letters.

in homo- or heterokaryotypic configuration. However, Van Delden and Kamping noticed that important life-history traits like development time, body weight (Van Delden and Kamping, 1991), male mating success and egg-laying distribution (W. Van Delden and A. Kamping, unpublished data) also varied among karyotypes both at high as well as at low temperatures. Development time of $In(2L)t/In(2L)t$ was always longer than that of $In(2L)t/ST$ or $ST/ST$, while the former karyotype invariably had the lowest body weight. The differences in development time were correlated with the time needed to attain critical weight for pupation. The shortest time (for temperatures not exceeding 25°C) was found for $In(2L)t/ST$ and the longest for $In(2L)t/In(2L)t$. The latter karyotype thus possesses the uncommon combination of slow development and low body weight. The observed differences among karyotypes give rise to pronounced fitness differences under appropriate (stress) conditions as witnessed by changes in inversion frequencies in polymorphic populations (Van Delden and Kamping, 1991). At temperatures exceeding 29°C $In(2L)t$ frequencies stay high, while they decline at 20 and 25°C. Apparently $ST$ possesses higher fitness at lower temperatures and $In(2L)t$ at high temperatures. Furthermore $In(2L)t$ is extremely sensitive to ethanol stress: the calculated fitness for $In(2L)t/In(2L)t$ was 0.17 compared with unity for the other karyotypes. The disadvantage of the slow development was exhibited by rapid decline of $In(2L)t$ from 0.50 to 0.10 in populations kept under a regime of high egg density and a transfer time of 14 days. Temperature and density effects were further shown by two populations both originating from flies caught in the Pyrenees, France, and both lacking the $Adh^S\alpha Gpdh^S$ haplotype, while the $Adh^S\alpha Gpdh^F$ haplotype was always associated with $In(2L)t$ (Kamping and Van Delden, 1997). These populations, Vernet-A and Vernet-B were replicated and thereafter continued both at 15 and 29°C for almost 5 years (about 42 and 130 generations respectively). Considerable divergence in the frequencies of $In(2L)t$, $Adh$ and $\alpha Gpdh$ occurred, including the elimination of the $Adh^F\alpha Gpdh^F$ haplotype from the Vernet-A population kept at 15°C and the $Adh^F\alpha Gpdh^S$ haplotye from the Vernet-B population kept at 29°C. In an additional experiment the effects on genetic composition of both low (50 larvae per vial) and high (250 larvae per vial) density for one generation were determined for all four populations. The outcome with respect to $In(2L)t$ frequencies is given in Table 2. The history of the populations, either kept at 15 or 29°C, is clearly reflected in the increase in $In(2L)t$ frequency at 29°C. Even a one-generation high-temperature effect is perceptible under uncrowding conditions. This confirms the hypothesis of inversion superiority at high-temperature stress. On the other hand, high crowding levels are supposed to decrease inversion frequencies, due to the slower development of inversion homokaryotypes and the resulting disadvantage with crowding. This is not clearly shown by the $In(2L)t$ frequencies in Table 2A. However, considering the heterokaryotype frequencies (Tab. 2B) shows generally higher heterokaryotype frequencies at high

Table 2. *In(2L)t* frequencies (A) and heterokaryotype frequencies (B) in four populations kept for one generation under low or high larval crowding conditions (see text for further explanation)

| A | | Test condition | | | |
| --- | --- | --- | --- | --- | --- |
| | | no crowding | | crowding | |
| population | history | 25°C | 29°C | 25°C | 29°C |
| Vernet A | 15°C | 0.17 | 0.27 | 0.27 | 0.28 |
| Vernet-A | 29°C | 0.38 | 0.45 | 0.35 | 0.30 |
| Vernet-B | 15°C | 0.02 | 0.04 | 0.04 | 0.01 |
| Vernet-B | 29°C | 0.20 | 0.24 | 0.25 | 0.29 |

| B | | Test condition | | | |
| --- | --- | --- | --- | --- | --- |
| | | no crowding | | crowding | |
| population | history | 25°C | 29°C | 25°C | 29°C |
| Vernet A | 15°C | 0.28 | 0.41 | 0.46 | 0.53 |
| Vernet-A | 29°C | 0.41 | 0.42 | 0.45 | 0.45 |
| Vernet-B | 15°C | 0.04 | 0.08 | 0.07 | 0.01 |
| Vernet-B | 29°C | 0.31 | 0.38 | 0.43 | 0.54 |

crowding levels among samples cultured at the same temperature but differing in crowding level. This reflects heterokaryotype superiority under stress crowding levels.

## Prospects

The similar latitudinal clines for allozymes, inversions, morphological and physiological traits observed for *D. melanogaster* on various continents, in both the Northern and Southern Hemispheres, strongly suggest the action of natural selection. The exact nature of the selection process is, however, still unclear. The situation is complicated first of all by uncertainty about the environmental stresses involved and their latitudinal variation. Temperature, or traits narrowly related to temperature, and perhaps humidity, seem obvious candidates. But differences in ecological situations, strongly opposite between tropical and temperate regions, may also cause differential selective forces. In this respect different food sources or the presumed stability of tropical areas and seasonal fluctuating conditions in temperate regions may be involved. Considerable differences in population size in time, and in recolonising events between both types of habitats will occur, which may give rise to either r- or K-selection (Parsons, 1983; Van Delden and Kamping, 1991). Another complication in the explanation of the latitudinal clines is formed by the interrelationships between the various traits and genetic polymorphisms involved. *In(2L)t*, for example, is nearly always associated with $Adh^S \alpha Gpdh^F$. $In(2L)t$ homokaryotypes have longer development times than the other two karyotypes, but $Adh^{SS}$ homozygotes also have slower development than the other two $Adh$ genotypes. Flies in tropical regions are smaller than in temperate regions, which may (partly) be associated with the high $In(2L)t$ frequencies in the tropics. Are tropical flies more sensitive to ethanol stress because they have lower body weight and ADH activity is allometrically correlated with body weight such that ADH activity per unit body weight decreases with decreasing body weight (Clarke et al., 1979)? Merely epistatic relations among allozymes, such as those that occur between $Adh$ and $\alpha Gpdh$, may give highly complex relationships which may vary with environmental factors as enunciated by Koehn et al. (1983). Are we in fact in the position described by Jones et al. (1977) in their review article on polymorphism in *Cepaea*, subtitled "A problem with too many solutions"?

An experimental approach aimed at deciphering the complex situation provided by variation at the various levels of single allozyme loci, polygenes and chromosomes should at least try to disentangle some of these factors. This is done in an experiment designed to separate the effects of the $In(2L)t/ST$ chromosomal polymorphism from the effects of the $Adh$ and/or the $\alpha Gpdh$ allozyme polymorphisms (Van Delden and Kamping, 1997). Seven population types were constructed by appropriate crosses between homozygous strains as indicated in Table 3. Crossing of the strain $Adh^{SS} \alpha Gpdh^{FF}$, homozygous for $In(2L)t$ for example, with $Adh^{FF} \alpha Gpdh^{SS}$, homozygous for $ST$ (abbreviated in Tab. 3 as SFI × FSN), for example, provided a population polymorphic for both $In(2L)t$, $Adh$ and $\alpha Gpdh$. When polymorphic, the initial frequency of the variants was 0.50, with maximum linkage disequilibrium. This experimental setup enables the analysis of the effects of various combinations such as the reaction of the inversion polymorphism in the absence of allozyme variation at the $Adh$ and $\alpha Gpdh$ loci

Table 3. Origin and composition of experimental populations, either polymorphic or monomorphic for $In(2L)t$, $Adh$ or $\alpha Gpdh$

| Cross* | Population type | Polymorphic(+) or monomorphic (−) | | |
|---|---|---|---|---|
| | | $In(2L)t$ | $Adh$ | $\alpha Gpdh$ |
| SFI × SFN | 1 | + | − | − |
| SFI × FSN | 2 | + | + | + |
| SFI × FFN | 3 | + | + | − |
| SFI × SSN | 4 | + | − | + |
| SFN × FSN | 5 | − | + | + |
| SFN × FFN | 6 | − | + | − |
| SFN × SSN | 7 | − | − | + |

* First letter of a strain stands for homozygous genotype of $Adh$, second letter for homozygous genotype of $\alpha Gpdh$ and third letter for homozygous karyotype of $In(2L)t$ (I) or $ST$ (N).

(population type 1), or the reaction of the inversion polymorphism when *Adh* is monomorphic but *αGpdh* is polymorphic (population type 4). Note that when *Adh* is monomorphic it is fixed for the $Adh^S$ allele, whereas when *αGpdh* is monomorphic it is fixed for the $αGpdh^F$ allele. The seven types of populations were maintained with two replicates at each of seven environmental conditions: 20, 25, 29 and 33°C (for part of the life cycle), ethanol-supplemented food, high egg densities with generation intervals of 14 days and high egg densities with generation intervals of 21 days (the latter three at 25°C).

A first principal conclusion from these experiments is that strongly different effects are exerted on the polymorphisms under study by the various stress conditions. An example is the strong decline in *In(2L)t* frequencies in populations kept on ethanol-supplemented food, while at high-temperature stress, for example, such a decline is moderate or absent. This confirms the previous assumption that distinct selection pressures associated with different stress factors are involved. A second conclusion is that the various population types may react quite differently to the same stress factor. An example is given in Figure 3, where frequency changes are shown for populations continuously kept on ethanol-supplemented food. Inversion frequencies decrease sharply in all polymorphic population types. $Adh^S$ frequencies also decrease drastically and, as it seems, quite independently from the presence or absence of *In(2L)t* in the population. For *αGpdh*, however, a sharp distinction can be made between population types 2 and 4 (with *In(2L)t*), which show a distinct rise in $αGpdh^S$ frequency and population types 5 and 7 (lacking *In(2L)t*), which show a very slight decrease. The explanation is that – as stated previously – severe selection is operating against *In(2L)t* in the presence of ethanol. This is clearly illustrated by population type 1, where no "disturbing" effect of allozyme polymorphism is present. Because of the coupling of $Adh^S$ with the inversion, $Adh^S$ will be carried along with *In(2L)t*, but in the inversion-free population types which are polymorphic for *Adh*, a similar decrease in $Adh^S$ occurs, demonstrating the well-known effect of selection against $Adh^S$ when ethanol is a stress factor. With respect to $αGpdh^S$ frequencies, there is a sharp distinction between population types 2 and 4, which exhibit considerable increases, and population types 5 and 7, which hardly appear influenced. This points to hitchhiking of $αGpdh^F$ with *In(2L)t* in the former population types, whereas the *αGpdh* polymorphism itself seems hardly to be affected by ethanol, as follows from the latter population types.

It is clear that the various karyotypes can differ considerably in fitness, especially under high-temperature and ethanol stress. However, the two allozyme loci studied also exhibit fitness differences on their own. It appears that *In(2L)t* – and probably other cosmopolitan inversions showing latitudinal clines as well – possesses a particular set of gene variants that provide high fitness to its carriers under conditions prevailing in tropical regions. Such coadapted gene complexes are to a great extent protected

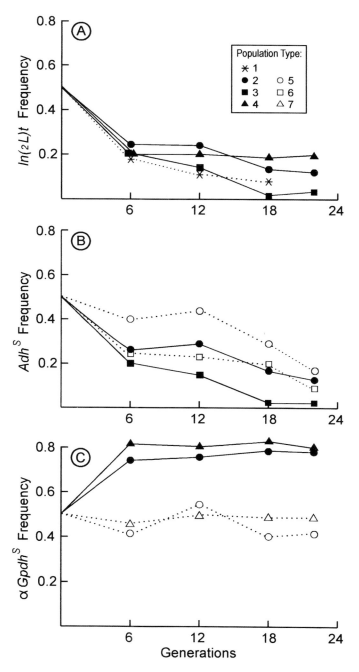

Figure 3. Changes in *In(2L)t* frequencies (A), in *Adh*[S] frequencies (B) and in *αGpdh*[S] frequencies (C) in populations kept on ethanol-supplemented food (12%). Population types are as described in Table 3.

against the destructive action of recombination. Dobzhansky and co-workers have provided ample evidence for such coadapted genes located in the third chromosome inversions in *D. pseudoobscura* (see Dobzhansky, 1970). These inversion polymorphisms appear often to be maintained by balancing selection predominantly by overdominance, though frequency-dependent selection has also been proposed as a potential mechanism (Dobzhansky, 1970; Lewontin, 1974).

In thus appears that the genetic system may take a key position in the genetic response to natural occurring stress factors as exemplified by the *In(2L)t* polymorphism. Recently it was discovered that in addition to Drosophilids, in Anopheline mosquitoes paracentric inversions are also involved in adaptation to environmental stresses. The sibling species of the *Anopheles gambiae* complex are characterised by a unique array of inversions and have a number of polymorphic inversions. Specific inversion types seem to be selected for under particular stress conditions generally involved with aridity and show clinal changes in frequencies (Mathiopoulos and Lanzaro, 1995). Differences among *A. gambiae* strains in susceptibility to *Plasmodium cynomolgi* B, a malaria parasite, were found to be linked with different forms of a polymorphic inversion *2La* (Crews-Oyen et al., 1993).

The important role of *In(2L)t* with respect to selective stress factors have lead us to search for the presumed complex of genes involved in coadaptation to stress factors. As a preliminary step the general levels of genetic variation in *In(2L)t* and the comparable region of *ST* are compared. This is to test the hypothesis that present-day inversions like *In(2L)t* originate from a unique mutation event in the past, which should be reflected by a relatively low level of variation, compared with *St* (see Lemeunier and Aulard, 1992). This was tested by means of RAPDs, RFLP and nucleotide-sequencing analysis. These tests all showed a confirmation of the assumption of lower genetic variability in the involved region. In the latter comparison, for example, a 500 bp fragment of the *αGpdh* gene was sequenced in three Dutch *In(2L)t* and three *ST* strains which possessed, except for the inverted region, a homogeneous background (Kamping and Van Delden, 1997). The results shown in Table 4 confirm the hypothesis of less varia-

Table 4. Comparison of the number of nucleotide substitutions in a 500-bp fragment of the *αGpdh* gene, within and between karyotypes

|           | St 1 | St 2 | St 3 | In(2L)t 1 | In(2L)t 2 | In(2L)t 3 |
|-----------|------|------|------|-----------|-----------|-----------|
| St 1      | –    | 5    | 4    | 8         | 9         | 6         |
| St 2      |      | –    | 5    | 9         | 10        | 9         |
| St 3      |      |      | –    | 8         | 9         | 8         |
| In(2L)t 1 |      |      |      | –         | 3         | 2         |
| In(2L)t 2 |      |      |      |           | –         | 3         |
| In(2L)t 3 |      |      |      |           |           | –         |

Three karyotypes, both of *St* and *In(2L)t* were compared.
Nucleotide diversity: within *In(2L)t*, .00533; within *St*, .00933; between *In(2L)t* and *St*, .01689.

tion in *In(2L)t*. Further comparisons of strains from the Netherlands and the French Pyrenees showed much greater resemblance among the *In(2L)t* strains from both regions than among the comparable *St* strains. Similar results were obtained for the *In(3L) Payne* inversion in *D. melanogaster* (Wesley and Eanes, 1994) and third chromosome inversions in *D. pseudo-obscura* (Aquadro et al., 1991). These assays confirm the preserved gene contents of the inverted regions, but of course they provide no information about the number, nature and interactions of the specific genes involved in adaptation to specific stresses. To elucidate the causes for the intriguing latitudinal clines in *D. melanogaster* and other organisms, a further elaboration of the functional approach for polymorphic single genes is needed. Such a study should include identification of the specific stress factors involved and should be extended to a study of gene interactions and the major gene effects on quantitative traits. With respect to the impact of the latter group of characters on stress resistance, a better understanding of their genetics and physiology is needed. With modern QTL (Quantitative Trait Loci) technology it should be possible, at least in a superb experimental organism like *D. melanogaster*, to localise and eventually characterise the genetic elements responsible for the various quantitative traits involved in stress resistance. Such an approach should be accompanied by identification of the relevant gene products. Finally, hypotheses with respect to coadaptation and clines in the context of stress resistance as derived from such studies could be tested in experiments as devised by Endler (1973, 1977). Such studies will, most probably, point to – often complex – interactions both at the phenotypic and the gene level.

## References

Aquadro, C.F., Weaver, A.L., Schaeffer, S.W., and Anderson, W.W. (1991) Molecular evolution of inversions in *Drosophila pseudoobscura*: The amylase gene region. *Proc. Natl. Acad. Sci. USA* 88:305–309.

Bénassi, V. and Veuille, M. (1995) Comparative population structuring of molecular and allozyme variation in *Drosophila melanogaster Adh* between Europe, West Africa and East Africa. *Genetical Research* 65:95–103.

Bijlsma-Meeles, E. and Bijlsma, R. (1988) The alcohol dehydrogenase polymorphism in *Drosophila melanogaster*: Fitness measurements and predictions under conditions with no alcohol stress. *Genetics* 120:743–753.

Bubli, O.A., Rakitskaya, T.A. and Imasheva, A.G. (1996) Variation of allozyme loci in populations of *Drosophila melanogaster* from the former USSR. *Heredity* 77:638–645.

Cavener, D.R. and Clegg, M.T. (1981) Multigenic adaptation to ethanol in *Drosophila melanogaster*. *Evolution* 35:1–10

Chambers, G.K. (1988) The *Drosophila* alcohol dehydrogenase gene-enzyme system. *Advances in Genetics* 25:39–197.

Chambers, G.K. (1991) Gene expression, adaptation and evolution in higher organisms: Evidence from studies of *Drosophila* alcohol dehydrogenase. *Comp. Biochem. Physiol.* 99B:723–730.

Clarke, B. (1975) The contribution of ecological genetics to evolutionary biology: Detecting the direct effects of natural selection on particular polymorphic loci. *Genetics* 79:101–113.

Clarke, B., Camfield, R.G., Galvin, A.M. and Pits, C.R. (1979) Environmental factors affecting the quantity of alcohol dehydrogenase in *Drosophila melanogaster*. *Nature* 280: 517–518.

Clausen, J., Keck, D.D. and Hiesey, W.M. (1940) Experimental studies in the nature of species. *Carnegie Inst. Wash. Publ.* 520: 1–452.

Crews-Oyen, A.E., Kumar, V. and Collins, F.H. (1993) Association of two esterase genes, a chromosomal inversion and susceptibility to *Plasmodium cynomolgi* in the African malaria vector *Anopheles gambiae*. *Am. J. Trop. Hyg.* 49: 341–347.

David, J.R. and Capy, P. (1988) Genetic variation of *Drosophila melanogaster* natural populations. *Trends in Genetics* 4: 106–111.

David, J.R., Bocquet, C. and De Sheemaeker-Louis, M. (1977) Genetic latitudinal adaptation of *Drosophila melanogaster*: New discriminative biometrical traits between European and equatorial African populations. *Genet. Res.* 30: 247–255.

David, J.R., Alonso-Moraga, A., Borai, F., Capy, P., Merçot, H., McEvey, S.F., Munoz-Serrano, A. and Tsakas, S. (1989) Latitudinal variation of *Adh* gene frequencies in *Drosophila melanogaster*: A Mediterranean instability. *Heredity* 62: 11–16.

Dobzhanksy, Th. (1970) *Genetics of the Evolutionary Process*. Columbia University Press, New York.

Endler, J.A. (1973) Gene flow and population differentiation. *Science* 179: 243–250.

Endler, J.A. (1977) *Geographic Variation, Speciation and Clines*. Princeton University Press, Princeton.

Geer, B.W., Langevin, M.L. and McKechnie, S.W. (1985) Dietary ethanol and lipid synthesis in *Drosophila melanogaster*. *Biochem. Genet.* 23: 607–622.

Grossman, A., Koreneva, L.G. and Lilitskaya, L.E. (1970) Variation of the alcohol dehydrogenase locus in natural populations of *Drosophila melanogaster*. *Genetika* 6: 91–96.

Hamrick, J.L. and Godt, M.J.W. (1990) Allozyme diversity in plant species. *In*: A.H.D. Brown, M.T. Clegg, A.L. Kahler and B.S. Weir (eds): *Plant Population Genetics, Breeding and Genetic Resources*. Sinauer, Sunderland, MA. pp. 43–63.

Heinstra, P.W.H. (1993) Evolutionary genetics of the *Drosophila* alcohol dehydrogenase gene-enzyme system. *Genetica* 92: 1–22.

Heinstra, P.W.H., Eisses, K.Th., Schoonen, W.G.E.J., Aben, W., De Winter, A.J., Van der Horst, O.J., Van Marrewyk, W.J.A., Beenakkers, A.M.Th., Scharloo, W. and Thörig, G.E.W. (1983) A dual function of alcohol dehydrogenase in *Drosophila*. *Genetica* 60: 129–137.

Hoffmann, A.A. and Parsons, P.A. (1991) *Evolutionary Genetics and Environmental Stress*. Oxford University Press, Oxford.

James, A.C., Azevedo, R.B.R. and Partridge, L. (1995) Cellular basis and developmental timing in a size cline of *Drosophila melanogaster*. *Genetics* 140: 659–666.

James, A.C. and Partridge, L. (1995) Thermal evolution of rate of larval development in *Drosophila melanogaster* in laboratory and field populations. *J. Evol. Biol.* 8: 315–330.

Johnson, F.M. and Powell, A. (1974) The alcohol dehydrogenase of *Drosophila melanogaster*. Frequency changes associated with heat and cold shock. *Proc. Natl. Acad. Sci. USA* 71: 1783–1784.

Jones, J.S., Leith, B. and Rawlings, P. (1977) Polymorphism in *Cepaea*: A problem with too many solutions? *Annu. Rev. Ecol. Syst.* 8: 109–143.

Kamping, A. and Van Delden, W. (1997) A comparison of Standard and *In(2L)t* karyotypes in *Drosophila melanogaster*: A molecular approach; *in press*.

Kimura, M. (1968) Genetic variability maintenance in a finite population due to mutational production of neutral and nearly neutral isoalleles. *Genet. Res.* 11: 247–269.

Kimura, M. (1983) *The neutral theory of molecular evolution*. Cambridge University Press, Cambridge.

Koehn, R.K., Zera, A.J. and Hall, J.G. (1983) Enzyme polymorphism and natural selection. *In*: M. Nei and R.K. Koehn (eds): *Evolution of Genes and Proteins*. Sinauer, Sunderland, MA, pp. 115–136.

Kreitman, M. (1983) Nucleotide polymorphism at the alcohl dehydrogenase locus of *Drosophila melanogaster*. *Nature* 304: 412–417.

Kreitman, M. and Hudson, P.R. (1991) Inferring the evolutionary histories of the *Adh* and *Adh-dup* loci in *Drosophila melanogaster* from patterns of polymorphism and divergence. *Genetics* 127: 565–582.

Lemeunier, F., and Aulard, S. (1992) Inversion polymorphism in Drosophila melanogaster. *In:* C.B. Krimbas and J.R. Powell (eds): *Drosophila Inversion Polymorphism*. CRC Press, Boca Raton, pp. 339–405.

Lemeunier, F., David, J.R., Tsacas, L. and Ashburner, M. (1986) *The melanogaster* species group. *In:* M. Ashburner, H.L. Carson and J.N. Thompson Jr. (eds): *The Genetics and Biology of Drosophila*, Volume 3e. Academic Press, London, pp. 147–256.

Lewontin, R.C. (1974) *The Genetic Basis of Evolutionary Change*. Columbia University Press, New York.

Mathiopoulos, K.D. and Lanzaro, G.C. (1995) Distribution of genetic diversity in relation to chromosomal inversion in the malaria mosquito *Anopheles gambiae. J. Mol. Evol.* 40: 578–584.

Mayr, E. (1963) *Animal Species and Evolution*. Belknap Press, Cambridge, MA.

Mitton, J.B. and Grant, M.C. (1984) Association among protein heterozygosity, growth rate and developmental homeostasis. *Annu. Rev. Ecol. Syst.* 15:479–499.

Nevo, E., Beiles, A. and Ben-Slomo, R. (1984) The evolutionary significance of genetic diversity: Ecological, demographic and life history correlates. *In:* G.S. Mani (ed): *Evolutionary Dynamics of Genetic Diversity*. Springer-Verlag, Berlin, pp. 13–213.

Oudman, L. (1993) *Adaptive aspects of the polymorphisms at the* Adh *and* αGpdh *loci in* Drosophila melanogaster. Ph.D. dissertation, University of Groningen.

Oudman, L., Van Delden, W., Kamping, A. and Bijlsma, R. (1991) Polymorphism at the *Adh* and αGpdh loci in *Drosophila melanogaster*: Effects of rearing temperature on developmental rate, body weight and some biochemical parameters. *Heredity* 67:103–115.

Oudman, L., Van Delden, W., Kamping, A. and Bijlsma, R. (1992) Interaction between the *Adh* and αGpdh loci in *Drosophila melanogaster*: Adult survival at high temperature. *Heredity* 68:286–297.

Parkash, R. and Shamina N. (1994) Geographical differentiation of allozyme variability in natural Indian populations of *Drosophila melanogaster. Biochem. Genet.* 32:63–73.

Parsons, P.A. (1983) *The Evolutionary Biology of Colonizing Species*. Cambridge University Press, New York.

Pipkin, S.B., Franklin-Springer, E., Law, S. and Lubega, S. (1976) New studies of the alcohol dehydrogenase cline in *Drosophila melanogaster* from Mexico. *J. Hered.* 67:258–266.

Powers, D.A., Lauerman, T., Crawford, D. and Dimichele, L. (1991) Genetic mechanisms for adapting to a changing environment. *Annu. Rev. Genet.* 25:629–659.

Prout, T. (1971) The relation between fitness components and population predition in *Drosophila*. I. The estimation of fitness components. *Genetics* 68:127–149.

Sperlich, D. and Pfriem, P. (1986) Chromosomal polymorphism in natural and experimental populations. *In:* M. Ashburner, H.L. Carson and J.N. Thompson Jr. (eds): *The Genetics and Biology of Drosophila*, Volume 3e. Academic Press, Oxford, pp. 257–309.

Van Delden, W. (1982) The alcohol dehydrogenase polymorphism in *Drosophila melanogaster. Evol. Biol.* 15:187–222.

Van Delden, W. and Kamping, A. (1979) The alcohol dehydrogenase polymorphism in populations of *Drosophila melanogaster*. III. Differences in developmental times. *Genet. Res.* 33:15–27.

Van Delden, W. and Kamping, A. (1980) The alcohol dehydrogenase polymorphism in *Drosophila melanogaster*. IV. Survival at high temperature. *Genetica* 51:179–185.

Van Delden, W. and Kamping, A. (1988) Selection against *Adh* null alleles in *Drosophila melanogaster. Heredity* 61:209–216.

Van Delden, W. and Kamping, A. (1989) The association between the polymorphism of the *Adh* and αGpdh loci and the *In(2L)t* inversion in *Drosophila melanogaster* in relation to temperature. *Evolution* 43:775–793.

Van Delden, W. and Kamping, A. (1991) Changes in relative fitness with temperature among second chromosome arrangements in *Drosophila melanogaster. Genetics* 127: 507–514.

Van Delden, W. and Kamping, A. (1997) Analysis of the association of allozyme and inversion polymorphisms in *Drosophila melanogaster; submitted*.

Van Delden, W., Boerema, A.C. and Kamping A. (1978) The alcohol dehydrogenase polymorphism in populations of *Drosophila melanogaster*. I. Selection in different environments. *Genetics* 99:161–191.

Van 't Land, J. (1997) *Latitudinal variation in wild populations of* Drosophila melanogaster. Ph.D. dissertation, University of Groningen.

Van 't Land, J., Van Delden, W. and Kamping, A. (1993) Variation in *In(2L)t* frequencies in relation to *Adh* and *alpha-Gpdh* polymorphisms in tropical populations of *D. melanogaster. Drosoph. Inf. Serv.* 72:102–104.

Van 't Land, J., Van Putten, P., Villarroel, H., Kamping, A. and Van Delden, W. (1995) Latitudinal variation in wing length and allele frequencies for *Adh* and *alpha-Gpdh* in populations of *Drosophila melanogaster* from Equador and Chile. *Drosoph. Inf. Serv.* 76:156.

Wesley, C.S. and Eanes, W.F. (1994) Isolation and analysis of the breakpoint sequences of chromosome inversion *In(3L)Payne* in *Drosophila melanogaster. Proc. Natl. Acad. Sci. USA* 91:3132–3136.

Environmental Stress, Adaptation and Evolution
ed. by R. Bijlsma and V. Loeschcke
© 1997 Birkhäuser Verlag Basel/Switzerland

# Stress and metabolic regulation in *Drosophila*

Andrew G. Clark

*Department of Biology, 208 Mueller Laboratory, Pennsylvania State University,
University Park, PA 16802, USA*

*Summary.* Environmental changes that result in stress (defined here as decreased absolute
viability and/or fecundity) result in extrinsic changes in metabolism that are to some extent
compensated by altered gene expression. The fact that different genotypes may respond differently to environmental stress may be of key importance to the maintenance of genetic variation in metabolic traits. Here we quantify a set of metabolic characters in genetically defined
lines of *Drosophila melanogaster* subjected to four stresses (3% acetic acid, 3% ethanol,
starvation and thermal stress) in order to assess the magnitude of environmental effects and
genotype × environment interactions. Genetic correlations were quantified, and many exhibit
significant heterogeneity across environments. Pleiotropically related traits may exhibit the
phenomenon of apparent selection, whose effects may be particularly strong in stressful
environments. This transient apparent selection may have a large consequence on the maintenance of genetic variation.

## Introduction

Three mechanisms that may serve to maintain quantitative genetic variation in populations are mutation-selection balance, pleiotropy and gene ×
environment interaction. Models of mutation-selection balance (Lande,
1975) run into problems with the correspondence between the per-locus
mutation rates and numbers of loci that affect a trait (Turelli, 1984). Pleiotropic effects of mutations on multiple traits may result in an apparent
selection on neutral traits that are pleiotropically related to traits associated
with fitness. Patterns of pleiotropy serve to increase the number of mutable
loci that can affect the variance in a trait, thereby getting around the mutation rate/number of loci problem faced by mutation-selection balance
models (Turelli, 1985). The problem with the model of pleiotropy is that it
is very difficult to design experiments that quantify the depth and complexity of pleiotropic interactions and to quantify the correlations of traits with
fitness. Without these, it is not really possible to test whether the theory
explains anything in nature.

   The study of genotype × environment interaction has a large literature
(reviewed in Via et al., 1995), but little of it is devoted to the question of
maintenance of quantitative genetic variation. More often, studies of norms
of reaction and phenotypic plasticity center on the question of how plasticity itself evolves. But changing the rank order of fitness of genotypes in
different environments can certainly maintain polymorphism in classical

population genetics models, and the effects of $G \times E$ on maintenance of quantitative genetic variation can be profound.

Gillespie and Turelli (1989) considered the importance of $G \times E$ on maintenance of variation with a pure additive model in which the genotypes had different fitnesses in different environments. Assuming that the most fit genotype may differ from one environment to another, they found that this simple model generates a situation in which genotypes with more heterozygous loci are generally more fit. Such a pattern maintains greater steady-state variation than a population in any one of the environments. It is possible that $G \times E$ is central to the problem of maintenance of variation, but again, empirical test of this hypothesis is not easy. Studies like those of Gupta and Lewontin (1982) routinely show that $G \times E$ interaction is very common in mapping genotypes to phenotypes, but abundance of $G \times E$ does not necessarily mean that the patterns of variation across environments promote polymorphism. As Gillespie and Turelli (1989) depressingly put it, there is no way to experimentally survey the full range of environments, and the artificial subset of environments that are chosen for experimental tests may not reflect what goes on in the full range of environments.

Stressful environments may change patterns of additive, dominant and epistatic genetic variance, although no consistent trends in such changes have been observed (reviewed in Hoffmann and Parsons, 1991). Blows and Sokolowski (1995) isolated six second chromosome isogenic lines of *Drosophila*, intercrossed the lines, and estimated developmental time and absolute viability in a series of environments. They found essentially no change in additive variance across environments, but dominance and epistasis increased at environmental extremes. Such patterns may reflect the past operation of selection on the population, and clearly affect the impact of stress on standing levels of genetic variation.

Environmental effects on metabolic traits are particularly amenable to study. Physiological responses to novel substances in the medium include altered expression of several enzymes. When the change is an increase in the expression of an enzyme, it is referred to as induction. Cases of induction in *Drosophila* include responses to elevated ethanol (McKechnie and Geer, 1984; Geer et al., 1988) and sucrose (Geer et al., 1981, 1983). Often the induction phenomenon results in a simple increased expression of the first enzyme involved in metabolism of the substrate, such as increased expression of alcohol dehydrogenase (ADH) in response to ethanol added to the medium (Geer et al., 1988). Metabolic consequences can be more complex, including in the case of an ethanol diet a shift toward lipid synthesis on elevated ethanol medium (Geer et al., 1985). In some cases, dietary changes are not as simple as first appears. Comparison of amylase activities on low and high starch diets shows that amylase expression is higher on the high starch diet, showing both an induction of amylase on the high starch food (Yamazaki and Matsuo, 1984) and a repression of amylase by glucose on the low starch food (Hickey et al., 1994).

Our primary interest initially is to describe the magnitude of variation in metabolic traits across environments, to quantify the heterogeneity in correlation patterns among them and to try to identify genetic variation responsible for the magnitude of environmental sensitivity of metabolic traits. Here we quantify the effects of four different post-eclosion metabolic stresses on 16 metabolic traits.

## Materials and Methods

### Drosophila *culturing*

With the use of balancer chromosomes, Chung-I Wu and colleagues generated a set of chromosome replacement lines of *Drosophila* which they kindly shared with us (Wu et al., 1995). The four founding lines were the marker stock *rucuca*, a stock from Zimbabwe (Z30), a stock from France (Fr) and a stock from Highgrove, California (Hg). The 19 lines have different combinations of chromosomes 1, 2, and 3, between the Z30 stock and the other two. In addition, three of the stocks are recombinants with the *rucuca* stock (which bears seven third-chromosome recessive markers) in order to partition the third chromosome. Females were allowed to lay eggs in bottles at standard density, and male progeny were split into five groups. The control group was placed on standard medium at 25°C. A second group was placed in vials with Whatman filter paper soaked in 3% acetic acid. A third group was placed in vials with Whatman paper soaked in 3% ethanol. A fourth group was placed on standard medium, but 45 h before homogenization, they were transferred to empty vials whose cotton ball was kept moist with water. A fifth group was placed on standard medium at 32°C. Henceforth, the environmental treatments go by the names Control, Acetic, EtOH, Starve and 32°C.

### *Enzyme kinetics*

On the sixth day post-eclosion, the flies from all treatments were collected, weighed and homogenized in groups of four flies. Procedures for preparing whole-fly tissue homogenates, dispensing into microtiter plates and scoring enzyme kinetics with a microtiter plate reader are described in detail in Wang and Clark (1994). Table 1 lists the metabolic traits that were assayed. Briefly, anesthetized flies were weighed in groups of four to the nearest 0.001 mg and were moved to a 4°C cold room where they were homogenized, centrifuged and distributed into 15 microtiter plates. Microtiter plates were frozen at − 70°C until kinetic assays were run. Kinetic tests were done colorimetrically. After adding appropriate reagents for each test, the optical density was recorded with a microtiter plate reader, which recorded the

Table 1. List of metabolic traits assayed

| | |
|---|---|
| WT: | live weight |
| PRO: | total protein |
| TRI: | triacylglycerol |
| GLY: | glycogen |
| ADH: | alcohol dehydrogenase |
| G6PD: | glucoe-6-phosphate dehydrogenase |
| GPDH: | glycerol-3-phosphate dehydrogenase |
| HEX: | hexokinase |
| ME: | malic enzyme |
| 6PGD: | 6-phosphogluconate dehydrogenase |
| PGI: | phosphoglucose isomerase |
| PGM: | phosphoglucomutase |
| TRE: | trehalase |

WT is in mg, PRO, TRI and GLY are µg/fly, and the remaining traits are all enzyme activities in units of nmoles of substrate turned over per fly per minute.

optical density of each well of the microtiter plate at a series of times. Enzyme activities were calculated from standards as the nmoles of substrate converted to product per fly per minute. Analysis was also done on a per milligram basis, and after removing allometric effects of weight and total protein, and the results did not differ appreciably from the simpler analysis of variance (ANOVA).

## Statistical methods

A few of the null hypotheses we wished to test include the following: (1) The genotypes were identical in metabolic traits. (2) The environmental treatments had no effect. (3) Differences in phenotype were independent of replaced chromosomes. (4) There were no genotype × environment interactions. These were all tested with standard methods of analysis of variance using the SAS procedure GLM. For the ANOVA, the model fitted was:

$$y_{ijkl} = \mu + G_i + E_j + (GE)_{ij} + V_{ijk} + e_{ijkl}$$

where $G_i$ is the effect of genotype $i$ ($i = 1-19$), $E_j$ is the effect of environmental treatment $j$ (where $j = 0, 1, 2, 3, 4$ for Control, Acetic, EtOH, Starve and 32°C, respectively), $(GE)_{ij}$ is the interaction of genotype $i$ with environment $j$, $V_{ijk}$ is the effect of vial $k$ in genotype $i$ and environment $j$, and the error term is $e_{ijkl}$. Environment and genotype are fixed effects, and the vial term is a random effect.

Finally, metabolic characters show extensive correlation structure, and we can ask whether the correlations are homogeneous across the posteclosion rearing media. Here we applied a test of homogeneity of correlations, testing the null hypothesis that correlation coefficients are the same for each pair of traits across the four experimental treatments (Sokal and Rohlf, 1985, p. 520).

## Results

*Changes in mean phenotypes*

The data consisted of 4 samples from each of the 19 genetic lines under each of five environments, for a total of 380 samples scored for weight, protein, lipid, glycogen and activities of nine enzymes. The means and standard errors of all characters in each experimental treatment are presented in Table 2. The among-line variance in the 13 traits was greater in the stressful environment than the control in 25 of the 52 comparisons that could be made. On the other hand, when traits were averaged over environments, the among-line variance of these mean phenotypes was less than the among-line variance in the control environment in 45 of the 52 comparisons. Changes in rank order of genotypes in different environments reduce the among-line variance of the average across environments and tend to retard the loss of variation.

From the phenotypic means of each line in each environment we calculated the changes in phenotype defined as the mean in environment minus the mean in the control treatment (Tab. 3). Changes larger than 20% of the mean were seen for TRI, ADH, ME, PGI and TRE. Many of the patterns of change make physiological sense. Ethanol, acetic acid and starvation generally have a negative effect on glucose metabolism, whereas elevated temperature tends to have a positive effect. Triacylglycerol storage, on the other hand, was reduced in all stressful environments.

Table 2. Means of all traits

| Trait | Control | Acetic | EtOH | Starve | 32°C |
|-------|---------|--------|------|--------|------|
| WT | 0.826 ± 0.013 | 0.858 ± 0.015 | 0.873 ± 0.015 | 0.710 ± 0.010 | 0.883 ± 0.018 |
| PRO | 52.894 ± 1.072 | 49.704 ± 0.599 | 51.570 ± 1.160 | 49.661 ± 1.100 | 53.990 ± 1.306 |
| TRI | 99.095 ± 11.979 | 67.240 ± 1.301 | 68.618 ± 1.822 | 60.206 ± 1.561 | 78.119 ± 2.869 |
| GLY | 14.109 ± 0.045 | 14.159 ± 0.053 | 14.281 ± 0.056 | 14.012 ± 0.040 | 14.224 ± 0.058 |
| ADH | 7.808 ± 0.083 | 8.999 ± 0.260 | 9.738 ± 0.296 | 8.745 ± 0.189 | 8.452 ± 0.110 |
| G6PD | 4.167 ± 0.135 | 4.069 ± 0.143 | 4.299 ± 0.152 | 3.321 ± 0.125 | 4.037 ± 0.105 |
| GPDH | 67.901 ± 2.684 | 62.173 ± 3.079 | 63.943 ± 3.283 | 55.180 ± 2.806 | 69.641 ± 2.972 |
| HEX | 3.441 ± 0.078 | 3.060 ± 0.073 | 3.000 ± 0.082 | 2.882 ± 0.079 | 3.088 ± 0.085 |
| ME | 10.434 ± 0.511 | 8.819 ± 0.379 | 8.895 ± 0.515 | 7.571 ± 0.301 | 9.199 ± 0.355 |
| PGD | 1.961 ± 0.020 | 1.920 ± 0.015 | 1.946 ± 0.016 | 1.894 ± 0.013 | 2.070 ± 0.020 |
| PGI | 14.048 ± 0.785 | 10.609 ± 0.408 | 10.130 ± 0.554 | 11.412 ± 0.540 | 14.394 ± 0.718 |
| PGM | 62.904 ± 0.539 | 65.903 ± 0.539 | 69.829 ± 0.552 | 70.040 ± 0.420 | 71.400 ± 0.572 |
| TRE | 21.270 ± 2.334 | 20.672 ± 2.025 | 11.632 ± 1.326 | 24.453 ± 2.702 | 24.714 ± 2.546 |

The sample size is $n = 76$ for each of the five treatments.

Table 3. Percent changes from control

| Trait | Acetic | Ethanol | Starve | 32°C |
|---|---|---|---|---|
| WT | 3.87 | 6.89 | − 14.06 | 6.89 |
| PRO | − 6.09 | − 2.56 | − 6.17 | 2.01 |
| TRI | − 32.15 | − 30.76 | − 39.24 | − 21.17 |
| GLY | 0.34 | 1.23 | − 0.68 | 0.82 |
| ADH | 15.25 | 24.72 | 2.00 | 8.24 |
| G6PD | 1.42 | 4.77 | − 16.99 | 0.57 |
| GPDH | − 8.08 | − 5.27 | − 17.82 | 3.52 |
| HEX | − 11.40 | − 13.76 | − 16.79 | − 10.58 |
| ME | − 15.48 | − 14.75 | − 27.44 | − 11.84 |
| 6PGD | − 2.08 | − 0.78 | − 3.44 | 5.57 |
| PGI | − 24.48 | − 27.89 | − 18.77 | 2.47 |
| PGM | 4.77 | 11.01 | 11.34 | 13.51 |
| TRE | − 2.81 | − 45.31 | 14.96 | 17.10 |

Table 4. Significance of effects of genotype and post-eclosion environment on metabolic characters

Mean squares

| Trait | Genotype | Environment | G × E | Error |
|---|---|---|---|---|
| WT | 0.194*** | 0.376*** | 0.008 | 1.721 |
| PRO | 115.8 | 279.5 | 79.29 | 87.73 |
| TRI | 9291.*** | 17954*** | 9329*** | 182.2 |
| GLY | 0.167 | 0.795*** | 0.360*** | 0.144 |
| ADH | 4.265 | 38.02*** | 3.213 | 3.145 |
| G6PD | 13.31*** | 10.97*** | 1.568*** | 0.425 |
| GPDH | 3743.*** | 2277.** | 475.38 | 477.5 |
| HEX | 3.052*** | 3.087*** | 0.304 | 0.259 |
| ME | 121.6*** | 79.84*** | 12.64*** | 6.196 |
| 6PGD | 0.081*** | 0.349*** | 0.018 | 0.017 |
| PGI | 64.31*** | 292.9*** | 43.382*** | 23.06 |
| PGM | 57.90*** | 936.6*** | 39.759*** | 14.03 |
| TRE | 386.5 | 2133.*** | 881.64*** | 252.6 |

* $p < 0.05$; ** $p < 0.01$; *** $p < 0.001$.
WT and PRO were analyzed by ANOVA, and the remaining variables were fitted with analysis of covariance with weight and total protein as covariates. The number of degrees of freedom are Environment (4), Genotype (18), G × E (72) and Error (285).

ANOVA was used to ascertain significance of the effects each stress treatment (Tab. 4). All 13 metabolic characters exhibited significant heterogeneity across environments, 9 of the 13 traits showed genotypic differences across lines, and 7 of the 13 exhibited significant genotype × environment interactions. The traits PRO, TRI, GLY, FAS, GPDH, GS and ME exhibited the most highly significant changes caused by the stress media. In addition to WT, the traits that were affected by both stress medium and genotype included G6PD, GP, HEX, PGM, TRE and ADH. Validity of the linear model was tested in part by verifying normality of the residuals with SAS procedure UNIVARIATE.

Table 4 reports the analyses of variance based on measures calculated on a per fly basis. This approach implicitly assumes that the relationship between live weight and each trait is the same across rearing media (e.g. a doubling in weight doubles enzyme activity). The one treatment that significantly affected weight was starvation. Table 4 looks very similar if this treatment is dropped, or if the analysis is repeated calculating each activity on a per milligram live weight basis. Total protein was assayed in order to standardize to a per milligram protein basis. Because total protein did not differ across environmental treatments or genotypes, it seemed unlikely that such standardizing would make a difference. As mentioned above, analysis of covariance treating total protein as a covariate, yielded a table remarkably like the simple ANOVA without weight and protein covariates.

### Genotype × environment interaction

Figure 1 displays 3 of the mean phenotypes for the complete set of 19 lines in five environments. In the case of live weight, it is clear that the genotypes differed in mean phenotype, because the rank order of genotypes remained fairly stable across environments. The effect of starvation is also readily apparent. The relative lack of crossing of lines in this norm of reaction plot

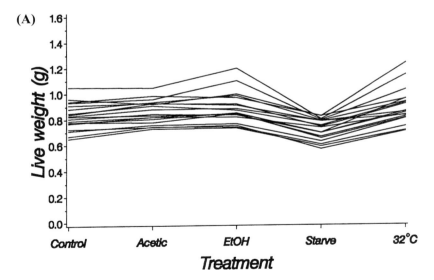

Figure 1. Norm of reaction plots showing the effects on metabolic traits of the five environmental treatments on post-eclosion male *Drosophila*. Each line represents one of the 19 genotypes. (A) Live weight showed little G × E interaction, with a clear depression in weight in the starvation treatment (B) PGM exhibited a clear environment effect, with somewhat more G × E interaction. (C) TRE showed strong G × E, as exhibited by the extent of crossing of lines.

is consistent with a lack of genotype × environment (G × E) interaction for weight. The figure in panel A is consistent with the statistics reported in Table 3: there are both significant genotype and treatment effects, but no interaction. Panel B shows a greater degree of line crossing for PGM, which did exhibit significant genotype × treatment interaction. Trehalase activity showed an extreme level of genotype × environment interaction, depicted by the chaotic pattern of lines in panel C. Figure 2 presents

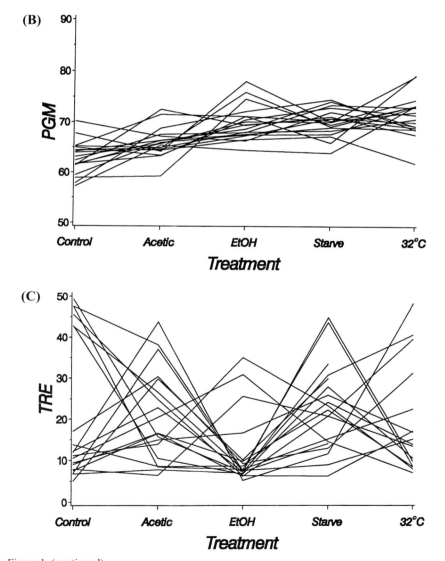

Figure 1 (continued)

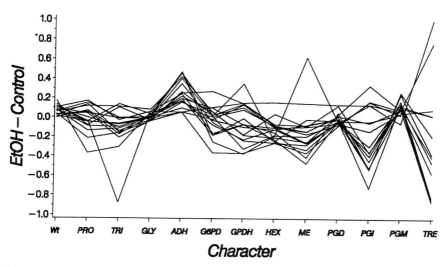

Figure 2. Changes in the 13 phenotypic characters caused by post-eclosion exposure to 3% ethanol for 5 days. Each line represents 1 of the 19 genotypes. Note that some genotypes appear fairly stable (change in phenotype close to zero) for all traits, and others are highly variable. Note also the opposite direction of change for some traits, particularly TRE.

another way to display the phenotypic means, illustrating the degree to which the lines differ in stability across environments. The plot shows that traits like GLY and PGD changed relatively little in comparing controls to ethanol-treated flies, but that ADH and ME, for example, changed to a greater extent and with greater variation among lines. TRI and ME show individual lines with very large effects, and TRE exhibits an exceptional level of heterogeneity in the direction of change among genotypes. Finally, this plot shows that some genotypes are relatively unperturbed in almost all metabolic traits by the ethanol assault, whereas other genotypes exhibit radical changes in several traits.

## Changes in patterns of genetic correlation

Many pairs of enzymes exhibit strong correlations in activity because of coordination of regulation. We can ask whether the stress medium or the selection resulted in altered correlations among metabolic traits by testing the null hypothesis that correlations are homogeneous across treatments. A simple function of the $z$-transformation of Pearson correlation coefficients is expected to have a central chi-square distribution (Sokal and Rohlf, 1985, p. 520), and we used this to provide a test (Tab. 5). Of the 55 test in Table 5, 10 were significant at the 5% level. Many of the correlations were stable across treatments, including the pair of enzymes leading into the

Table 5. Homogeneity of correlations across environments

| C1 | C2 | Cont | Acetic | EtOH | Starve | 32°C | Chi-square |
|----|----|------|--------|------|--------|------|------------|
| TRI | GLY | 0.504 | 0.297 | 0.333 | 0.301 | 0.508 | 1.14 |
| | ADH | 0.329 | − 0.350 | 0.029 | 0.151 | 0.292 | 4.99 |
| | G6PD | − 0.722 | 0.161 | 0.307 | 0.562 | − 0.005 | 21.61 |
| | GPDH | − 0.068 | 0.126 | 0.401 | 0.602 | 0.698 | 6.20 |
| | HEX | − 0.088 | 0.477 | 0.689 | 0.553 | 0.414 | 3.29 |
| | ME | − 0.611 | 0.265 | − 0.054 | − 0.012 | 0.270 | 10.34 |
| | PGD | − 0.662 | − 0.232 | 0.674 | 0.384 | 0.696 | 31.82 |
| | PGI | − 0.255 | 0.214 | 0.014 | 0.250 | 0.058 | 2.62 |
| | PGM | 0.560 | 0.047 | 0.212 | − 0.247 | 0.508 | 7.44 |
| | TRE | − 0.204 | − 0.259 | 0.185 | 0.021 | − 0.462 | 4.03 |
| GLY | ADH | 0.131 | 0.243 | 0.288 | 0.566 | 0.565 | 0.47 |
| | G6PD | 0.377 | 0.054 | 0.319 | 0.152 | 0.024 | 1.02 |
| | GPDH | − 0.075 | 0.074 | 0.298 | 0.242 | 0.555 | 3.35 |
| | HEX | − 0.093 | − 0.036 | 0.251 | 0.122 | 0.432 | 2.84 |
| | ME | 0.148 | − 0.075 | − 0.047 | − 0.346 | 0.601 | 9.76 |
| | PGD | 0.105 | 0.274 | 0.696 | 0.308 | 0.413 | 1.90 |
| | PGI | − 0.226 | − 0.323 | − 0.128 | − 0.171 | 0.345 | 4.43 |
| | PGM | − 0.093 | 0.475 | 0.378 | 0.155 | 0.348 | 2.26 |
| | TRE | 0.304 | 0.182 | 0.053 | 0.455 | − 0.247 | 4.54 |
| ADH | G6PD | − 0.153 | − 0.026 | − 0.093 | 0.426 | 0.234 | 4.05 |
| | GPDH | 0.233 | − 0.165 | 0.209 | 0.169 | 0.588 | 4.71 |
| | HEX | 0.022 | − 0.399 | − 0.192 | 0.084 | 0.553 | 9.70 |
| | ME | − 0.492 | − 0.211 | − 0.154 | − 0.173 | 0.205 | 3.88 |
| | PGD | − 0.368 | 0.332 | 0.339 | 0.499 | 0.306 | 7.16 |
| | PGI | − 0.366 | − 0.472 | − 0.731 | − 0.410 | 0.175 | 6.55 |
| | PGM | 0.390 | 0.542 | 0.649 | 0.631 | 0.300 | 3.83 |
| | TRE | 0.454 | − 0.002 | − 0.391 | 0.121 | 0.006 | 6.64 |
| G6PD | GPDH | 0.431 | 0.184 | 0.461 | 0.533 | 0.197 | 0.83 |
| | HEX | 0.396 | 0.425 | 0.622 | 0.575 | 0.085 | 0.34 |
| | ME | 0.591 | 0.418 | 0.201 | 0.343 | − 0.018 | 2.14 |
| | PGD | 0.735 | 0.472 | 0.513 | 0.611 | 0.141 | 1.10 |
| | PGI | 0.189 | 0.077 | 0.104 | − 0.082 | 0.002 | 0.62 |
| | PGM | − 0.599 | 0.312 | − 0.269 | 0.155 | 0.059 | 10.23 |
| | TRE | 0.273 | − 0.283 | − 0.112 | 0.130 | − 0.234 | 3.78 |
| GPDH | HEX | 0.688 | 0.545 | 0.510 | 0.634 | 0.727 | 9.41 |
| | ME | 0.243 | 0.193 | 0.079 | 0.155 | 0.469 | 0.58 |
| | PGD | 0.333 | 0.301 | 0.371 | 0.654 | 0.793 | 0.52 |
| | PGI | 0.045 | 0.287 | − 0.201 | 0.277 | 0.336 | 2.97 |
| | PGM | 0.024 | 0.102 | 0.206 | − 0.143 | 0.236 | 1.38 |
| | TRE | 0.241 | 0.052 | − 0.132 | 0.425 | − 0.604 | 12.41 |
| HEX | ME | 0.429 | 0.283 | − 0.007 | 0.477 | 0.242 | 0.90 |
| | PGD | 0.430 | − 0.093 | 0.589 | 0.571 | 0.679 | 3.14 |
| | PGI | 0.315 | 0.368 | 0.268 | 0.367 | 0.450 | 2.36 |
| | PGM | − 0.309 | − 0.127 | − 0.115 | − 0.273 | − 0.089 | 0.02 |
| | TRE | − 0.038 | − 0.425 | − 0.012 | 0.372 | − 0.361 | 7.09 |
| ME | PGD | 0.538 | 0.341 | − 0.172 | 0.040 | 0.341 | 4.78 |
| | PGI | 0.575 | 0.384 | − 0.014 | 0.505 | 0.252 | 1.67 |
| | PGM | − 0.488 | 0.167 | − 0.131 | − 0.324 | 0.094 | 5.05 |
| | TRE | − 0.094 | − 0.107 | 0.155 | 0.000 | − 0.217 | 1.21 |
| PGD | PGI | 0.136 | − 0.234 | − 0.162 | − 0.103 | 0.397 | 4.62 |
| | PGM | − 0.422 | 0.590 | 0.377 | 0.405 | 0.215 | 10.35 |
| | TRE | − 0.130 | 0.041 | − 0.217 | 0.438 | − 0.667 | 13.31 |
| PGI | PGM | − 0.651 | − 0.089 | − 0.651 | − 0.706 | − 0.534 | 1.40 |
| | TRE | − 0.030 | 0.140 | 0.250 | 0.029 | − 0.524 | 6.67 |
| PGM | TRE | 0.063 | 0.085 | − 0.285 | 0.044 | 0.080 | 1.69 |

Partial correlations of pairs of characters in each of the five treatments and the chi-square for the test of homogeneity of the correlation coefficients. The chi-square has 3 d.f., and is significant at the 5% level if it exceeds 7.815 (but see text about the issue of multiple tests).

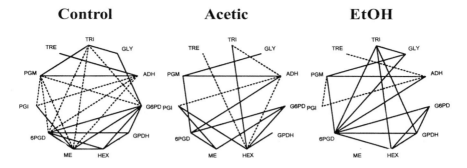

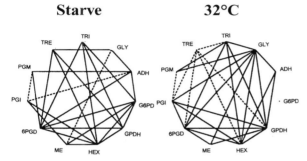

Figure 3. Significant correlations among line means. Positive correlations between pairs of traits are indicated by solid lines, and negative correlations are indicated by dashed lines. Differences among environments are readily apparent.

pentose phosphate shunt, G6PD and 6PGD, but Figure 3 gives the overall impression of rather substantial changes across environments in the pattern of correlations. The similarity of the pattern of correlations in the acetic acid and ethanol treatments is consistent with the close relationship of the entrance of these two compounds into metabolism (two steps apart).

*Ascribing effects to individual chromosomes*

The genotypes that were tested include two sets of balanced chromosome replacements. One set had all possible combinations of the three major chromosomes from a French line and a Zimbabwean line. Designating the first, second and third chromosomes by F or M, these eight lines can be written FFF, FFZ, FZF, FZZ, ZFF, ZFZ, ZZF and ZZZ. ANOVA of the phenotypes from these lines allows tests for differences between each of the F and Z chromosomes on each trait (Tab. 6). A significant chromosome × environment interaction term implies the existence of a gene or

Table 6. Significance of effects of each chromosome on each metabolic trait and on the sensitivity to the environment

| Trait | Environment | I | II | III | I × E | II × E | III × E |
|---|---|---|---|---|---|---|---|
| WT | | | | *** | | | |
| PRO | | | | | | | |
| TRI | | | | | | | |
| GLY | | ** | | | *** | | |
| ADH (2) | | | | | | | |
| G6PD (1) | | * | | *** | | | |
| GPDH (2) | | | | * | | | |
| HEX (1) | * | | | *** | | | |
| ME (3) | ** | * | | | | | |
| PGD (1) | *** | | | | | | |
| PGI (2) | | * | ** | | | ** | |
| PGM (3) | *** | | | | *** | | * |
| TRE (2) | | * | * | * | * | * | ** |

Numbers in parentheses after each enzyme indicate the chromosomal location of the structural gene for the respective enzyme. "Environment" refers to differences among the stress treatments; I, II and III are effects of each major chromosome; and I × E and so on refer to interactions between each chromosome and the environmental stresses. $* p < 0.05$, $** p < 0.01$, $*** p < 0.0001$.

genes on the given chromosome which mediate the degree of phenotypic change when comparing one environment with another. Such genes mediate the environmental lability of the trait. Table 6 shows that significant individual chromosome effects were detected for 8 of the 13 traits, and 4 of the traits showed evidence for genes that modified environmental lability. Table 6 also indicates the chromosomal location of the structural loci for each enzyme. Molecular scoring of alleles at the structural genes has not been done, but note that at least some of the effects of G6PD, PGI and TRE may be due to direct effects and interactions with the structural gene. Figure 4 shows an attempt to illustrate the effect of the chromosome × environment interaction graphically. If a chromosome had no effect on a trait, this figure would appear as a horizontal line. If the different chromosomes had an effect on the trait that was independent of environment, this figure would appear as a set of parallel lines. Crossing lines, as seen in Figure 4, indicate a genotype × environment interaction mediated by the one chromosome. In the example of Figure 4, the different third chromosomes examined in the study differed in the response of TRE activity of the different environments.

## Discussion

Stressful environments often cause immediate physiological changes which are often related to the changes seen in long-term selections to the stress. Selection for resistance to knockdown by ethanol fumes, for exam-

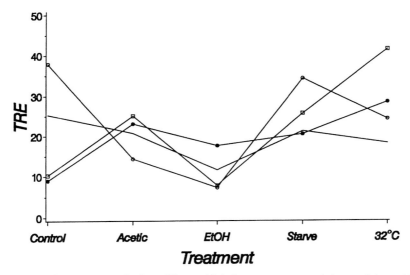

Figure 4. Difference among the four different third chromosomes on trehalase activity and its sensitivity to post-eclosion environmental stress. Symbols are Dot: France, Circle: *rucuca*, Square: Highgrove, Point: Zimbabwe30.

ple, favors higher-activity ADH alleles in both long and short term (Hoffmann and Cohan, 1987). Desiccation acclimation also is probably mediated by the same genetic variation responsible for long-term desiccation resistance (Hoffmann, 1990, 1993). Lines of *Drosophila* reared at 28°C are more tolerant of heat shock than are lines reared at lower temperatures (Cavicchi et al., 1995). Chen (1993) showed a similar relationship between cold pre-treatment and cold-shock tolerance. The Chateau Tahbilk winery population of *D. melanogaster* was shown to have increased resistance to ethanol vapor than flies caught outside the winery, again giving a response that reinforced the short-term physiological response (Hoffmann and McKechnie, 1991). Artificial selection on different concentrations of ethanol resulted in a direct response in increased knockdown resistance, consistent with the winery population result (Oakeshott et al., 1985).

Unfortunately, the ADH story is more complicated, as Bijlsma-Meeles and Bijlsma (1988) observed significant heterozygote advantage even in the absence of ethanol stress. Breaking the difference into selection components, they saw no larval viability difference among ADH genotypes, but they did find overdominance in the adult components fecundity and virility. Being able to identify a simple mechanism (as in the ADH story) does not by itself prove anything.

A close tie between environmental stresses and the mechanism of metabolic regulation adds insight to the cause of correlated and pleiotropic changes. The ability to actively regulate expression of enzymes in a way that adapts to changes in the available food would appear to be advantageous. Effects of dietary carbohydrate and ethanol on enzyme activities have been widely studied in *Drosophila*, and many aspects of the regulation are beginning to be understood. Many of the changes appear to be associated with NADH:$NAD^+$ ratio, including GPDH (Geer et al., 1983), ADH and FAS (Geer et al., 1985). In some cases, the evidence is very clear that dietary changes affect transcript levels. For example, ethanol in the medium clearly induces elevated transcription of *Adh* in *Drosophila* larvae (Geer et al., 1988). Interpretation of increased enzyme activities after a challenge with synthetic medium as induction should be made with caution. High-protein diet causes an increase in xanthine dehydrogenase activity but no change in the amount of cross-reacting material to XDH antibodies (Collins et al., 1970). The change in activity is caused by a post-translational change caused by co-factor binding. Large correlated responses to selection on metabolic traits can also sometimes be understood in terms of simple mechanisms. The 100-fold increase in fitness differences on maltose medium between controls and the 2000-generation selected *Escherichia coli* of Richard Lenski's lab (Leroi et al., 1994; Travisano et al., 1995) is almost certainly due to pleiotropic changes mediated by changes in glucose uptake, since the metabolic steps subsequent to maltose hydrolysis are identical (Bennett and Lenski, 1996).

A minority of correlations between metabolic traits differed significantly across environmental treatments. In another study, directional selection on thorax size in *D. melanogaster* for 23 generations resulted in significant direct effects, and a likelihood ratio test revealed significant changes in the genetic covariance matrices among morphological traits (Wilkinson et al., 1990). Testing correlations on an element-by-element basis provides the advantage in seeing just which terms changed. Geer and Laurie-Ahlberg (1984) compared correlations across media that differed in composition, and they found the genetic correlations to be very stable. In an analysis of larval phenotypes, Geer et al. (1991) saw strong positive correlation between ADH activity, flux from ethanol to fatty acid, and survival on 4.5% ethanol. We did a short-term stress on post-eclosion adults, with the idea that most of the metabolic traits examined are mediated by the fat body, which undergoes a complete turnover from eclosion (when the larval fat body tissues partially remain) to around 5 days of age. We saw reduced triacylglycerol under ethanol stress, probably for the simple reason that the flies were feeding less.

Associations between metabolic changes and fitness are even more difficult to establish accurately. Geer et al. (1991) provided data that show that lines of flies differ in ADH activity, fat storage and survival on ethanol. The lines with the highest ADH activity also deposit the most fat and have the highest survival rate on ethanol. While one might suspect that the higher ADH activity better enables these lines to detoxify the substrate, Geer et al. (1991) give evidence that it is the elevated lipid that provided the survival advantage. Chakir et al. (1996) found strong correlation between tolerance of ethanol and tolerance of acetic acid in the medium, an effect that was most strongly mediated by a gene on the third chromosome (the *Adh* structural gene is on chromosome 2). *D. melanogaster* exhibits striking response to artificial selection on starvation resistance (Chippendale et al., 1996) mediated largely through increased lipid storage during the larval phase. These examples and the results reported here support the idea that genotype × environment interactions are prevalent, and that the pattern of apparent selection on a trait mediated through webs of pleiotropic effects can be radically altered by stressful environments.

*Acknowledgments*
I thank Michael Abraham, Sarah Blass, John Masly, Carrie Tupper and Lei Wang for technical assistance. This work was supported by grant DEB-9419631 from the U.S. National Science Foundation.

# References

Bennett, A.F. and Lenski, R.E. (1996) Evolutionary adaptation to temperature. V. Adaptive mechanisms and correlated responses in experimental lines of *E. coli*. *Evolution* 50: 493–503.

Bijlsma-Meeles, E. and Bijlsma, R. (1988) The alcohol dehydrogenase polymorphism in *Drosophila melanogaster*: Fitness measurements and predictions under conditions with no alcohol stress. *Genetics* 120:743–753.

Blows, M.W. and Sokolowski, M.B. (1995) The expression of additive and nonadditive genetic variation under stress. *Genetics* 140:1149–1159.

Cavicchi, S., Guerra, D., La Torre, V. and Huey, R.B. (1995) Chromosomal analysis of heat shock tolerance in *D. melanogaster* evolving at different temperatures in the laboratory. *Evolution* 49:676–684.

Chakir, M., Capy, P., Genermont, J., Pla, E. and David, J.R. (1996) Adaptation to fermenting resource in *D. melanogaster*: Ethanol and acetic acid tolerances share a common genetic basis. *Evolution* 50:767–776.

Chen, C.-P. (1993) Increase in cold-shock tolerance by selection of cold resistant lines in *Drosophila melanogaster Ecol. Ent.* 18:184–196.

Chippendale, A.K., Chu, T.J.F. and Rose, M.R. (1996) Complex trade-offs and the evolution of starvation resistance in *D. melanogaster*. *Evolution* 50:753–766.

Collins, J.F., Duke, E.J. and Glassman, E. (1970) Nutritional control of xanthine dehydrogenase. I. The effect in adult *Drosophila melanogaster* of feeding a high protein diet to larvae. *Biochimica et Biophysica Act* 208:294–303.

Geer, B.W. and Laurie-Ahlberg, C.C. (1984) Genetic variation in the dietary sucrose modulation of enzyme activities in *Drosophila melanogaster*. *Genet. Res.* 43:307–321.

Geer, B.W., Williamson, J.H., Cavener, D.R. and Cochrane, B.J. (1981) Dietary modulation of glucose-6-phosphate dehydrogenase (EC 1.1.1.4.9) and 6-phosphogluconate dehydrogenase (EC 1.1.1.44) in *Drosophila melanogaster*. In: G. Bhaskaran, S. Friedman and J.G. Rodriguez (eds) *Current Topics in Insect Endocrinology and Nutrition*. pp. 253–282.

Geer, B.W., McKechnie, S.W. and Langevin, M.L. (1983) Regulation of *sn*-glycerol-3-phosphate dehydrogenase in *Drosophila melanogaster* larvae by dietary ethanol and sucrose. *J. Nutr.* 113:1632–1642.

Geer, B.W., Langevin, M.L. and McKechnie, S.W. (1985) Dietary ethanol and lipid synthesis in *Drosophila melanogaster*. *Biochem. Genet.* 23:607–622.

Geer, B.W., McKechnie, S.W., Bentley, M.M., Oakeshott, J.G., Quinn, E.M. and Langevin, M.L. (1988) Induction of alcohol dehydrogenase by ethanol in *Drosophila melanogaster*. *J. Nutr.* 118:398–407.

Geer, B.W., McKechnie, S.W., Heinstra, P.W.H. and Pyka, M.J. (1991) Heritable variation in ethanol tolerance and its association with biochemical traits in *Drosophila melanogaster*. *Evolution* 45:1107–1119.

Gillespie, J.H. and Turelli, M. (1989) Genotype environment interactions and the maintenance of polygenic variation. *Genetics* 121:129–138.

Gupta, A.P. and Lewontin, R.C. (1982) A study of reaction norms in natural populations of *D. pseudoobscura*. *Evolution* 36:934–948.

Hickey, D.A., Benkel, B.F., Fong, Y. and Benkel, B.F. (1994) A *Drosophila* gene promoter is subject to glucose repression in yeast cells. *Proc. Natl. Acad. Sci. USA* 91:11109–11112.

Hoffmann, A.A. (1990) Acclimation for desiccation resistance in *D. melanogaster* and the association between acclimation responses and genetic variation. *J. Insect. Physiol.* 36:885–891.

Hoffmann, A.A. (1993) Selection for adult desiccation resistance in *Drosophila melanogaster*: Fitness components, larval resistance and stress correlations. *Biol. J. Linn. Soc.* 48:43–55.

Hoffmann, A.A. and Cohan, F.M. (1987) Genetic divergence under uniform selection. III. Selection for knockdown resistance to ethanol in *Drosophila melanogaster* populations and their replicate lines. *Heredity* 58:425–443.

Hoffmann, A.A. and McKechnie, S.W. (1991) Heritable variation in resource utilization and response in a winery population of *Drosophila melanogaster*. *Evolution* 45:1000–1015.

Hoffmann, A.A. and Parsons, P.A. (1991) *Evolutionary Genetics and Environmental Stress*. Oxford University Press, Oxford.

Lande, R. (1975) The maintenance of genetic variability by mutation in a polygenic character with linked loci. *Genet. Res.* 26:221–234.

Leroi, A.M., Lenski, R.E. and Bennett, A.F. (1994) Evolutionary adaptation to temperature. III. Adaptation of *E. coli* to a temporally varying environment. *Evolution* 48:1222–1229.

McKechnie S.W. and Geer, B.W. (1984) Regulation of alcohol dehydrogenase in *Drosophila melanogaster* by dietary alcohol and carbohydrate. *Insect Biochem.* 14:231–242.

Oakeshott, J.G., Cohan, F.M. and Gibson, J.B. (1985) Ethanol tolerances of *Drosophila melanogaster* populations selected on different concentrations of ethanol supplemented medium. *Theor. Appl. Genet.* 69:603–608.

Sokal, R.R. and Rohlf, F.J. (1985) *Biometry,* Second Edition. Freeman, San Francisco.

Travisano, M., Vasi, F. and Lenski, R.E. (1995) Long-term experimental evolution in *E. coli*. III. Variation among replication populations in correlated responses to novel environments. *Evolution* 49:189–200.

Turelli, M. (1984) Heritable genetic variation via mutation-selection balance: Lerch's zeta meets the abdominal bristle. *Theor. Pop. Biol.* 25:138–193.

Turelli, M. (1985) Effects of pleiotropy on predictions concerning mutation-selection balance for polygenic traits. *Genetics* 111:165–195.

Via, S., Gomulkiewicz, R., De Jong, G., Scheiner, S.M., Schlichting, C.D. and van Tienderen, P.H. (1995) Adaptive phenotypic plasticity: Consensus and controversy. *Trends Ecol. Evol.* 10:212–218.

Wang, L. and Clark, A.G. (1994) Physiological genetics of the response to a high-sucrose diet by *Drosophila melanogaster*. *Biochem. Genet.* 33:149–164.

Wilkinson, G.S., Fowler, K. and Partridge, L. (1990) Resistance of genetic correlation structure to directional selection in *Drosophila melanogaster*. *Evolution* 44:1190–2003.

Wu, C.-I., Hollocher, H., Begun, D.J., Aquadro, C.F., Xu, Y. and Wu, M.-L. (1995) Sexual isolation in *D. melanogaster*. A possible case of incipient speciation. *Proc. Natl. Acad. Sci. USA* 92:2519–2523.

Yamazaki, T. and Matsuo, Y. (1984) Genetic analysis of natural populations of *D. melanogaster* in Japan. III. Genetic variability of inducing factors of amylase and fitness. *Genetics* 108:223–235.

# Acclimation and response to thermal stress

Environmental Stress, Adaptation and Evolution
ed. by R. Bijlsma and V. Loeschcke
© 1997 Birkhäuser Verlag Basel/Switzerland

# Phenotypic and evolutionary adaptation of a model bacterial system to stressful thermal environments

Albert F. Bennett[1] and Richard E. Lenski[2]

[1] *Department of Ecology and Evolutionary Biology, University of California, Irvine, CA 92697, USA*
[2] *Center for Microbial Ecology, Michigan State University, East Lansing, MI 48824, USA*

*Summary.* We studied both phenotypic and evolutionary adaptation to various thermal environments using the bacterium *Escherichia coli* as an experimental model system. We determined that 42°C was stressful to a bacterial clone adapted to 37°C, based on reductions in both absolute and competitive fitness, as well as induction of a heat stress response. This clone was also used to found replicated populations that were propagated for thousands of generations under several different thermal regimes, including 42°C. Evolutionary adaptation of the populations to 42°C resulted in an increase in both absolute and relative fitness at that temperature, measured respectively as an increase in the number of descendants (and their biovolume) and in competitive ability relative to the ancestral clone. The replicated experimental lineages achieved their evolutionary improvement by several distinct pathways, which produced differential preadaptation to a non-stressful nutrient environment. Adaptation to this stressful temperature entailed neither a change in the ancestral thermal niche nor any pronounced trade-offs in fitness within the thermal niche, contrary to *a priori* predictions. This study system has several important advantages for evaluating hypotheses concerning the effects of stress on phenotypic and evolutionary adaptation, including the ability to obtain lineages that have evolved in controlled and defined environments, to make direct measurements of fitness and to quantify the degree of stress imposed by different environments.

## Introduction

Organisms have the ability to tolerate and survive in a range of different thermal environments, which defines their thermal niche. The niche may be very broad in eurythermic species from temperate regions or quite narrow in stenotherms from polar environments (Precht et al., 1973; Cossins and Bowler, 1987). Regardless of the width of the thermal niche, there is typically a pronounced asymmetry in performance and functional capacity associated with its upper and lower extremes (Huey and Stevenson, 1979; Huey and Kingsolver, 1989). Exposure to progressively lower temperatures lowers performance and slows rate processes, often to the point of immobility, but these decrements may not be lethal in themselves. In fact, many organisms, even some vertebrates, can tolerate freezing (Cossins and Bowler, 1987; Storey, 1992). In contrast, exposure to progressively higher temperatures accelerates performance up to a maximal level, and then further temperature increments cause functional capacity to decline precipitously and result in rapid death. Thus, acute exposure to temperatures at the upper extreme of the thermal niche is liable to be more stressful and damaging than exposure to low temperatures. Since heat stress may become lethal so rapidly, the ability to respond phenotypically to high temperatures may be critical to survival. Over long time periods, the ability of a population to respond genetically and evolve adaptive responses to thermally stressful environments may be crucial to its persistence, particularly in a period of warming climates (Holt, 1990; Karieva et al., 1993). The phenotypic and evolutionary responses of organisms and populations to heat stress are thus of considerable theoretical and practical importance.

Studies of phenotypic and evolutionary responses to heat stress are clearly very desirable. However, discussions and experimental investigations that deal with stress often become bogged down in problems of defining and quantifying the topic of interest (Hoffmann and Parsons, 1991). Many commonsense definitions of stress are so vague that they are not very useful in structuring a quantitative study of its implications for organismal function and survival. A major achievement in studies of stress has been the recognition of the induction of a variety of different gene products (heat shock or stress proteins) in organisms exposed to potentially damaging environments (Morimoto et al., 1990). The production of these proteins may allow an investigator to define operationally whether an environment is indeed stressful and to quantify to some extent the degree of stress imposed. Even so, the organismal consequences of the induction of stress proteins are poorly understood. It is through the differential functioning, survival and reproduction of individual organisms that natural selection operates, and it is the impact of stress on these factors that is ultimately important for evolution. We believe that an analysis of stress, including both its phenotypic and evolutionary consequences, should ultimately be based on these organismal- and population-level consequences. Therefore, we and others (Koehn and Bayne, 1989; Sibly and Calow, 1989; Hoffmann and Parsons, 1991; Lenski and Bennett, 1993; Forbes and Calow, this volume) favor the definition of stress as an environmental factor that causes a reduction in fitness, i.e. the reproductive potential of the individual organism or an entire population. An evolutionary adaptation to stress is one that increases the fitness of a population in the stressful environment.

While quantitative measurements of the impact of stress on fitness are desirable, they are also very difficult to obtain in most kinds of organisms. It may even be difficult to measure the effects of stress on presumptive fitness components, and these difficulties have inhibited progress in investigations of the phenotypic and evolutionary responses to stressful environments, including high-temperature environments. In contrast, it is relatively easy to measure fitness in populations of bacteria. This ability to quantify the impact of stress on fitness in these organisms, along with a suite of other features of their laboratory culture and population structure, makes them ideal subjects for investigating general principles of both phenotypic and evolutionary adaptation to stress. In this chapter, we report the results of our experimental investigations of the phenotypic and evolutionary responses of a model bacterial system to heat stress. This work is an extension of our previous essay on this topic (Lenski and Bennett, 1993). We discuss below the suite of properties that make bacteria such valuable subject species for investigations of stress, and we outline the general features of our experimental system. We next present our findings on the phenotypic (non-genetic) responses of the ancestral bacterial strain, which is adapted to 37°C, to high-temperature stress in the form of acute exposure to 41.5–42°C. We then discuss the evolutionary adaptation of replicated lineages of this strain to persistent high-temperature stress. Finally, we discuss some general conclusions from our research.

## Why bacteria?

It is obvious that bacteria offer many advantages as experimental subjects to a laboratory scientist. They are easy to maintain and propagate in large numbers, and a great wealth of information is available about their molecular and cell biology and genetics. In addition, they offer many special advantages for the experimental evolutionist, in particular someone interested in adaptations to stressful environments. These special features include the following:

1) *Fitness can be readily quantified.* If the best criterion to evaluate the direct phenotypic consequences of stress is fitness reduction (see "Introduction"), then it is necessary to be able to measure fitness accurately and rapidly. For bacterial populations, such measurements are straighforward and can be made in a variety of ways. "Absolute fitness" may be measured by growing replicate cultures in both benign and putatively stressful environments (e.g. moderate and high temperatures) and comparing the impact of the environments on total population production. A reduction in the total number or size (biovolume or biomass) of cells in a population is evidence of stress. Absolute fitness may also be measured simply as the ability of a population to sustain itself in a particular set of environmental conditions. Absolute fitness tests the performance ability of a population of like organisms, in the absence of competition. "Relative fitness" can be measured by experiments involving direct competition between two different populations for a common pool of resources; it is expressed as the differential rate of offspring production during competition. For instance, one can quantify the relative fitness of two different phenotypic (acclimation) states in a defined environment, or one can obtain the relative fitness of genotypes with different evolutionary derivations. If a population is cultured for a long period in a novel environment (e.g. at high temperature), then evolu-

tionary adaptation may lead to increased fitness. This adaptation can be best quantified by direct competition experiments between the ancestral strain and the derived population that was selected in the novel environment. An increase in the rate of offspring production of the derived population relative to that of its ancestor is evidence of evolutionary adaptation, and its mechanistic bases and correlated consequences are then open to further investigation. All these types of fitness measurements are feasible and can be made quickly with bacterial populations.

2) *Experimental evolutionary studies are feasible over relatively short periods of time.* Bacteria may be conveniently propagated in such large numbers that even rare beneficial mutations may occur and be subject to natural selecton (since the frequency of the appearance of a mutation is the product of mutaton rate and population size). It is therefore possible to propagate bacteria in a novel environment (e.g. at high temperature) and to observe genetic change and evolutionary adaptation arising via selection for spontaneous mutations within the population. Because of the large population sizes and rapidity of reproduction, evolutionary changes may occur quite rapidly. For example, we observed evolutionary adaptation in the bacterium *Escherichia coli* to high and stressful temperatures within only 200 generations, which took only 30 days in the experimental regime that we employed (Bennett et al., 1990).

3) *Clonal reproduction facilitates experimental replication and control.* True experiments require rigorus regulation of the variables of interest, replication of treatments, and appropriate controls. These features are all possible in evolutionary experiments using bacteria. The asexual nature of bacterial reproduction facilitates the creation of genetically identical populations, many replicates of which can be cultured simultaneously in a novel (e.g. stressful) environment as well as in the ancestral environment, the latter serving as experimental controls. Since each population is physically separated from the others after its creation, it is an independent replicate of the experiment, and measurements of the performance of replicate cultures can be analyzed statistically to determine whether significant directional change has occurred in the variables of interest. Moreover, all genetic changes arise *de novo* in each population, owing to its clonal origin. Thus, unlike in experiments with most higher organisms, one can be sure with bacteria that parallel responses to selective regimes must have arisen by independent genetic events. Populations propagated in the ancestral environment serve as a control for evolutionary changes that may be associated with features of the culture regime other than those of direct experiment interest to the investigators. Further, the fact that the founding ancestral clone can be frozen and later revived permits it to be used as a comparative basis for measurements of evolutionary change in both the experimental and control lineages. For instance, a direct assessment of evolutionary

adaptation to a stressful environment can be obtained by measuring the reproductive success of an experimentally derived lineage relative to that of its ancestor as they compete with one another in the stressful environment.

4) *Adaptation can be analyzed at all levels of biological organization.* So far in this discussion, we have concentrated on the utility of bacteria for observations at the population and organismal levels of biological organization. In terms of wealth of information, however, it is at the cellular, molecular and genetic levels that bacteria are exemplary experimental organisms. It is no exaggeration to say that in these areas the bacterium *E. coli* is the best investigated and understood organism of all species on earth (Neidhardt et al., 1987). Further, phenotypic responses to heat stress, that is, activation of heat shock proteins, in bacteria are homologous to those in eukaryotes (Neidhardt and VanBogelen, 1987; Gross et al., 1990), indicating that responses to stress are widely shared among very different kinds of organisms. Bacteria have already played an important role in elucidating our understanding of the mechanistic bases of the heat shock response. It should also be possible to use bacteria to investigate the mechanisms underlying the evolutionary response to stressful environments. Clones of bacteria that have been shown to have adapted to stressful environments (in the types of evolutionary experiments discussed above) can be subjected to detailed genetic and molecular analyses to determine the mechanistic bases of that adaptation. The availability of the ancestral clone permits the precise identification of the genetic changes that have occurred in the experimental lineages, even when these changes are few in number. Replicate experimental lineages can be analyzed for the variety of alternative forms and mechanisms of adaptation that may have occurred during the experiment, providing evidence for the diversity (or lack of it) of mechanistic solutions to a common environmental challenge.

5) *Natural diversity in thermotolerance.* The diversity of natural thermal environments occupied by the Bacteria and Archaea is unparalleled. They occur in every environment on earth. Bacteria found in permafrost can be revived and grow when thawed. Psychrophilic bacteria that grow at nearly freezing temperatures occur in Antarctic seas (Straka and Stokes, 1960; Baross and Morita, 1978). Mesophilic bacteria with intermediate temperature optima abound in all temperate and tropical environments, including commensals and pathogens of mesophilic higher organisms. Thermophilic bacteria and archaea from hot springs and hydrothermal vents grow optimally at temperatures of 75 to 105°C (Brock, 1986; Stetter, 1995). Each of these groups has a limited range of growth temperatures: optimal temperatures for mesophiles may be lethally hot for psychophiles and yet too cold to permit growth of thermophiles. Presumably all of these groups have evolved different mechanisms of coping with these incredibly diverse thermal environ-

ments. The comparative potential provided by these groups for the analysis of natural evolutionary adaptation to thermally diverse and stressful environments is unmatched in any other group of organisms.

For a wide variety of reasons, therefore, bacteria provide exceptional opportunities for the investigation of adaptation to thermal stress.

## The model study system

We have studied the phenotypic and evolutionary adaptation of lineages of the mesophilic bacterium *E. coli* B to different thermal environments, including thermally stressful environments (Bennett et al., 1990, 1992; Bennett and Lenski, 1993, 1996, 1997; Leroi et al., 1994a, b; Mongold et al., 1996). Here we summarize briefly some important features of the design of these experiments; readers are referred to the original publications for more details. We founded our study system from a single clone (designated here as the "ancestor") of a bacterial lineage that had previously been propagated at 37°C by serial dilution in minimal glucose (25 μg/ml) medium for 2000 generations (Lenski et al., 1991). During this initial period, that lineage underwent extensive evolutionary adaptation, increasing its relative fitness by over 30% (Lenski et al., 1991). With continued propagation under the same experimental conditions, however, subsequent fitness improvement was relatively slight (<3% during the next 2000 generations, Bennett and Lenski (1996)), indicating that further evolutionary adaptation to the basic culture conditions was becoming increasingly difficult. The use of an ancestor that was already well adapted to a defined set of culture conditions increased the likelihood that temperature-specific adaptations, rather than general adaptations to culture conditions, would be seen when experimental lines were propagated in novel environments. A genetic marker (± ability to metabolize arabinose) was incorporated into the ancestral strain that permitted identification of lines in competition experiments, and the selective neutrality of the marker was verified over a wide range of experimental temperatures.

The thermal niche of the ancestral strain was determined to be between 19 and 42°C (Fig. 1), the range of temperatures over which it can persist in culture with daily 100-fold serial dilution (requiring a growth rate of 6.64 generations per day). Temperatures below 19°C are not lethal but sufficiently retard growth rate so that the bacteria are unable to produce 6.64 generations daily and are therefore eventually diluted to extinction by the serial transfer regime. In contrast, high temperatures inhibit growth completely or kill the bacteria. A population of the ancestor is able to sustain itself indefinitely at 42.1°C, but it is unable to grow at all at 42.3°C and is therefore rapidly diluted to extinction. At 43°C, the ancestor not only cannot grow but dies at a significant rate. These data illustrate the pronounced

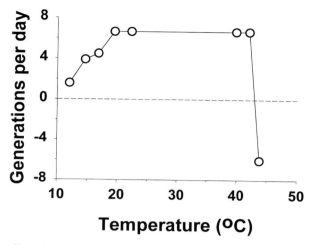

Figure 1. The effect of temperature on population growth rates of the ancestral bacterial clone. Each day, bacterial populations were diluted 100-fold into fresh medium, and the populations grew until nutrients were exhausted. The maximum number of cell divisions possible under this regime is 6.64 ($2^{6.64} = 100$). The thermal niche is defined as the range of temperatures in which this rate of growth occurs. At slower rates of growth, the populations are progressively diluted to extinction. A value of zero indicates no net growth, and negative values indicate cell death as well as dilution. (Data recalculated from Bennett and Lenski, 1993.)

asymmetry of heat and cold stress discussed in the "Introduction": cold temperatures may be inhibiting, but hot temperatures may be rapidly lethal. The thermal niche of this clone is undoubtedly affected by our use of a minimal glucose medium: growth rates are slower and thermal niches consequently narrower in minimal as compared with rich medium (Herendeen et al., 1979). The phenotypic responses of the ancestor to temperatures near its upper lethal limits – but still within its thermal niche – are discussed in the section on "Phenotypic responses to heat stress".

Clones of the ancestral strain were used to found an evolutionary experiment involving groups of populations propagated in four novel thermal environments (constant 20, 32 and 42°C and 32–42°C, the latter being a daily alteration between these two temperatures), as well as continuing propagation in the ancestral environment of 37°C. The ancestral strain lacks plasmids and is strictly asexual, so all groups began with an identical genetic background and all changes in fitness are attributable to *de novo* mutations and subsequent selection. Six experimental lines (3 each of the two genetic marker states for arabinose utilization) were founded for each group (= temperature regime). All 30 lines (five groups with sixfold replication) were propagated by serial dilution in minimal glucose medium for 2000 generations (300 days) under their defined temperature regimes. The absence of cross-contamination among populations and of external contamination by other bacteria was verified by testing appropriate genetic markers. Adaptation of the experimental and control lines to their respective thermal regimes was measured every 200 generations by direct competition experiments against the ancestral form that possessed the opposite genetic marker state. Ancestral and derived lines were preconditioned separately at the experimental temperature for 1 day before being mixed together in fresh medium, so that both competitors were phenotypically acclimated to the experimental conditions. The fitness of a derived line relative to its ancestor was calculated as the ratio of their growth rates according to the formula

$$W = \log(E_f/E_i)/\log(A_f/A_i)$$

where subscripts $i$ and $f$ denote initial and final values, respectively, for the population densities of the evolutionarily derived ($E$) and ancestral ($A$) genotypes. Of particular interest in the context of evolutionary adaptation to heat stress are changes in the lines of the 42°C group, which was propagated at a temperature within 1°C of the upper limit of the ancestral thermal niche. The properties of this group are discussed in the section on "Evolutionary responses to heat stress".

## Phenotypic responses to heat stress

What are the short-term (phenotypic) and long-term (evolutionary) effects of exposure to high temperature on the functioning and reproductive performance of *E. coli*? Can high temperatures be demonstrated to be stressful to this organism? What are the correlated consequences of that stress for organismal performance in other environments? We begin to answer these questions by examining first the phenotypic responses of the ancestral bacterial strain to exposure to temperatures near the upper limit of its thermal niche. The ancestral strain is well adapted to 37°C, and the upper limit of its niche is approximately 42.3°C. What are the proximate phenotypic effects of exposure to temperatures around 42°C?

1) *Depression of absolute fitness.* The effect of temperature on absolute fitness, the net production of cells, can be measured directly from the number and size of descendants produced at different temperatures given a constant nutrient base. The ancestral strain was grown at several different temperatures, in a medium containing a fixed amount of nutrient (25 µg glucose/ml), for 1 day, during which time the supply of glucose was exhausted. The number of descendent bacteria and their average size was then measured with a Coulter counter, and the total biovolume yield was calculated (Fig. 2; A. F. Bennett and R. E. Lenski, unpublished data). The number of cells produced declines markedly as temperature rises much above 37°C. It is halved at 40°C; at 42.0°C, it is only 11% that at 37°C. Mean cell size increases over this temperature

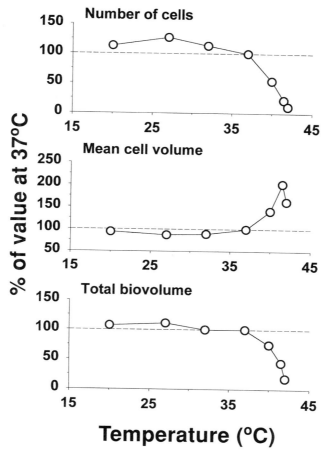

Figure 2. The effect of temperature on total reproductive output of populations of the ancestral clone, given a constant amount of nutrient. Total number of cells produced, the mean cell volume, and their product – the total biovolume produced – were measured with a Coulter cell counter. All values are normalized to values measured at 37°C. (Unpublished data of the authors.)

range (see also Neidhardt and VanBogelen, 1987), at least to some extent because some daughter cells do not separate after replication and thereby form chains of two or more cells (Tsuchido et al., 1986). While the latter effect partially offsets the former effect, overall yield – measured as the total volume of offspring produced – declines, being only 75% at 40°C and 18% at 42.0°C of the biovolume produced at 37°C. These results demonstrate unambiguously that high temperatures are indeed stressful to the ancestral bacterial strain, using our definition of a stressful environment as one that reduces fitness (see Introduction).

2) *Heat shock response.* Growth of the ancestral strain at 41.5°C results in increased thermotolerance at lethally higher temperatures (Leroi et al., 1994a): the death rate of cells at 50°C is only half that of cells that were grown at 32°C. This increased thermotolerance is the original definition of the heat shock response (Neidhardt and VanBogelen, 1987). Exposure to 42°C activates numerous heat shock genes and induces the expression of many heat shock (stress) proteins in *E. coli* (Neidhardt and VanBogelen, 1981, 1987; Gross et al., 1990). The latter are presumably involved in increased tolerance to heat and other forms of stress, but exact causal and mechanistic associations among these factors are still poorly understood (Neidhardt and VanBogelen, 1987; Watson, 1990).

3) *Decreased competitive ability.* It is generally believed that phenotypic adjustments to new environments improve function in those environments (e.g. Hoffmann and Parsons, 1991; Rome et al., 1992). In the case of exposure to different temperatures, a host of different physiological adjustments, termed "acclimation" or "acclimatization", occurs (Hochachka and Somero, 1984; Cossins and Bowler, 1987); at high temperatures, the heat shock response is one of these. According to the "beneficial acclimation assumption" (Leroi et al., 1994a), these resulting phenotypic changes benefit the organism and improve its functional capacities at that temperature in comparison with phenotypes that would result from prior exposure to other temperatures. Contrary to that assumption, however, we found that exposure to heat stress actually handicaps the performance of the ancestral bacterial strain at high but nonlethal temperatures, in comparison with identical bacteria without prior heat exposure (Leroi et al., 1994a). We acclimated cultures of the ancestral strain for one day at either 32 or 41.5°C, then reciprocally cross-competed them the next day and measured their relative fitness according to their acclimation state (Tab. 1). The bacteria acclimated to the competition temperature had prior exposure and underwent various physiological adjustments to that temperature, and so we anticipated that they should be competitively superior to non-acclimated forms. This was indeed the case in competition at 32°C, where the 32°C-acclimated form had higher fitness than the 41.5°C-acclimated form. But at 41.5°C, the 32°C-acclimated form was again competitively superor, with a relative fitness advantage of about 17%. Prior exposure to heat stress in fact disabled the bacteria and made them less able to compete, rather than improving their performance in the acclimation environment. The reduced competitive fitness caused by acclimation to 41.5°C might reflect some physiological handicap associated with expression of stress proteins when they are not actually needed to prevent heat damage (Leroi et al., 1994a). For example, synthesis of most other proteins is repressed by the heat shock response (Neidhardt and VanBogelen, 1987; Watson, 1990). However, further studies at non-

Table 1. The effect of thermal acclimation on relative fitness of the ancestral strain at high temperature

| Acclimation temperatures | | Competition temperature | Mean fitness | p (W = 1) |
|---|---|---|---|---|
| 32°C | 32°C | 32°C | 1.006* | 0.6 |
| 41.5°C | 41.5°C | 41.5°C | 0.993* | 0.6 |
| 32°C | 41.5°C | 32°C | 0.921+ | < 0.001 |
| 32°C | 41.5°C | 41.5°C | 0.827+ | < 0.001 |

Populations of the ancestral strain expressing reciprocal genetic markers were acclimated to either 32 or 41.5°C for 1 day and then allowed to compete against one another during a second day. Mean fitness (W) of the two different forms is calculated either as the fitness of the Ara+ form relative to that of the Ara− form (*), or as the fitness of the 41.5°C-acclimated form relative to that of the 32°C-acclimated form (+). Differences in fitness between the acclimation states are indicated by statistically significant departures from a mean fitness value of 1. Bacteria acclimated to 32°C are superior to those acclimated to 41.5°C during subsequent competition at either 32 or 41.5°C. (Data from Leroi et al., 1994a.)

stressful temperatures (Bennett and Lenski, 1997) have found other examples of competitive inferiority associated with acclimation, so the mechanistic basis of this fitness depression remains unknown.

## Evolutionary responses to heat stress

The ancestral bacterial strain, which had evolved at and adapted to 37°C, is clearly stressed at 41–42°C. As outlined in the previous section, exposure to these temperatures depresses both net production and competitive fitness. What then is the effect of long-term exposure to these thermally stressful temperatures? Do evolutionary changes occur that better adapt the lineage to function at these temperatures? Are there any general patterns in the direct responses to these temperatures or in the correlated responses at other temperatures? What are the mechanisms underlying adaptation to heat stress? We have begun to investigate these questions experimentally by examining the properties of the 42°C group, comprising six replicated experimental lines derived from the ancestral strain that were propagated at 41–42°C for 2000 generations. In this group, we can observe directly the evolution of adaptation to heat stress. The following are some of the main findings from this evolution experiment.

1) *Improvement in absolute fitness.* After 2000 generations of culture at 41–42°C, the absolute fitness, or net productivity, of the lines of the 42°C group had improved substantially in the thermally stressful environment (Fig. 3). At 41.5°C, the number of descendants produced from an equal amount of nutrient is twice that produced by the ancestral strain. Average cell size did not change greatly, so the biovolume yield at 41.5°C is double that of the ancestor. In fact, the mean biovolume of

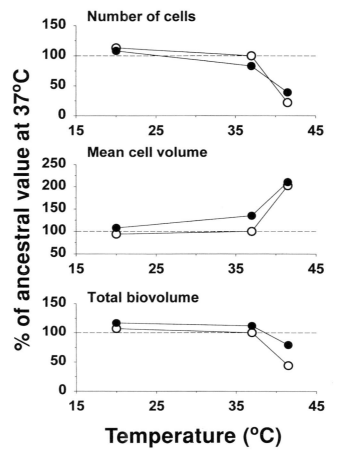

Figure 3. The effect of temperature on reproductive output of the ancestral clone (open circles) and the 42°C experimental group (filled circles). Mean values of the six experimental lines of the latter group are shown. Total number of cells produced, the mean cell volume, and their product – the total biovolume produced – were measured with a Coulter cell counter. All values are normalized to those of the ancestral clone at 37°C. (Unpublished data of the authors.)

the selected lines at 41.5°C is now 80% that of the ancestral clone at is own evolutionary temperature of 37°C. Yield efficiency, the conversion of energy into biomass, has consequently improved substantially during evolution in the thermally stressful environment. This dramatic improvement in productivity at 41.5°C is correlated with more modest but also significant gains at other temperatures in the thermal niche. In the 32–42°C group as well, improvement in biovolume yield at 41.5°C, the thermally stressful component of the environment, was greater than that at 32°C, the non-stressful component (A. F. Bennett and R. E. Lenski, unpublished data). Clearly, evolution resulted in extensive improvement under heat stress in these experimental groups.

2) *Rapid and extensive improvement in competitive fitness.* Within 200 generations (30 days), the mean fitness of the six experimental lines of the 42°C group had increased about 8.5% at that temperature (Bennett et al., 1990). This 8.5% advantage accrues each and every generation, and so after only 1 day of competition ($\sim$ 6.6 generations of binary fission), the average derived cell has almost 50% more descendants than does a cell of the ancestral strain [exp $(0.085 * 6.6 * \ln(2)) = 1.48$]. No change in fitness at 37°C had occurred in this time in the 37°C group, indicating that the fitness improvement in the 42°C group was probably the result of specific adaptation to temperature and not general adaptation to other aspects of culture conditions. In further support of that conclusion, even after 400 generations, no increase in fitness was observed in any experimental or control group at its selective temperature besides the 42°C group. After 200 generations, the mean fitness of the lines of the 42°C group had increased $\sim$33.5% when assayed at 42°C, a substantially greater improvement than occurred in any other group at its corresponding selective temperature (Tab. 2). Compounding this 33.5% fitness difference as above indicates that the average individual cell in the 42°C group has almost fivefold more descendants than does the ancestor after only 1 day of competition at that temperature. Interestingly, the fitness improvement of the 32−42°C group was also far greater at 42°C than it was at 32°C, even though the group spent the same amount of time (and had nearly equal numbers of generations) at each temperature. Consequently, we conclude that in our study system evolutionary adaptation to environments involving heat stress was far more rapid and extensive than it was to any other thermal environments, including those near the lower extreme of the thermal niche. This finding thus supports the assertion that stressful environments should promote rapid evolutionary change (e.g. Parsons, 1987, this volume; Hoffmann and Parsons, 1991).

Table 2. Fitness of different experimental groups, relative to that of their common ancestor, after 2000 generations in their respective selective environments

| Experimental group | Assay temperature | Mean fitness ± 95% C. L. |
|---|---|---|
| 20°C | 20°C | 1.087 ± 0.027 |
| 32°C | 32°C | 1.107 ± 0.028 |
| 37°C | 37°C | 1.025 ± 0.020 |
| 42°C | 42°C | 1.335 ± 0.168 |
| 32−42°C | 32−42°C | 1.174 ± 0.068 |
| 32−42°C | 32°C | 1.049 ± 0.016 |
| 32−42°C | 42°C | 1.267 ± 0.079 |

All fitness values are significantly greater than 1, indicating evolutionary adaptation to the selective environments (Data from Bennett and Lenski, 1996.); C. L.: Confidence Limits.

3) *Evolution of acclimation response.* Did the phenotypic acclimation response of the 42°C group also evolve during adaptation to high temperature? The previous measurements on competitive fitness of this group involved both acclimation and competition at high temperature. Part of the observed fitness improvement may therefore have been due to improvements in the phenotypic acclimation response at high temperature. If the magnitude of the observed fitness benefit were reduced by acclimation to the ancestral temperature prior to competition at high temperature, such a difference would indicate that the acclimation effect itself had evolved in a way that was beneficial during the constant exposure to high temperature for 2000 generations. We ran paired competition experiments between the lines of the 42°C group and their common ancestor (A. F. Bennett and R. E. Lenski, unpublished data). All of the competition experiments were performed at 41.5°C, but in one half of the experiment both competitors were acclimated to 41.5°C and in the other half both were acclimated to 37°C. The difference in relative fitness between the two treatment pairs provides a measure of the evolutionary change in the effect of acclimation (Bennett and Lenski, 1997). Averaging over all six lines, the mean fitness relative to the common ancestor was 1.40 following acclimation at 41.5°C but only 1.24 following acclimation to 37°C (A. F. Bennett and R. E. Lenski, unpublished data). This difference is only marginally significant ($t = 2.161$, 5 d. f., $p = 0.083$) due to considerable heterogeneity among the lines in the effect of acclimation. However, at least two of the lines had much higher fitnesses following acclimation to 41.5°C than they did following acclimation to 37°C. Therefore, in at least some of the lines, a substantial portion of their improved fitness at high temperature depends on acclimation to high temperature. Hence, not only can acclimation influence fitness directly (as shown in the section on phenotypic responses of the ancestor to heat stress), but the acclimation benefit itself can improve during evolution in a thermally stressful environment.

4) *Diversity of adaptive pathways.* Within the 42°C group are six independently derived experimental lines, each of which adapted significantly to this stressful environment. They all began with a common genetic background, and all changes were the result of natural selection acting on mutations that arose *de novo* within the independent lineages. Was there a common adaptive mechanism in all six lines, or did the evolving bacteria "discover" a diversity of adaptive solutions to the problem of heat stress? We do not yet know the physiological mechanisms by which these experimental lines adapted to heat stress, although temperature-specific alterations in glucose uptake appear to be involved in some cases (Bennett and Lenski, 1996). By analyzing heterogeneity in patterns of performance (e.g. fitness in different environments), we can derive minimal estimates for the number of phenotypically distinct lines

within the selected group. For instance, analysis of variance of relative fitness in the selective environment (42°C, minimal glucose medium) indicates two distinct clusters of lines which are statistically different from each other (Bennett and Lenski, 1996). Competition in another nutrient (42°C, minimal maltose medium) also indicates two statistically distinct clusters of lines (Bennett and Lenski, 1996). Analysis of the number of descendant cells and average cell size in the selective environment also indicates two significantly different clusters (A. F. Bennett and R. E. Lenski, unpublished data). Therefore, we conclude that multiple, two or more, adaptive pathways were followed by these lines in their evolutionary adjustment to this stressful environment.

5) *Little change in the thermal niche.* The lines of the 42°C group were maintained within 1°C of their upper limit for persistence in serial dilution culture. During 2000 generations, they experienced considerable adaptive evolution at that temperature. In the course of that evolution, was there an increase in heat tolerance and a shift in the upper boundary of the thermal niche to even higher temperatures? And was there a loss of cold tolerance associated with adaptation to heat stress? Surprisingly, the thermal niche of the 42°C did not change much during its adaptation to high temperature (Bennett and Lenski, 1993). Only one of the six experimental lines increased its upper thermal limit, and that increment was less than 1°C; the mean upper limit of the thermal niche for the group as a whole did not change significantly from that of the ancestor. Likewise, the lower boundary of the thermal niche did not change significantly from its ancestral condition, about 19°C. In spite of fitness gains in the high-temperature environment, the broader thermotolerance of this experimental group was essentially unaltered. In contrast, the 20°C group did experience a downward shift in its thermal niche, decreasing both its lower and upper thermal limits by 1−2°C (Mongold et al., 1996).

Although no thermophilic forms evolved in our main evolution experiments, further experiments at temperatures in excess of 43°C yielded a small number of mutant forms capable of surviving at these temperatures (Bennett and Lenski, 1993; J. Mongold, unpublished observations). Their thermophily persists even after storage and growth at lower temperatures, indicating that the effect is genetic (and not merely delayed phenotypic acclimation to high temperature). We call them "Lazarus" mutants, because they emerge from otherwise dying populations. The upper boundary of their thermal niche is not greatly extended (none can persist in minimal glucose medium above 45°C), but they are capable of growing at temperatures 1−2°C higher than can the ancestral strain. Thus, while it is possible for more thermophilic forms to emerge from the genetic background of the ancestral bacterial strain, none became fixed in our main experiment, even among those lines that experienced 42°C. Preliminary studies indicate that, although the Lazarus mutants can grow at higher temperatures, their fitness within the ancestral thermal niche is somewhat compromised (J. Mongold, personal communication). Consequently, if one had arisen in an experimental population, it would have been outcompeted by more mesophilic genotypes. These Lazarus mutants can apparently prosper only under the hard selective regime of temperatures that are lethal to the rest of the population.

6) *No tradeoffs in fitness at other temperatures.* Many evolutionary models assume that adaptation to one environment necessarily entails a loss of fitness in some other environments; in the case of thermal environments, improvement in function at one temperature is assumed to be accompanied by poorer performance at temperatures at the opposite end of the thermal niche (e.g. Levins, 1968; Lynch and Gabriel, 1987; Pease et al., 1989). The 42°C group experienced considerable improvement in its absolute and competitive performance at 41–42°C. Although there was essentially no change in its thermal niche, did it lose competitive ability at lower temperatures within that niche? That is, was there a tradeoff in performance, and ultimately fitness, associated with evolutionary adaptation to high temperature? Surprisingly again, no general tradeoff was detected. All six lines in the 42°C group had higher fitness than the ancestor between about 40 and 42°C. One of the six lines in this group showed a modest reduction in fitness at low temperatures, but the mean fitness of the group as a whole did not differ significantly from that of the ancestor between about 19 and 37°C (Fig. 4). Therefore, it is evidently possible for these bacteria to undergo

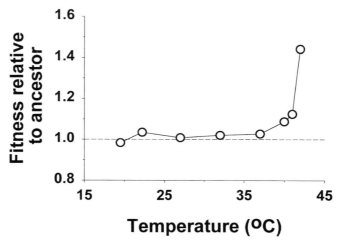

Figure 4. The effect of temperature on the fitness of the 42°C group relative to the ancestral clone, measured during direct competition. Mean values of the six experimental lines are reported. (Data recalculated from Bennett and Lenski, 1993.)

extensive evolutionary adaptation to a stressfully high temperature without simultaneously experiencing a loss of fitness at lower temperatures.

7) *Unexpected preadaptation to other environments.* Expression of heat shock proteins enhances not only thermotolerance but also resistance to other forms of stress (e.g. ethanol exposure, Neidhardt and VanBogelen (1987)). In fact, they are sometimes termed "stress proteins" in recognition of this cross-resistance. Thus, phenotypic responses to heat stress are already known to enhance function in a variety of different stressful environment. What we found in our experiments was that evolutionary responses to heat stress pleiotropically enhanced fitness in other environments. Specifically, during their evolution in minimal glucose medium under thermal stress, five of the six lines in the 42°C group became even more fit in maltose than in glucose, indicating a hyper-preadaptation to this novel nutrient environment (Bennett and Lenski, 1996). Averaging over all six lines, this group had a mean fitness relative to the ancestor of 1.55 in maltose at 42°C, as compared with a mean fitness of 1.34 in glucose at 42°C. The fitness differential in maltose implies that a derived cell has about 12 times as many descendants as an ancestral cell after only 1 day of competition between them. Maltose is a glucose dimer, but it is transported into the bacterial cell through a different pathway from that used for glucose transport (Nikaido and Saier, 1992). Evidently, adaptation to heat stress in one nutrient environment had the correlated consequence of a spectacular inprovement in fitness in another nutrient environment, even though this nutrient was never present during the selective history. This and other correlated responses are presumed to arise as pleiotropic side-effects of the selected mutations, and they may reveal important information about the physiological mechanisms of evolutionary adaptation to heat stress.

## Conclusions

The significance of our findings does not lie in the elaboration of the precise patterns and mechanisms by which our experimental organisms adapted to different thermal environments. In the most narrow sense, our results apply only to the single ancestral genotype of *E. coli* that we used to found these experiments. Each experimental organism will have its own particular features and, moreover, be subject to historical constraints and stochastic effects (Travisano et al., 1995). Only comparative studies of widely different kinds of organisms will reveal truly general features and patterns of adaptation to different environments, such as thermal stress.

However, what our experiments can do is to provide an extraordinarily comprehensive view and rigorous analysis of the effects of heat stress in one mesophilic organism and the evolutionary responses to that stress. Our results are useful in evaluating general assertions about the effects of stress and patterns of evolutionary response to stressful environments. As the first necessary step, we have demonstrated unambiguously that the high-temperature environment is stressful to the ancestral organism by a variety of criteria: activating a stress response, depressing net productivity and reducing competitive fitness. By observing how the experimental lines adapt phenotypically and evolutionarily to this stress, we can evaluate the generality of various assertions about adaptation to stressful environments. We found, for instance, evidence to support the prediction that evolutionary rates in stressful environments should exceed those in non-stressful environments. Contrary to *a priori* expectations, however, we found that an increase in heat tolerance and an upward displacement of thermal niche is not a necessary correlate of evolutionary adaptation to a stressfully high temperature. We also found that tradeoffs (that is, performance decrements in other protions of the niche) are not necessary correlates of adaptation to heat stress: increments in fitness at one temperature extreme were often not accompanied by a loss of fitness at the other temperature extreme. Also contrary to expectations, prior exposure and acclimation to heat stress was shown not to be beneficial in the stressful environment, but rather actually to decrease competitive ability and reproductive success. Thus, many widely held expectations about general patterns of evolution in stressful environments were contradicted by this experimental system. Some of these expectations are explicitly built into models that seek to predict evolutionary responses to environmental change, such as global warming (e.g. (Lynch and Gabriel, 1987; Pease et al., 1989; Bürger and Lynch, this volume). It seems to us that these assumptions have to be re-examined, considering that a rigorous experimental test has failed to lend support to them.

Our study also produced other intriguing results for which there were no particular *a priori* expectations. For instance, the increment in yield efficiency associated with evolutionary adaptation to high temperature is not necessarily anticipated by evolutionary theory (Vasi et al., 1994): differential reproduction is anticipated to be rewarded, but faster growth may occur by either increased or decreased efficiency. How general is the increase in bioenergetic efficiency in adaptive evolution? Another interesting result is the hyper-preadaptation to the novel maltose environment. We had no *a priori* expectation of this result. Are there many other unexpected correlated consequences of adaptation to stressful environments? Are such extreme cases of preadaptation more common during adaptation to stressful than to non-stressful environments? Thus, as with any good experimental system, we are able not only to test existing hypotheses but also to generate new questions and hypotheses of general interest for further examination. The great utility of our system is that it has the flexibility to test both phenotypic and evolutionary hypotheses, generate new ones and then perform further evolutionary experiments to test these too.

*Acknowledgments*
Many people have contributed to the ideas and experiments in our studies. In particular we thank P. McDonald and S. Simpson for their excellent technical assistance with this work and our co-authors K. Dao, A. Leroi, J. Mittler and J. Mongold. This research has been supported the U.S. National Science Foundation, currently grants to the authors (IBN-9507416) and to the Center for Microbial Ecology at Michigan State University (BIR-9120006).

# References

Baross, J.A. and Morita, R.Y. (1978) Microbial life at low temperature: Ecological aspects. *In*: D.J. Kushner (ed.): *Microbial Life in Extreme Environments*. Academic Press, New York, pp. 9–71.

Bennett, A.F., Dao, K.M. and Lenski, R.E. (1990) Rapid evolution in response to high temperature selection. *Nature* 346:79–81.

Bennett, A.F. and Lenski, R.E. (1993) Evolutionary adaptation to temperature. II. Thermal niches of experimental lines of *Escherichia coli*. *Evolution* 47:1–12.

Bennett, A.F. and Lenski, R.E. (1996) Evolutionary adaptation to temperature. V. Adaptive mechanisms and correlated responses in experimental lines of *Escherichia coli*. *Evolution* 50:493–503.

Bennett, A.F. and Lenski, R.E. (1997) Evolutionary adaptation to temperature. VI. Phenotypic acclimation and its evolution in *Escherichia coli*. *Evolution* 51:36–44

Bennett, A.F., Lenski, R.E. and Mittler, J.E. (1992) Evolutionary adaptation to temperature. I. Fitness responses of *Escherichia coli* to changes in its thermal environment. *Evolution* 46:16–30.

Brock, T.D. (1986) *Thermophilic Microorganisms and Life at High Temperatures*. Springer-Verlag, New York.

Cossins, A.R. and Bowler, K. (1987) *Temperature Biology of Animals*. Chapman and Hall, New York.

Gross, C., Straus, D.B., Erickson, J.W. and Yura, T. (1990) The function and regulation of heat shock proteins in *Escherichia coli*. *In*: R.I. Morimoto, A. Tissieres and C. Georgopoulos, (eds). *Stress Proteins in Biology and Medicine*. Cold Spring Harbor Laboratory Press, Cold Spring Harbor, New York, pp. 167–189.

Herendeen, S.L., VanBogelen, R.A. and Neidhardt, F.C. (1979) Levels of major proteins of *Escherichia coli* during growth at different temperatures. *J. Bacteriol.* 139:185–194.

Hochachka, P.W. and Somero, G.N. (1984) *Biochemical Adaptation*. Princeton University Press, Princeton.

Hoffmann, A.A. and Parsons, P.A. (1991) *Evolutionary Genetics and Environmental Stress*. Oxford University Press, Oxford.

Holt, R. (1990) The microevolutionary consequences of climate change. *Trends Ecol. Evol.* 5:311–315.

Huey, R.B. and Kingsolver, J.G. (1989) Evolution of thermal sensitivity of ectotherm performance. *Trends Ecol. Evol.* 4:131–135.

Huey, R.B. and Stevenson, R.D. (1979) Integrating thermal physiology and ecology of ectotherms: A discussion of approaches. *Am. Zool.* 19:367–384.

Karieva, P.J., Kingsolver, J.G. and Huey, R.B. (1993) *Biotic Interactions and Global Change*. Sinauer Associates, Sunderland, MA.

Koehn, R.K. and Bayne, B.L. (1989) Towards a physiological and genetical understanding of the energetics of the stress response. *Biol. J. Linn. Soc.* 37:157–171.

Lenski, R.E. and Bennett, A.F. (1993) Evolutionary response of *Escherichia coli* to thermal stress. *Am. Nat.* 142:S47–S64.

Lenski, R.E., Rose, M.R., Simpson, S.C. and Tadler, S.C. (1991) Long-term experimental evolution in *Escherichia coli*. I. Adaptation and divergence during 2000 generations. *Am. Nat.* 138:1315–1341.

Leroi, A.M., Bennett, A.F. and Lenski, R.E. (1994a) Temperature acclimation and competitive fitness: An experimental test of the Beneficial Acclimation Assumption. *Proc. Nat. Acad. Sci. USA* 91:1917–1921.

Leroi, A.M., Lenski, R.E. and Bennett, A.F. (1994b) Evolutionary adaptation to tempera-
    ture. III. Adaptation of *Escherichia coli* to a temporally varying environment. *Evolution*
    48:1222–1229.
Levins, R. (1968) *Evolution in Changing Environments*. Princeton University Press, Princeton.
Lynch, M. and Gabriel, W. (1987) Environmental tolerance. *Am. Natl.* 129:283–303.
Mongold, J.A., Bennett, A.F. and Lenski, R.E. (1996) Evolutionary adaptation to temperature.
    IV. Adaptation of *Escherichia coli* at a niche boundary. *Evolution* 50:35–43.
Morimoto, R., Tissieres, A. and Georgopoulos, C. (eds) (1990) *Stress Proteins in Biology and
    Medicine*. Cold Spring Harbor Laboratory Press, Cold Spring Harbor, New York.
Neidhardt, F.C., Ingraham, J.L., Low, K.B., Magasanik, B., Schaechter, M. and Umbarger, H.E.
    (eds) (1987) *Escherichia coli and Salmonella typhimurium: Cellular and Molecular Biology*.
    American Society for Microbiology, Washington, D.C.
Neidhardt, F.C. and VanBogelen, R.A. (1981) Positive regulatory gene for temperature-control-
    led proteins in *Escherichia coli*. *Biochem. Biophys. Res. Commun.* 100:894–900.
Neidhardt, F.C. and VanBogelen, R.A. (1987) Heat shock response. *In*: F.C. Neidhardt, J.L.
    Ingraham, K.B. Low, B. Magasanik, M. Schaechter and H.E. Umbarger (eds): *Escherichia
    coli and Salmonella typhimurium: Cellular and Molecular Biology*. American Society for
    Microbiology, Washington, D.C., pp. 1334–1345.
Nikaido, H. and Saier, M.H. Jr. (1992) Transport proteins in bacteria: Common themes in their
    design. *Science* 258:936–942.
Parsons, P.A. (1987) Evolutionary rates under environmental stress. *Evol. Biol.* 21:311–347.
Pease, C.M., Lande, R. and Bull, J.J. (1989) A model of population growth, dispersal and evo-
    lution in a changing environment. *Ecology* 70:1657–1664.
Precht, H., Christofersen, J., Hensel, H. and Larcher, W. (1973) *Temperature and Life*. Springer-
    Verlag, Berlin.
Rome, L.C., Stevens, E.D. and John-Alder, H.B. (1992) The influence of temperature and
    thermal acclimation on physiological function. *In*: M.E. Feder and W.W. Burggren (eds):
    *Environmental Physiology of the Amphibians*. University of Chicago Press, Chicago, pp.
    183–205.
Sibly, R.M. and Calow, P. (1989) A life-cycle theory of responses to stress. *Biol. J. Linn. Soc.*
    37:101–116.
Stetter, K.O. (1995) Microbial life in hyperthermal environments. *Am. Soc. Micro. News* 61:
    285–290.
Storey, K.B. (1992) Natural freeze tolerance in ectothermic vertebrates. *Annu. Rev. Physiol.* 54:
    619–637.
Straka, R.D. and Stokes, J.L. (1960) Psychrophilic bacteria from Antarctica. *J. Bacteriol.*
    80:622–625.
Travisano, M., Mongold, J.A., Bennett, A.F. and Lenski, R.E. (1995) Experimental tests of the
    roles of adaptation, chance and history in evolution. *Science* 267:87–90.
Tsuchido, T., VanBogelen, R.A. and Neidhardt, F.C. (1986) Heat-shock response in *Escherichia
    coli* concerns cell division. *Proc. Natl. Acad. Sci. USA* 83:6959–6963.
Vasi, F., Travisano, M. and Lenski, R.E. (1994) Long-term experimental evolution in *Escheri-
    chia coli*. II. Changes in life-history traits during adaptation to a seasonal environment. *Am.
    Nat.* 144:432–456.
Watson, K. (1990) Microbial stress proteins. *Adv. Microb. Physiol.* 31:183–223.

Environmental Stress, Adaptation and Evolution
ed. by R. Bijlsma and V. Loeschcke
© 1997 Birkhäuser Verlag Basel/Switzerland

# Ecological and evolutionary physiology of heat shock proteins and the stress response in *Drosophila*: Complementary insights from genetic engineering and natural variation

Martin E. Feder[*,+] and Robert A. Krebs[*]

*Department of Organismal Biology and Anatomy,[*] The Committee on Evolutionary Biology[+], and The College,[+] The University of Chicago, 1027 East 57th Street, Chicago, IL 60637, USA*

*Summary.* Classical adaptational and genetic engineering approaches offer complementary insights to understanding biological variation: the former elucidates the origins, magnitude and ecological context of natural variation, while the latter establishes which genes can underlie natural variation. Studies of the stress or heat shock response in *Drosophila* illustrate this point. At the cellular level, heat shock proteins (Hsps) function as molecular chaperones, minimizing aggregation of peptides in non-native conformations. To understand the adaptive significance of Hsps, we have characterized thermal stress that *Drosophila* experience in nature, which can be substantial. We used these findings to design ecologically relevant experiments with engineered *Drosophila* strains generated by unequal site-specific homologous recombination; these strains differ in *hsp70* copy number but share sites of transgene integration. *hsp70* copy number markedly affects Hsp70 levels in intact *Drosophila*, and strains with extra *hsp70* copies exhibit corresponding differences in inducible thermotolerance and reactivation of a key enzyme after thermal stress. Elevated Hsp70 levels, however, are not without penalty; these levels retard growth and increase mortality. Transgenic variation in *hsp70* copy number has counterparts in nature: isofemale lines from nature vary significantly in Hsp70 expression, and this variation is also correlated with both inducible thermotolerance and mortality in the absence of stress.

## Evolutionary adaptational biology and genetic engineering as complementary approaches

The major issues concerning stress, adaptation and evolution include how organisms vary in their tolerance of stress; the biochemical, physiological, morphological and behavioral mechanisms that underlie stress tolerance; and the evolutionary processes that create and sustain variation in stress tolerance. Investigations of these questions can proceed from at least two alternative research strategies (Fig. 1). In the evolutionary adaptationist program, investigators either locate existing populations that differ in stress tolerance or create study populations through experimental selection on pre-existing variation, choose an underlying trait for analysis and hypothesize its effect on stress tolerance, and perform experiments or observations to test for the predicted covariation between the trait and stress tolerance. In the molecular genetic approach, by contrast, investigators choose a gene for analysis, create variation in that gene *de novo*, and observe any gain or

## TWO COMPLEMENTARY APPROACHES

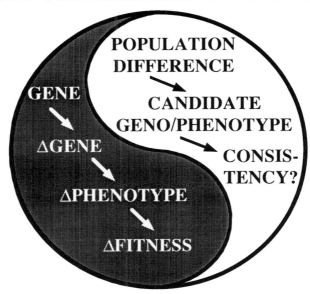

Figure 1. The evolutionary adaptationist program and the molecular genetic program: two approaches for understanding stress, adaptation and evolution. The former approach often proceeds by demonstrating that variation in performance and/or fitness is consistent with a hypothesized effect of a trait, but often cannot unambiguously attribute such effects to the gene or trait under study. By manipulating a gene of interest directly, the latter approach can often unambiguously demonstrate the consequences of that gene for performance and/or fitness, but may create genetic variation that will seldom if ever occur in nature. When combined, these two approaches may complement one another.

loss of function that may result. Thus, the evolutionary approach focusses on genetic variants that may arise through natural or semi-natural genetic and evolutionary processes, while the molecular genetic approach may be free to create any desired variant. These alternative research strategies each have strengths and weaknesses. Our thesis, which we will illustrate with our work on the heat shock proteins of *Drosophila*, is that these strategies are complementary; that is, if deployed in combination, the resulting insights may exceed those yielded by either strategy alone. Neither strategy may be sufficient; both may be essential.

The complementarity of these research strategies arises from the source and magnitude of the variation they analyze: (1) The evolutionary approach can elucidate what variants have arisen and might arise through the natural processes that create and constrain genetic variation, whereas the genetic engineering approach can yield variants that might never occur naturally. Accordingly, the evolutionary approach can help define the "evolutionary feasibility" of engineered variants. (2) By creating greater variation than may be seen in nature, genetic engineering can facilitate experimental and comparative studies where natural variation is too limited for current analytical techniques. (3) The critical potential contribution of the genetic engineering approach is that it permits a single gene to be manipulated in isolation against a constant genetic background (Feder, 1996). Organisms comprise multiple traits, and manipulating a single trait while all else is held constant is a near-impossible undertaking by conventional means. Even the most extraordinarily detailed and comprehensive research programs of natural genetic variants have conceded that they cannot exclude that their findings are due to linkage between a gene under study and some "mystery gene" (Powers et al., 1993). Moreover, whereas in natural variation the suite of genetic changes responsible for a given phenotypic variant may be impossible to characterize and may differ in multiple evolutionary events (i.e. parallelism and convergence), the molecular genetic approach enables an investigator to specify which gene(s) will undergo manipulation.

Both the evolutionary adaptationist approach and the genetic engineering approach must surmount the same difficulty: that of providing an ecological context for the model under study. Ideally, practitioners of both approaches should examine organisms undergoing the same intensities, frequencies and kinetics of stress that these organisms experience in the wild. Both specifying experimental conditions that correspond to nature and inferring the implications of laboratory study for fitness in nature can be problematic when an ecological context is absent. Evolutionary adaptationists often study organisms for which a rich background of natural history already exists, which can provide an adequate ecological context. Genetic engineers, by contrast, often choose model systems for their experimental tractability and convenience of laboratory culture. In many cases, these favored models have not been studied in the wild; basic ecological information can be lacking. Whatever the cause of inadequate ecological data, its provision is essential for meaningful experimentation and interpretation. In our case, the availability of an especially powerful system for controlled transgenic manipulation of *Drosophila* (Welte et al., 1993) led us to begin an evolutionary adaptationist analysis of thermal stress in the species, and then to alternate recursively between evolutionary and molecular genetic studies. Through this work, we hope to illustrate the contribution that each approach provides to understanding stress resistance in this species.

## The stress response and heat shock proteins

Since their discovery, the heat shock response and heat shock proteins have intrigued evolutionary and molecular biologists alike. In response to high temperature and nearly every other stress known, almost every organism studied expresses a characteristic suite of proteins, termed heat shock proteins (Hsps) or stress proteins (Lindquist and Craig, 1988; Morimoto et al., 1994; Feder et al., 1995; Feder, 1996). The Hsps are highly conserved and occur in several well-characterized families, most named by molecular weight (e.g. the Hsp70 family). These proteins are now known to be only part of a larger machinery that manages unfolded proteins within the cell (Hartl, 1996). During initial synthesis, intracellular transport and under the influence of cell stress, proteins may not be in their native conformation. In non-native proteins, residues that would ordinarily be internalized and isolated from one another in the native state can be exposed and interact, which can lead non-native proteins to aggregate. Such aggregations can be harmful if not lethal to the cell. Hsps and their constitutively expressed cognates function as molecular chaperones; they recognize and bind to non-native proteins, thus impeding aggregation (Hartl, 1996). Molecular chaperones then either release bound proteins, which can then fold, or target bound proteins for degradation and removal from the cell (Parsell and Lindquist, 1993). At least one Hsp, Hsp104 of yeast, can remove proteins from aggregations (Parsell et al., 1994). Because of their stress-inducible expression and their role in coping with stress-damaged proteins, the Hsps have long been considered important for stress tolerance. Unequivocal demonstrations of the stress tolerance of Hsps have been few, however, due especially to the difficulty of distinguishing the effects of Hsps from those of the many other physiological and biochemical mechanisms that contribute to thermotolerance. Molecular genetic manipulations now provide these demonstrations in a growing number of eukaryotic systems. With respect to Hsp70 family members, for example, overexpression and introduction of exogenous Hsp70 improves the thermotolerance of various mammalian cell types in cell culture (Li et al., 1991; Liu et al., 1992; Lee et al., 1993; Heads et al., 1994; Mailhos et al., 1994; Li et al., 1995), protects cells against ultraviolet radiation (Simon et al., 1995), protects whole mammalian hearts against post-ischemic trauma (Marber et al., 1995; Plumier et al., 1995) and improves the inducible thermotolerance of *Drosophila* cells in culture (Solomon et al., 1991). Introduction of anti-Hsp70 antibodies disrupts transcription (Moreau et al., 1994) and the development of tolerance to ischemia (Nakata et al., 1993) and heat (Riabowol et al., 1988; Lee et al., 1993).

The present study concerns the major Hsp of *D. melanogaster*, Hsp70. In *Drosophila* cells in culture, Hsp70 is virtually absent from the unstressed cell and undergoes massive expression during and after heat shock, at its peak accounting for 1% of total cellular protein (Lindquist, 1980; Velazquez et al., 1980; Velazquez et al., 1983; Velazquez and Lindquist, 1984). Although Hsp70 of *Drosophila* has not been characterized as a molecular chaperone, its counterparts in other species can clearly perform as molecular chaperones both *in vitro* and in the cell (Hartl, 1996). Typically, 10 nearly identical genes encode Hsp70 in the wild-type genome (Ish-Horowicz et al., 1979). Other family members in *Drosophila* include Hsp68 and a variety of constitutively expressed cognates (Palter et al., 1986). In transformed *Drosophila* cells in culture, thermotolerance is highly correlated with Hsp70 levels (Solomon et al., 1991). These facts led us to investigate whether Hsp70 was a significant evolutionary adaptation to stress.

## Natural thermal stress: A contribution of the evolutionary adaptationist approach

Understanding how Hsp70 contributes to thermotolerance in *Drosophila* requires information on whether *Drosophila* encounters thermal stress in nature, if such stress is sufficient to induce Hsp70 expression and whether this induced expression varies genetically within and/or among *Drosophila* populations. The effects of temperature on maintenance, growth and reproduction in *Drosophila* are well known (Fig. 2). At temperatures above approximately 27°C, various body functions fail in direct proportion to the magnitude and duration of heat shock. Parsons (1978) has cited 30°C as the highest constant temperature at which *D. melanogaster* populations can

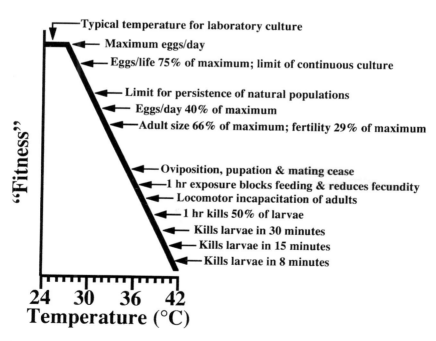

Figure 2. Effect of temperature on traits related to fitness in *D. melanogaster*. Data are from Parsons, 1978; David et al., 1983; Ashburner, 1989; Huey et al., 1992; Krebs and Loeschcke, 1994a; Feder et al., 1997a.

persist in the wild. Individual *Drosophila* can tolerate higher temperatures temporarily, but such temperatures eventually compromise reproduction.

Do *Drosophila* encounter such temperatures in the wild? Surprisingly, despite the massive study of *Drosophila*, numerous investigations of the effects of temperature upon it and its extensive characterization along climatic gradients, few studies report actual body temperatures of *D. melanogaster* in the field (Feder, 1996). Such direct measurements are necessary to document thermal stress in adults, because their great mobility and small size potentially enable them to escape thermal stress by behavioral thermoregulation (Krebs and Bean, 1991; E. Toolson, personal communication). By contrast, the preadult developmental stages, eggs, larvae, and sometimes pupae, reside on or within necrotic fruit. For these stages, the thermal regime of the host fruit becomes that of the *Drosophila* within it. From studies of necrotic fruit and *Drosophila* outside the laboratory, the following points emerge:

– If necrotic fruit (e.g. apples, peaches) is in the sun, it can readily attain temperatures that would be harmful if not lethal to any indwelling *Drosophila*. On warm summer days, temperatures of fruit containing larvae can be as warm as 44°C, and frequently exceed 35°C (Feder et al., 1997a). Fruit temperatures are largely a function of size, as the greater thermal inertia and evaporative cooling capacity of very large host plants (e.g. columnar cacti) apparently result in more moderate temperatures (Krebs and Loeschcke, 1994b; E. Toolson, personal communication).

– In *D. melanogaster*, behaviors that might mitigate such stress are either ineffective or non-existent. Adult females can avoid fruit that is stressfully warm at the time of oviposition (Fogleman, 1979; Schnebel and Grossfield, 1986). The limited thermal inertia of small fruits, however, means that fruit temperatures at the time of oviposition are poor indicators of future thermal conditions. Although the physical-chemical-biological "signature" of past heating could be used as an indicator of a future thermal stress, ovipositing females do not avoid fruit that has previously been heated (Feder et al., 1997b). Once deposited on an apple or peach, larvae are restricted to that environment. There larval *Drosophila* have little opportunity to thermoregulate behaviorally, as these fruits tend to be thermally homogeneous (Feder et al., 1997a).

– Larvae and pupae undergoing such natural thermal stress express Hsps in response. For example, Hsp70 levels in wild larvae equal or exceed the highest levels detected in wild-type strains undergoing heat shock in the laboratory (Feder et al., 1997a).

Previous molecular genetic investigations of the heat shock response in *Drosophila* had proceeded without any reference to temperatures that *Drosophila* naturally experience, prompting the criticism that Hsp expression in this species might have no bearing on fitness in nature and therefore be irrelevant as an adaptation to stress. The characterization of natural thermal stress in *Drosophila* creates an ecological context for the interpretation of molecular data and enhances the design of evolutionary and genetic experiments to come. Elsewhere in this volume, Loeschcke et al. review natural phenotypic and genetic variation among populations of *Drosophila*. Below we consider variation in Hsp70 expression and thermotolerance within *Drosophila* populations.

## Controlled variation against a constant genetic background: A contribution of the molecular genetics approach

Much of our knowledge of the traits that contribute to stress tolerance has come from comparisons of individuals, populations, species and/or higher taxa in contrasting environments, in which differences in traits are found to be consistent with differences in stress tolerance. Correlations of stress tolerance and underlying traits have readily led to reductionistic studies, which have elucidated numerous candidate mechanisms of stress tolerance. In addition to Hsps, these tolerance-related traits include fatty acid saturation, quantitative and qualitative changes in numerous enzymes, cell membrane stability, energy storage, alternative metabolic strategies, osmotic mediators of protein stability and a host of others (Precht et al., 1973; Cossins and Bowler, 1987). The multiplicity of candidate mechanisms of stress tolerance, however, creates a problem in moving in the opposite direction, i.e. inferring functional and/or evolutionary significance from an underlying trait. So many potential mechanisms of stress tolerance differ among individuals, populations, species and/or higher taxa that attributing functional and/or adaptive significance to any one of them can be inherently equivocal. In many cases, investigators must content themselves with assuming that the trait they are studying (and not other correlated traits) is actually responsible for an observed functional or evolutionary difference, all else equal (Lewontin, 1978). The key contribution of the genetic engineering approach is that it can hold other traits genetically constant while manipulating a gene of interest. With proper controls, such manipulation can establish that a gene of interest is necessary, sufficient, neither necessary nor sufficient, or both necessary and sufficient to yield a predicted phenotype. Similarly, some whole-organism traits with complex genetic bases (e.g. body size, reproductive status) are amenable to "phenotypic engineering" in studies of adaptation (Sinervo and Basolo, 1996).

The execution of the genetic engineering approach with respect to Hsp70 in *Drosophila* is problematic, however, because the wild-type genome already includes 10 copies of the gene (Ish-Horowicz et al., 1979). Thus, the effects of adding a single transgene or knocking out an existing copy might well be undetectable against this multi-copy background. Susan Lindquist and colleagues have contributed greatly to the resolution of this problem by developing a technique (Golic and Lindquist, 1989; Welte et al., 1993) wherein flies are transformed with a transgene construct bearing multiple copies of a gene of interest flanked by yeast recombination targets (Fig. 3). In thus-transformed flies, expression of FLP recombinase, a protein that enables homologous recombination in yeast, results in unequal homologous recombination such that two daughter strains eventually result: an extra-copy strain bearing two or more copies of the transgene construct, and an excision strain bearing only the construct's flanking

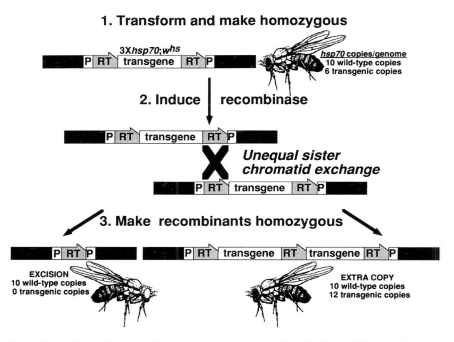

Figure 3. A scheme for controlled mutagenesis in *Drosophila* via site-specific recombination (Golic and Lindquist, 1989; Welte et al., 1993). A transgene construct is created that includes the transgene of interest (in this case, three copies of the *Drosophila hsp70* gene) and an eye-color marker ($w^{hs}$), flanked by yeast FLP recombination target sequences (RT) and P element sequences (P). This construct is transformed into the germline of *D. melanogaster*, where it integrates at random into a chromosome (shown in black); transformants are made homozygous (step 1). Male transformants are mated to females transformed with the *FLP recombinase* gene under control of the *hsp70* promoter. Heat-shocking their offspring induces FLP recombinase, which sometimes catalyzes unequal recombination among sister chromatids (step 2). Such recombination yields one chromatid with two copies of the sequence flanked by the recombination targets, and a second chromatid from which this sequence has been excised. Subsequent crosses then isolate the recombinants, eliminate the chromosome-bearing *FLP recombinase*, and make the recombinants homozygous. These crosses result in two strains (step 3), an extra-copy strain homozygous for two copies of the RT-flanked transgene construct (in this case, 12 copies of *hsp70*), and an excision strain with an identical site of transgene integration into the genome but lacking any copies of the transgene of interest. All strains bear 10 wild-type copies of *hsp70* at other locations on the same chromosome or on other chromosomes. Strains can be identified by eye color.

elements, but not the transgene itself. In addition, both the extra-copy strain and the excision strain share the site of transgene integration; thus, the excision strain controls for positional or insertional mutagenesis (i.e. a phenotype due solely to disruption of the host genome rather than to the transgene). In rare cases, a difference in the size of the remaining construct in the extra-copy and excision strains can itself result in a phenotype due to unusual genetic circumstances. Examination of multiple pairs of extra-copy and excision strains can effectively rule out this possibility, because the likelihood that two independent chromosomal interruptions would yield the same phenotype in relation to the size of the transgene insert is negligibly small (Welte et al., 1993).

Is the difference in *hsp70* copy number between extra-copy and excision

This entire procedure has now been used to create extra-copy strains of *Drosophila* bearing 12 *hsp70* transgenes in addition to the 10 wild-type copies, and matched excision strains with only the wild-type copies (Welte et al., 1993). Under diverse conditions and at multiple developmental stages, extra-copy *Drosophila* accumulate Hsp70 to higher levels than do their excision counterparts, and they generally express Hsp70 more rapidly as well (Figs. 4, 5) (Welte et al., 1993; Feder et al., 1996; Krebs and Feder, 1997a; M. Tatar and J. Curtsinger, personal communication). Although both the genes for other Hsps and the wild-type Hsp70s continue to express upon heat shock in the extra-copy *Drosophila*, the transgenic mRNA is specifically elevated in the extra-copy strains (Welte et al., 1993).

Is the difference in *hsp70* copy number between extra-copy and excision strains sufficient to affect tolerance of ecologically relevant heat stress? To answer the question, we compared thermotolerance of extra-copy and ex-

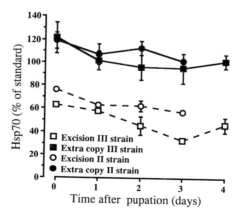

Figure 4. Effect of age and *hsp70* copy number on Hsp70 levels in pupae of two pairs of extra-copy and excision transformants. Hsp70 levels were determined for lysates of whole pupae with an Hsp70-specific ELISA and are expressed as a percentage of a standard lysate of *Drosophila* S2 cells (Feder et al., 1996; Feder et al., 1997a). Means are plotted ± one standard error. II and III refer to the chromosomes bearing the transgenes in each pair of strains.

cision *Drosophila* with and without exposure to a Hsp70-inducing pre-treatment, typically 36°C. Differences in thermotolerance between pre-treated and control (i.e. unpretreated) *Drosophila* are attributable to the entire suite of heat-inducible responses, whereas differences between extra-copy and excision strains are attributable specifically to *hsp70* copy number. As has been previously reported for most ectotherms in which it has been studied, pretreatment dramatically improves the thermotolerance of both extra-copy and excision *Drosophila*, except in early embryos. The improvement of thermotolerance, however, is greater in extra-copy than in excision *Drosophila* in several circumstances (Welte et al., 1993; Feder et al., 1996). In wandering third-instar larvae, Hsp70 levels differ most greatly between extra-copy and excision strains at 1 h after pretreatment. At this time, the thermotolerance of excision larvae has increased to approximately 150% of control levels, whereas the thermotolerance of extra-copy larvae has increased to approximately 350%. Pretreatment of 6-h embryos also results in much greater hatching success in extra-copy strains than in excision strains.

Although candidate mechanisms for this Hsp70-enhanced thermotolerance are numerous, an appropriate target of investigation is the unfolding of proteins, as is consistent with the role of Hsps as molecular chaperones. We chose alcohol dehydrogenase (ADH) as a model protein for several reasons: ADH denatures at temperatures experienced by *Drosophila* in the field, ADH alleles vary in thermal stability and this variation is correlated with climatic gradients (Sampsell and Sims, 1982; Sampsell and Barnette, 1985; Chambers, 1988). Moreover, ADH has been the subject of numerous evolutionary investigations (see above references and elsewhere in this volume). To examine the effects of Hsp70 level on thermal inactivation and recovery of ADH, we exposed second-instar larvae to 40°C for 15 min, which reduced ADH activity by 40% (Fig. 5). If larvae of the extra-copy strain were pretreated at 36°C before the 40°C exposure, Hsp70 was present at high levels during the 40°C heat shock. In these larvae, ADH activity recovered to near-initial levels during the 2 h after heat shock. No such recovery occurred in extra-copy larvae that were not pretreated before heat shock, nor in excision larvae with or without pretreatment. The greater levels of Hsp70 in pretreated larvae of the extra-copy strain could have yielded this pattern by reactivating pre-existing ADH unfolded by the heat shock and/or by protecting *de novo* synthesis of ADH after the heat shock. Nonetheless, this is apparently the first example of Hsp-mediated restoration of thermally inactivated protein function in an intact multicellular eukaryote.

In summary, these data show that alteration in *hsp70* copy number is sufficient to enhance inducible tolerance of temperatures that might well be encountered in the field by *Drosophila* larvae and pupae. Furthermore, the manner in which the experimental strains were created establishes the participation of Hsp70, rather than some linked trait, in the production of thermotolerance. Whether Hsp70 exerts these effects alone or in combina-

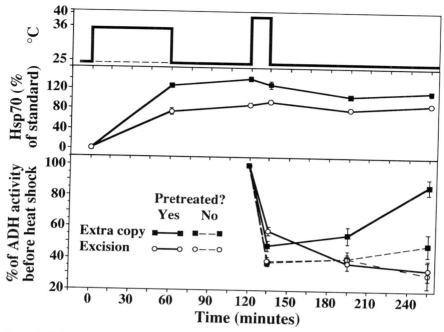

Figure 5. Effect of *hsp70* copy number on reactivation of alcohol dehydrogenase in whole second-instar larvae (chromosome III transformants). Top: Groups of larvae were exposed to 40°C for 15 min after either being exposed to 36°C for 1 h (pretreated) or being kept at 25°C (unpretreated). Middle: Hsp70 levels in pretreated larvae of the extra-copy and excision strains. Data are plotted as in Figure 4. Unpretreated larvae expressed no Hsp70 before or during heat shock. Bottom: Alcohol dehydrogenase (ADH) activity of whole larvae. ADH activity was determined from crude homogenates according to McKechnie and Geer (1984), standardized by μg of protein in the homogenate and expressed as a percentage of the ADH activity determined before heat shock. Means are plotted ± one standard error.

tion with other chaperones (e.g. the *Drosophila* DnaJ homologue (Hartl, 1996)), and whether Hsp70 itself promotes thermotolerance, protects some other thermotolerance factor against heat damage, signals the induction of other thermotolerance factors and so on, remain for future investigation.

## Molecular and evolutionary approaches intersect: Hsp70 as a trade-off

Diverse molecular investigations suggest that Hsp70 may sometimes harm cells in addition to protecting cells from stress and/or aiding cells' recovery from stress. Hsp70 is undetectable in *Drosophila* cells and tissues in the absence of stress (Velazquez et al., 1983). A large array of transcriptional, translational and post-translational regulatory mechanisms enforce both this absence of Hsp70 and its rapid degradation after recovery from stress is complete (Lindquist, 1993). If these controls are overridden by experimentally expressing Hsp70 in the absence of stress, the growth of cells in culture slows until the Hsp70 is sequestered in intracellular granules (Feder et al., 1992). Coleman et al. (1995) have suggested that these phenomena occur because massive Hsp expression depletes the cell of resources that are needed for other cellular processes. Alternatively, ectopic Hsp70 expression may be toxic to cells. For example, it may bind inappropriately to its cellular targets and/or fail to release them at appropriate times. Indeed, Dorner et al. (1988, 1992) have reported that overexpression of the mammalian Hsp70 family member Grp78 inhibits the secretion of certain proteins from the cell because it retains these proteins within the Grp78-dependent secretory pathway; proteins not dependent on Grp78 for secretion are unaffected. These putative mechanisms are not mutually exclusive.

Evolutionary biology can supply a theoretical framework for interpreting such findings: that of a trade-off, in which a trait that increases fitness is linked to a trait that decreases fitness (Stearns and Kaiser, 1996). Attempts to understand trade-offs, however, have often been frustrated by lack of a detailed understanding of the traits in question and how they actually interact to yield the trade-off (Stearns and Kaiser, 1996). Hsp70, for which the positive and negative effects may share a common and increasingly well-understood mechanism (binding to unfolded proteins), may be a useful model for the study of trade-offs. This possibility manifested itself in our work with *Drosophila* when excision strains outperformed extra-copy strains in several tests of thermal tolerance (Krebs and Feder, 1997a). For example, in studies of first-instar larvae that began several hours after hatching, pretreated extra-copy larvae exhibited significantly lower survival to adulthood than excision larvae. This assay tests not only acute survival of a heat shock but also the ability to grow, metamorphose and eclose afterwards. Pretreatment and/or heat shock initially disrupted growth more severely in the extra-copy strain than in the excision strain, although extra-copy larvae made up this difference within 72 h. Comparisons of thermotolerance in an independently derived pair of extra-copy and excision strains and in early third-instar larvae of both pairs of strains revealed essentially the same pattern. Interestingly, even in the control treatment, which involved neither pretreatment nor heat shock, extra-copy larvae showed higher mortality than excision larvae.

Thermal injury of feeding and protection against it may have special ecological importance, because intense larval feeding provides the nutrients and energy for metamorphosis and initial stages of reproduction. Our ongoing analyses of the pathogenesis of heat damage single out the gut as a target for these effects. One day after a 38.5°C heat shock without pretreatment, cells die massively throughout the foregut, gastric caeca, midgut and hindgut. Hsp70-specific immunofluorescence remains elevated during

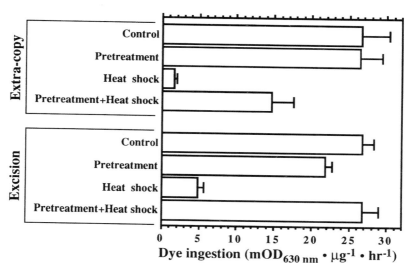

Figure 6. Effect of heat shock, pretreatment and *hsp70* copy number on feeding rates of third-instar larvae (chromosome II transformants). Treatments: Control, sham transfer at constant 25°C; pretreatment, 36°C for 50 min and 25°C for 1 h; heat shock, 38.5°C for 50 min. Larvae were fed yeast paste for 16 h after the end of treatment, and then transferred to a yeast slurry made with 1% FD & C Blue #1 dye. This dye does not transit the gut wall, and thus serves as a marker for ingestion (Edgecomb et al., 1994). After 75 min in the dyed yeast, groups of larvae were homogenized and analyzed for dye-specific absorbance at 630 nm and total protein content. Dye ingestion per hour was standardized by µg of protein in the homogenate. Means are plotted ± one standard error.

this time, unlike in other tissues. Whether these patterns differ between extra-copy and excision strains is not yet known. These strains differ in recovery of feeding after heat shock, however. A 38.5°C heat shock dramatically depresses feeding for at least 16 h afterwards (Fig. 6). Pretreatment alone has little effect on feeding at this time, but significantly ameliorates the effect of the 38.5°C heat shock if administered beforehand (Fig. 6). The improvement due to pretreatment is significantly greater in the excision strain than in the extra-copy strain, which expresses higher levels of Hsp70.

These findings prompted an experiment in which extra-copy and excision larvae underwent multiple pretreatments, during which Hsp70 levels were chronically elevated (but to the different levels characteristic of each strain). In the excision larvae, larva-to-adult survival was unrelated to the number of pretreatments (Fig. 7). In the extra-copy larvae, by contrast, larva-to-adult survival was inversely correlated with the number of heat shocks (Krebs and Feder, 1997a). Thus, engineered variation in *hsp70* copy number can amplify aspects of a prospective trade-off that may be more difficult to recognize in natural variants.

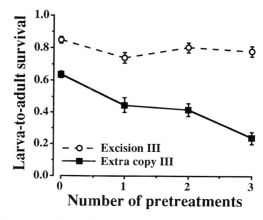

Figure 7. Effect of number of Hsp70-inducing treatments on survival to adulthood in extra-copy and excision larvae (Krebs and Feder, 1997a). For larvae given one or more treatments, the first pretreatment was 2–4 h after hatching, the second 24 h later and the third another 24 h later. Each treatment exposed larvae to 36°C for 1 h. Means are plotted ± one standard error. Pretreatment number significantly reduced survival in the extra-copy strain but had no effect on the excision strain (p < 0.001 for strain X pretreatment number interaction; analysis of variance on arcsine-square root transformed data).

## Natural variation in Hsp70: A final contribution of the evolutionary approach

The preceding section suggests that stabilizing selection may maintain *hsp70* copy number and Hsp70 expression at present levels in natural populations; that is, the disadvantages of any increase in these traits would outweigh the corresponding advantages. Alternatively, the present *hsp70* copy number and levels of Hsp70 expression could have become fixed long ago and bear no relationship to ongoing stabilizing selection. Transgenic experimentation cannot resolve this issue, which requires return to the evolutionary approach.

The conditions necessary for ongoing selection on Hsp70 in nature are that individuals vary in expression within populations, that this variation has a genetic basis and that these genetic differences affect fitness. Natural variation in *hsp70* copy number in *D. melanogaster* has never been examined thoroughly, although studies with laboratory stocks and cell lines suggest a considerable potential for natural variation (Mirault et al., 1979; Lis et al., 1981b, c). To characterize variation in Hsp70 expression, we measured Hsp70 accumulation in 20 isofemale lines derived from a summer collection of *D. melanogaster* from a population in an Indiana Orchard (Krebs and Feder, 1997b). These lines varied approximately two-fold in Hsp70 expression in response to a standard heat shock (Fig. 8). Although comparable data are not yet available for other populations of *Drosophila*

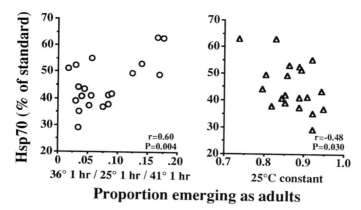

Figure 8. Natural inter-individual variation in Hsp70 expression and thermotolerance, as inferred from 20 isofemale lines derived from a single natural population (Krebs and Feder, 1997b). Hsp70 levels were determined and standardized as in Figure 4. These levels are plotted against first-instar larva-to-adult survival after sequential exposure to 36°C for 1 h, 25°C for 1 h and 41°C for 1 h (left graph), and survival at constant 25°C. Significance levels are for individual correlation coefficients.

or most other species, diIorio et al. (1996) have demonstrated substantial genetic variation in the expression of two Hsp70 family members among hemiclones of *Poeciliopsis*. Next, we examined the potential relationship of this variation in Hsp70 to fitness by comparing the tolerance of the 20 iso-female lines to heat shock with their Hsp70 expression. Hsp70 level was positively correlated with tolerance of high temperatures, as assessed by larva-to-adult survival. McColl et al. (1996) provide additional evidence that Hsps can be subject to ongoing natural selection. They selected *Drosophila* for resistance to paralysis at high temperatures, analyzed *hsp68* and *hsr-omega* via denaturing gradient gel electrophoresis, and discovered that the experimental selection regime had brought about changes in both genes. While the nature and the functional consequences of these changes await characterization, clearly the genes for Hsps are not refractory to selection.

Unexpectedly, Hsp70 level and tolerance of high temperatures were each negatively correlated with larva-to-adult survival at 25°C (Fig. 8). Thus, the isofemale lines derived from a natural population manifest the same inverse relationship between Hsp70 level upon heat shock and mortality in the absence of stress as do the genetically engineered strains. Here the evolutionary and molecular approaches corroborate one another. Both approaches raise the same concern, however: How can variation in a gene expressed only in the presence of stress affect survival in the absence of stress? Tests of several possible explanations are presently underway.

## Conclusion

The complementarity of genetic and phenotypic analyses and of laboratory and field approaches within the evolutionary adaptationist approach is by no means a novel concept (Koehn, 1987; Feder and Watt, 1993; Powers et al., 1993; Partridge, 1994), nor is the exploitation of transgenic variation in the analysis of evolutionary adaptation (Feder and Block, 1991). The challenge is to incorporate all of these in a single research program. Often, the limiting factor will be the availability of a suitable model system (Feder and Watt, 1993). In *Drosophila*, the incidence of natural thermal stress and the amenability to genetic engineering are key features lacking in many other candidate models.

The complementary contributions of the evolutionary adaptationist approach and the molecular genetic approach deserve re-emphasis. The field data establish that *Drosophila* naturally encounter thermal stress, to which they respond by expressing Hsps, and provide assurance that the transgenic variation has a counterpart in nature. Without these data, the analyses of the transgenic strains risk becoming an exercise in mechanistic pharmacology with no clear relevance to the function and fitness of organisms in nature. Experimentation with the transgenic strains unequivocally attributes variation in high-temperature tolerance to Hsp70. Without such a comparison, hundreds of other traits known to affect thermotolerance must be accounted for before making a similar attribution. Genetic engineering can be an incredibly powerful addition to the tool kit of evolutionary adaptationists. However, realizing the full potential of transgenic manipulation as an analytical tool in adaptational biology will require its deployment in concert with the classical tools of evolutionary and functional biology.

*Acknowledgments*
We thank Susan Lindquist for continuing encouragement and insight, and Ray Huey for comments on the manuscript. Evelyn Tin (Knox College) gathered data on alcohol dehydrogenase activity while supported by a summer research fellowship from the Howard Hughes Medical Institute. Figure 3 is modified from original artwork by Michael Welte. The Warner-Jenkinson Corp., St. Louis, MO, donated FD & C Blue #1 dye. Research was supported by grants from the National Science Foundation (IBN94-08216 and BIR94-19545) and the Louis Block Fund of the University of Chicago.

## References

Ashburner, M. (1989) Drosophila: *A Laboratory Manual*. Cold Spring Harbor Laboratory Press, Cold Spring Harbor, NY.
Chambers, G.K. (1988) The *Drosophila* alcohol dehydrogenase gene-enzyme system. *Adv. Genet.* 25:39–108.
Coleman, J.S., Heckathorn, S.A. and Hallberg, R.L. (1995) Heat-shock proteins and thermotolerance: Linking ecological and molecular perspectives. *Trend Ecol. Evolut.* 10:305–306.
Cossins, A.R. and Bowler, K. (1987) *Temperature Biology of Animals*. Chapman and Hall, London.

David, J.R., Allemand, R., Van Herrewege, J. and Cohet, Y. (1983) Ecophysiology: Abiotic factors. *In*: M. Ashburner, H.L. Carson and J.N. Thompson (eds): *The Genetics and Biology of* Drosophila, third Edition. Academic Press, London, pp. 105–170.

diIorio, P.J., Holsinger, K., Schultz, R.J. and Hightower, L.E. (1996) Quantitative evidence that both Hsc70 and Hsp70 contribute to thermal adaptation in hybrids of the livebearing fishes *Poeciliopsis. Cell Stress & Chaperones* 1:139–147.

Dorner, A.J., Krane, M.G. and Kaufman, R.J. (1988) Reduction of endogenous GRP78 levels improves secretion of a heterologous protein in CHO cells. *Mol. Cell. Biol.* 8:4063–4070.

Dorner, A.J., Wasley, L.C. and Kaufman, R.J. (1992) Overexpression of GRP78 mitigates stress induction of glucose regulated proteins and blocks secretion of selective proteins in Chinese hamster ovary cells. *EMBO J.* 11:1563–1571.

Edgecomb, R.S., Harth, C.E. and Schneiderman, A.M. (1994) Regulation of feeding behaviour in adult *Drosophila melanogaster* varies with feeding regime and nutritional state. *J. Exp. Biol.* 197:215–236.

Feder, J.H., Rossi, J.M., Solomon, J., Solomon, N. and Lindquist, S. (1992) The consequences of expressing hsp70 in *Drosophila* cells at normal temperatures. *Genes & Develop.* 6:1402–1413.

Feder, M.E. (1996) Ecological and evolutionary physiology of stress proteins and the stress response: The *Drosophila melanogaster* model. *In*: I.A. Johnston and A.F. Bennett (eds): *Animals and Temperature: Phenotypic and Evolutionary Adaptation.* Cambridge University Press, Cambridge, pp. 79–102.

Feder, M.E. and Block, B.A. (1991) On the future of physiological ecology. *Funct. Ecol.* 5:136–144.

Feder, M.E. and Watt, W.B. (1993). Functional biology of adaptation. *In*: R.J. Berry, T.J. Crawford and G.M. Hewitt (eds): *Genes in Ecology.* Blackwell Scientific Publications, Oxford, pp. 365–391.

Feder, M.E., Parsell, D.A. and Lindquist, S.L. (1995). The stress response and stress proteins. *In*: J.J. Lemasters and C. Oliver (eds): *Cell Biology of Trauma.* CRC Press, Boca Raton, FL, pp. 177–191.

Feder, M.E., Cartaño, N.V., Milos, L., Krebs, R.A. and Lindquist, S.L. (1996) Effect of engineering *hsp70* copy number on Hsp70 expression and tolerance of ecologically relevant heat shock in larvae and pupae of *Drosophila melanogaster. J. Exp. Biol.* 199:1837–1844.

Feder, M.E., Blair, N. and Figueras, H. (1997a) Natural thermal stress and heat-shock protein expression in *Drosophila* larvae and pupae. *Funct. Ecol.* 11:90–100.

Feder, M.E., Blair, N. and Figueras, H. (1997b) Oviposition site selection and temperature: Unresponsiveness of ovipositing *Drosophila* to cues of potential thermal stress on offspring. *Anim. Behav.* 53:585–588.

Fogleman, J. (1979) Oviposition site preference for temperature in *D. melanogaster. Behav. Genet.* 9:407–412.

Golic, K.G. and Lindquist, S. (1989) The FLP recombinase of yeast catalyzes site-specific recombination in the *Drosophila* genome. *Cell* 59:499–509.

Hartl, F.U. (1996) Molecular chaperones in cellular protein folding. *Nature* 381:571–580.

Heads, R.J., Latchman, D.S. and Yellon, D.M. (1994) Stable high level expression of a transfected human HSP70 gene protects a heart-derived muscle cell line against thermal stress. J. *Mol. Cell. Cardiol.* 26:695–699.

Huey, R.B., Crill, W.D., Kingsolver, J.G. and Weber, K.E. (1992) A method for rapid measurement of heat or cold resistance of small insects. *Funct. Ecol.* 6:489–494.

Ish-Horowicz, D., Pinchin, S.M., Schedl, P., Artavanis, T.S. and Mirault, M.E. (1979) Genetic and molecular analysis of the 87A7 and 87C1 heat-inducible loci of *D. melanogaster. Cell* 18:1351–1358.

Koehn, R.K. (1987) The importance of genetics to physiological ecology. *In*: M.E. Feder, A.F. Bennett, W.W. Burggren and R.B. Huey (eds): *New Directions in Ecological Physiology.* Cambridge University Press, Cambridge, pp. 170–185.

Krebs, R.A. and Bean, K.L. (1991) The mating behavior of *Drosophila mojavensis* on organ pipe and agria cactus. *Psyche* 98:101–109.

Krebs, R.A. and Feder, M.E. (1997a) Deleterious consequences of Hsp70 overexpression in *Drosophila melanogaster* larvae. *Cell Stress & Chaperones* 2:60–71.

Krebs, R.A. and Feder, M.E. (1997b) Natural variation in the expression of the heat-shock protein Hsp70 in a population of *Drosophila melanogaster*, and its correlation with tolerance of ecologically relevant thermal stress. *Evolution* 51:173–179.

Krebs, R.A. and Loeschcke, V. (1994a) Costs and benefits of activation of the heat-shock response in *Drosophila melanogaster*. *Funct. Ecol.* 8:730–737.

Krebs, R.A. and Loeschcke, V. (1994b) Response to environmental change: Genetic variation and fitness in *Drosophila buzzatii* following temperature stress. *In*: V. Loeschcke, J. Tomiuk and S.K. Jain (eds): *Conservation Genetics*. Birkhäuser, Basel, pp. 309–321.

Lee, Y.J., Kim, D., Hou, Z.Z., Curetty, L., Borrelli, M.J. and Corry, P.M. (1993) Alteration of heat sensitivity by introduction of hsp70 or anti-hsp70 in CHO cells. *J. Therm. Biol.* 18:229–236.

Lewontin, R.C. (1978) Adaptation. *Sci. Amer.* 239:156–169.

Li, G.C., Li, L.G., Liu, Y.K., Mak, J.Y., Chen, L.L. and Lee, W.M. (1991) Thermal response of rat fibroblasts stably transfected with the human 70-kDa heat shock protein-encoding gene. *Proc. Natl. Acad. Sci. USA* 88:1681–1685.

Li, L., Shen, G. and Li, G.C. (1995) Effects of expressing human Hsp70 and its deletion derivatives on heat killing and on RNA and protein synthesis. *Exp. Cell Res.* 217:460–468.

Lindquist, S. (1980) Translational efficiency of heat-induced messages in *Drosophila melanogaster* cells. *J. Mol. Biol.* 137:151–158.

Lindquist, S. (1993) Autoregulation of the heat-shock response. *In*: J. Ilan (ed.) *Translational Regulation of Gene Expression 2*. Plenum Press, New York, pp. 279–320.

Lindquist, S. and Craig, E.A. (1988) The heat-shock proteins. *Annu. Rev. Genet.* 22:631–677.

Lis, J., Neckameyer, W., Mirault, M.E., Artavanis, T.S., Lall, P., Martin, G. and Schedl, P. (1981a) DNA sequences flanking the starts of the hsp70 and alpha beta heat shock genes are homologous. *Dev. Biol.* 83:291–300.

Lis, J.T., Ish, H.D. and Pinchin, S.M. (1981b) Genomic organization and transcription of the alpha beta heat shock DNA in *Drosophila melanogaster*. *Nucleic Acids Res.* 9:5297–5310.

Lis, J.T., Neckameyer, W., Dubensky, R. and Costlow, N. (1981c) Cloning and characterization of nine heat-shock-induced mRNAs of *Drosophila melanogaster*. *Gene* 15:67–80.

Liu, R.Y., Li, X., Li, L. and Li, G.C. (1992) Expression of human hsp70 in rat fibroblasts enhances cell survival and facilitates recovery from translational and transcriptional inhibition following heat shock. *Cancer Res.* 52:3667–3673.

Mailhos, C., Howard, M.K. and Latchman, D.S. (1994) Heat shock proteins hsp90 and hsp70 protect neuronal cells from thermal stress but not from programmed cell death. *J. Neurochem.* 63:1787–1795.

Marber, M.S., Mestril, R., Chi, S.H., Sayen, M.R., Yellon, D.M. and Dillmann, W.H. (1995) Overexpression of the rat inducible 70-kD heat stress protein in a transgenic mouse increases the resistance of the heart to ischemic injury. *J. Clin. Invest.* 95:1446–1456.

McColl, G., Hoffmann, A.A. and McKechnie, S.W. (1996) Response of two heat shock genes to selection for knockdown heat resistance in *Drosophila melanogaster*. *Genetics* 143:1615–1627.

McKechnie, S.W. and Geer, B.W. (1984) Regulation of alcohol dehydrogenase in *Drosophila melanogaster* by dietary alcohol and carbohydrate. *Insect Biochem.* 14, 231–242.

Mirault, M.E., Goldschmidt, C.M., Artavanis, T.S. and Schedl, P. (1979) Organization of the multiple genes for the 70,000-dalton heat-shock protein in *Drosophila melanogaster*. *Proc. Natl. Acad. Sci. USA* 76:5254–5258.

Moreau, N., Laine, M.C., Billoud, B. and Angelier, N. (1994) Transcription of amphibian lampbrush chromosomes is disturbed by microinjection of HSP70 monoclonal antibodies. *Exp. Cell Res.* 211:108–114.

Morimoto, R.I., Tissieres, A. and Georgopoulos, C. (eds) (1994) *The Biology of Heat Shock Proteins and Molecular Chaperones*. Cold Spring Harbor Laboratory Press, Cold Spring Harbor, NY.

Nakata, N., Kato, H. and Kogure, K. (1993) Inhibition of ischaemic tolerance in the gerbil hippocampus by quercetin and anti-heat shock protein-70 antibody. *Neuroreport* 4:695–698.

Palter, K.B., Watanabe, M., Stinson, L., Mahowald, A.P. and Craig, E.A. (1986) Expression and localization of *Drosophila melanogaster* hsp70 cognate proteins. *Mol. Cell. Biol.* 6:1187–1203.

Parsell, D.A. and Lindqust, S. (1993) The function of heat-shock proteins in stress tolerance: Degradation and reactivation of damaged proteins. *Annu. Rev. Genet.* 27:437–496.

Parsell, D.A., Kowal, A.S., Singer, M.A. and Lindquist, S. (1994) Protein disaggregation mediated by heat-shock protein Hsp104. *Nature* 372:475–478.

Parsons, P. (1978) Boundary conditions for *Drosophila* resource utilization in temperate regions, especially at low temperatures. *Am. Nat.* 112, 1063–1074.

Partridge, L. (1994) Genetic and nongenetic approaches to questions about sexual selection. *In:* C.R.B. Boake (ed): *Quantitative Genetic Studies of Behavioral Evolution.* University of Chicago Press, Chicago, pp. 126–141.

Plumier, J.C., Ross, B.M., Currie, R.W., Angelidis, C.E., Kazlaris, H., Kollias, G. and Pagoulatos, G.N. (1995) Transgenic mice expressing the human heat shock protein 70 have improved post-ischemic myocardial recovery. *J. Clin. Invest.* 95:1854–1860.

Powers, D.A., Smith, M., Gonzalez-Villasenor, I., DiMichele, L., Crawford, D.L., Bernardi, G. and Lauerman, T. (1993) A multidisciplinary approach to the selectionist/neutralist controversy using the model teleost *Fundulus heteroclitus. In:* D. Futuyma and J. Antonovics (eds): *Oxford Surveys in Evolutionary Biology* 9, pp. 43–107.

Precht, H., Christophersen, J., Hensel, H. and Larcher, W. (eds) (1973) *Temperature and Life.* Springer-Verlag, Berlin.

Riabowol, K.T., Mizzen, L.A. and Welch, W.J. (1988) Heat shock is lethal to fibroblasts microinjected with antibodies against hsp70. *Science* 242:433–436.

Sampsell, B.M. and Barnette, V.C. (1985) Effects of environment temperatures on alcohol dehydrogenase activity levels in *Drosophila melanogaster. Biochem. Genet.* 23:53–59.

Sampsell, B.M. and Sims, S. (1982) Effect of *Adh* genotype and heat stress on alcohol tolerance in *Drosophila melanogaster. Nature* 296:853–855.

Schnebel, E.M. and Grossfield, J. (1986) Oviposition temperature range in four *Drosophila* species triads from different ecological backgrounds. *Am. Midl. Nat.* 116:25–35.

Simon, M.M., Reikerstorfer, A., Schwarz, A., Krone, C., Luger, T.A., Jaattela, M. and Schwarz, T. (1995) Heat shock protein 70 overexpression affects the response to ultraviolet light in murine fibroblasts: Evidence for increased cell viability and suppression of cytokine release. *J. Clin. Invest.* 95:926–933.

Sinervo, B. and Basolo, A. (1996). Testing adaptation using phenotypic manipulations. *In:* M.R. Rose and G.V. Lauder (eds): *Adaptation.* Academic Press, New York, pp. 149–185.

Solomon, J.M., Rossi, J.M., Golic, K., McGarry, T. and Lindquist, S. (1991) Changes in csp70 alter thermotolerance and heat-shock regulation in *Drosophila. The New Biologist* 3:1106–1120.

Stearns, S.C. and Kaiser, M. (1996) Effects on fitness components of P-element inserts in *Drosophila melanogaster:* Analysis of tradeoffs. *Evolution* 50:795–806.

Velazquez, J.M., DiDomenico, B.J. and Lindquist, S. (1980) Intracellular localization of heat shock proteins in *Drosophila. Cell* 20:679–689.

Velazquez, J.M. and Lindquist, S. (1984) Hsp70: Nuclear concentration during environmental stress and cytoplasmic storage during recovery. *Cell* 36:655–662.

Velazquez, J.M., Sonoda, S., Bugaisky, G. and Lindquist, S. (1983) Is the major *Drosophila* heat shock protein present in cells that have not been heat shocked? *J. Cell. Biol.* 96:286–290.

Welte, M.A., Tetrault, J.M., Dellavalle, R.P. and Lindquist, S.L. (1993) A new method for manipulating transgenes: Engineering heat tolerance in a complex, multicellular organism. *Curr. Biol.* 3:842–853.

Environmental Stress, Adaptation and Evolution
ed. by R. Bijlsma and V. Loeschcke
© 1997 Birkhäuser Verlag Basel/Switzerland

# High-temperature stress and the evolution of thermal resistance in *Drosophila*

Volker Loeschcke[1], Robert A. Krebs[2], Jesper Dahlgaard[1]
and Pawel Michalak[1, 3]

[1] *Department of Ecology and Genetics, University of Aarhus, Ny Munkegade,
DK-8000 Aarhus C, Denmark*
[2] *Department of Organismal Biology and Anatomy, The University of Chicago,
1027 East 57th Street, Chicago, IL 60637, USA*
[3] *Present address: Department of Zoology, Jagiellonian University, Krakow, Poland*

*Summary.* The evolution of thermal resistance and acclimation is reviewed at the population level using populations and isofemale lines of *Drosophila buzzatii* and *D. melanogaster* originating from different climatic regions. In general, ample genetic variation for thermal resistance was found within and among populations. A rough correlation between the climate of origin and thermal resistance was apparent. Acclimation at a non-lethal temperature led to a significant increase in survival after heat shock, and recurrent acclimation events generally increased survival even further. Acclimation effects lasted over several days, but this effect decreased gradually with time since acclimation. Protein studies showed that the concentration of Hsp70 in adult flies is greatly increased by acclimation and thereafter gradually decreases with time. For populations with relatively high survival at one life stage, survival often was low at other life stages. Furthermore, selection on different life stages showed that a selection response in one life stage did not necessarily result in a correlated response in another. These observations indicate that different mechanisms or genes at least in part are responsible for or are expressed at different developmental stages. Selection for increased resistance was successful despite low heritabilities for the trait. Survival and fertility were compared between acclimated and non-acclimated flies, and a cost of expressing the "heat shock response" was identified in that increased survival after acclimation was accompanied by reduced fertility. The relative costs increased under nutritional stress. Metabolic rate was genetically variable but did not correlate with temperature resistance. The more resistant lines, however, often had shorter developmental time. Inbreeding reduced thermal stress tolerance of adult flies, but it did not reduce tolerance of embryos that possibly are exposed to strong natural selection for thermal stress resistance. In general, inbreeding may reduce stress resistance, and thus multiple stressful events may account for increased inbreeding depression in harsh environments.

## Introduction

Genetic variation in thermal tolerance is abundant, providing the raw material for evolutionary responses in natural and laboratory populations (Hoffmann and Parsons, 1991). Our interest is to comprehend factors that affect this variation in natural populations, and to use identified variants to further our understanding of the mechanistic basis of thermotolerance. Genetic variants that enhance thermotolerance are expected to benefit individuals under thermal extremes. However, as fitness effects of genes often differ across environments, stress-resistant genotypes may have reduced fitness in conditions favourable to rapid growth. Even the response to stress may have costs, causing selection for a physiological balance between the consequences of failing to respond sufficiently to environmental change and those incurred by responding unnecessarily (Krebs and Loeschcke, 1994a).

We would also like to know if one primary genetic system provides the armament by which individuals tolerate perturbations to the physiological environment, or whether many systems of similar importance determine the thermal phenotype. Here we review research on genetic variation in thermotolerance and the fitness consequences of inducing a thermotolerant phenotype and ask whether the preponderance of evidence favours a single stress-defense system or whether thermotolerance is a product of many interactive genes. By answering this question with respect to natural variation in thermotolerance, investigations into the mechanistic basis of thermotolerance can be directed to particular target loci and/or to isolating and mapping phenotypic variants to identify as yet unknown players in the phenotypic change.

## Genetic variation in stress tolerance

Natural variation for resistance to high temperature has been known for many years. Stress tolerance correlates positively with variation in the thermal environment from which populations originate, as first shown by Timofeef-Ressovsky (Dobzhansky, 1937, pp. 154–155). He showed that egg-to-adult survival of strains of *D. funebris* from southern Russia exceeds that of northern strains at 29°C, while viability of the northern strains exceeds that of southern ones at 15°C. Coyne et al. (1987) observed a similar relationship for pupal thermotolerance in *D. pseudoobscura*, but not of adults. Numerous attempts to compare stress tolerance and climate of origin also have been performed in the cosmopolitan *D. melanogaster*. Where extreme differences in environmental conditions are found between populations, variation in thermotolerance occurs in the expected directon (Parsons, 1980), but where climatic variation is small, patterns are inconsistent (Krebs et al., 1996). Evidence for thermal adaptation to stressful environments has been clearly demonstrated in other organisms, including many lizard species (Huey and Bennett, 1990; Ulmasov et al., 1992), which probably are much more philopatric than *D. melanogaster*. However, even in *Drosophila* species constrained to ephemeral resources, like the cactophilic *D. buzzatii*, patterns to natural variation in thermotolerance are present but are less obvious than expected based on climatic data alone (Loeschcke et al., 1994; Krebs and Loeschcke, 1995a, b).

Historically, survival has been the most commonly used character for measuring stress effects on fitness of Drosophilids, but organisms respond to thermal stresses in so many ways that evolutionary consequences from periodic stress exposures cannot be predicted easily (Hoffmann and Parsons, 1991). Reproductive performance of males and females declines before conditions are life threatening (Krebs and Loeschcke, 1994a), and stress markedly reduces fitness of survivors (Krebs and Loeschcke 1994b, c). Evolutionarily, death and sterility generally are equivalent in non-social organisms. Changes in other fitness-related traits also may vary genetically in their response to thermal stress, although few have been examined in detail. We tested for genetic variation in survival and in offspring production after exposing individuals to 40°C for 90 min, using 10 isofemale lines of *D. melanogaster* from a Danish locality (Krebs and Loeschcke, 1994b). Effects on survival were not significantly different among lines, but male fertility varied significantly among survivors from the different lines.

## Acclimation to high temperatures

Insects and almost all other organisms can increase their resistance to future high-temperature stress by acclimation, i.e. responding to a short pretreatment at non-lethal high temperature. In *D. buzzatii* and *D. melanogaster*, preteated individuals remain more heat-resistant than non-treated individuals for several days, and repeated pretreatment may increase the level of resistance (Loeschcke et al., 1994; Krebs and Loeschcke, 1994a). Maximum resistance is achieved between 8 and 32 h after pretreatment (Krebs and Loeschcke, 1994a; J. Dahlgaard, V. Loeschcke, P. Michalak and J. Justesen, *submitted*), although the precise optimum probably depends on the temperature and duration of the acclimation treatment. Both in *D. melanogaster* (Fig. 1) and in *D. buzzatii* (Loeschcke et al., 1994), isofemale lines that are relatively more resistant than others to a direct heat stress also generally are the more resistant after acclimation. In *D. melanogaster*, females usually are more heat-resistant than males, and in isofemale lines where the relative performance of the sexes was opposite, i.e. males better than females without acclimation, the common female-superior performance relationship holds with acclimation (Fig. 1). Similarly in *D. buzzatii*, males and females may respond very differently to different stress treatments, although males normally are the more resistant to heat (Loeschcke et al., 1994; Krebs and Loeschcke, 1996).

## Temperature stress and selection

A clear example of stress temperatures are those at which a population cannot maintain growth or which disrupt the reproductive capability of individuals. After heat stress, however, reproduction usually recovers after individuals are returned to lower temperatures for some days. In *D. melanogaster*, constant 30–31°C is about the maximum for rearing eggs to adult (Parsons, 1978). Upper limits for the cactophilic *D. buzzatii*, which

**Not conditioned**

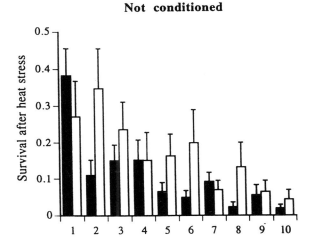

**Conditioned**

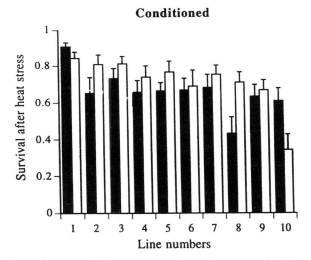

Figure 1. Relative resistance to high temperatures of *D. melanogaster* adults either pretreated or not pretreated. Ten isofemale lines originating from the Canary Islands were pretreated for 75 min at 37°C then kept at 25°C for 16 h and heat-shocked at 39°C for 90 min, or heat-shocked 80 min at 39.3°C without pretreatment. Flies were allowed to recover at 25°C for 22 h before survival was scored as the ability to walk. Males are presented as filled bars and females as open bars.

may encounter high temperatures more frequently than *D. melanogaster* in nature, are only slightly higher. Fitness declines rapidly as upper limits are approached, while lower limits taper off slowly (Huey and Kingsolver, 1989; Huey and Bennett, 1990). Perhaps not coincidentally, temperatures just above this range are those that clearly induce the heat shock response (DiDomenico et al., 1982; Lindquist and Craig, 1988). However, some physiological acclimation, or heat hardening, which increases performance at higher temperatures, may occur at lesser temperatures (Crill et al., 1996).

Furthermore, rearing at different thermal regimes leads to adaptive differences with respect to relative performance under a variety of temperature treatments. For instance, experiments on flies evolving in a constant thermal environment indicate that lines perform better for many traits at their own rearing temperature than do lines reared at cooler or warmer temperatures (Mourad, 1965; Cavicchi et al., 1989; Huey et al., 1991). The higher temperature lines also tolerate high-stress temperatures better than those reared at lower temperatures (Stephanou et al., 1983; Kilias and Alahiotis, 1985; Lints and Bourgois, 1987; Cavicchi et al., 1995). However, it is not known whether differences result from reduced tolerance to heat in lines evolving in a cold environment or to increased tolerance of those reared under warmer regimes. The highest-temperature lines used in continuous rearing are those at 28°C by S. Cavicchi, yet lines may be reared with daily pulses to much higher temperatures. *D. melanogaster* may be reared with temperatures climbing to at least 35°C for 6 h a day, with lines at 25°C for 18 h (V. Loeschcke, unpublished data), and similarly *D. buzzatii* can be reared at 38.2°C for 6 h if adults are kept for a few days at 25°C for egg laying. In this environment, lines evolve increased thermotolerance (Loeschcke and Krebs, 1996).

Greater differences in thermal tolerance can be produced using truncated selection for higher survival at stress temperatures (e.g. White et al., 1970; Morrison and Milkman, 1978; Stephanou et al., 1983; Quintana and Prevosti, 1990; Huey et al., 1992; Krebs and Loeschcke, 1996; McColl et al., 1996). Responses to artificial selection are also more rapid, with selection responses in thermotolerance usually increasing gradually and leveling off slowly, as expected from quantitative inheritance. Crosses between *D. melanogaster* lines selected for knockdown resistance and their respective controls produced progeny of intermediate resistance, a result that is compatible with an additive genetic architecture of knockdown resistance (McColl et al., 1996). As genetic variation for stress resistance is likely complex, and involves many traits, an additive basis for this variation is predicted (Feder, 1996). However, reciprocal crosses between selected lines in *D. buzzatii* (Krebs and Loeschcke, 1996), and among divergent populations of *D. melanogaster* and *D. buzzatii* (Krebs et al., 1996), provide a contrasting picture of genetic variation in the stress resistance between these two species. For *D. melanogaster*, we also obtained results for which explanation required interactions among many genes. Some high-tolerance effects that were attributable to the X chromosome clearly were dominant in inheritance, and epistatic interactions were suggestive in specific crosses. Explaining variation in survival of *D. buzzatii* adults after heat shock was simpler, requiring as few as three loci, but of these three, one locus expresses much more influence on males than on females (Krebs et al., 1996). This effect also contributed to a greater selection response for increased thermotolerance in males (Krebs and Loeschcke, 1996).

While both constant rearing and selection protocols can verify substantial genetic variation in thermal sensitivity or tolerance within populations, in nature no populations encounter a persistent shift to a constant temperature environment nor will truncated selection be frequently repeated on the whole population (Cavicchi et al., 1995). Indisputably, *Drosophila* have the capacity to evolve stress tolerance, but unspecified are the physiological characters actually affected in natural and laboratory populations, or how structural or regulatory changes in genes may enhance tolerance.

## Variation among life stages in thermal resistance

By focusing upon specific loci involved in stress responses, research may address developmental patterns of protein expression, variation among tissues for sensitivity to stress and/or developmental time points of higher or lower whole-body thermotolerance. *Drosophila* provide an important model system for characterizing the genetic basis of stress resistance, because as a complex eukaryote changes in stress responses and selection coefficients on those responses vary across its complex life cycle. For example, as an alternative to changes in physiology, survival may increase through behavioural avoidance of high temperatures, thereby reducing the importance of physiologically plastic responses to changing environments. However, within the lifetime of any individual, behavioural avoidance is largely limited to the adult stage in *D. melanogaster*. Eggs and pupae are immobile, and larvae are confined to the thermal environment of the fruit they inhabit, in which temperatures may become high, while thermal variation within the fruit is low (Feder, 1996; Feder et al., 1997). Such restrictions, however, cannot be fully generalized across species, as the necrotic prickly pear inhabited by *D. buzzatii* may vary extensively in temperature between shaded and sun-exposed portions of the cladode (Krebs and Loeschcke, 1995b), possibly enabling larvae to avoid more extreme temperatures. Comprehending the environment is a prerequisite for understanding natural selection pressures on tolerance to extremes. How successfully a population responds to conditions of stress and which traits respond will depend greatly on the heritability of tolerance as a whole, and on the periodicity and degree of stress exposures (Krebs and Loeschcke, 1994c).

Differences in thermal environments across development are relevant to predictions of evolutionary responses, because genetic variation in survival may differ among life stages of *Drosophila*. For seven populations of *D. buzzatii* where survival after thermal stress was measured from eggs, larvae, pupae and adults, high tolerance at the different life stages was unrelated or poorly correlated among the various stages (Krebs and Loeschcke, 1995a, b).

For a strain to be tolerant in one, but not other life stages, indicates that different mechanisms control variation in thermotolerance along development, or that key components of thermotolerance are shut down at some time points. Large differences in thermotolerance of the different stages also suggests that stress-defense systems may be developmentally specific or limited in their time of expression. Pupae withstand much higher temperatures than do eggs, larvae or adults in *Drosophila* (Krebs and Loeschcke, 1995b; Feder et al., 1996, 1997) and in blowflies (Tiwari et al., 1995), but variation among populations is low for pupae (Coyne et al., 1987; Krebs and Loeschcke, 1995b). Lack of genetic variation in this stage limits the opportunities for investigating mechanisms of thermotolerance in pupae. Divergent lines are necessary for studying mechanistic variation, but for *D. buzzatii* adults, in which quantitative variation for thermal resistance is extensive, patterns of induction and repression of the thermal phenotype appear very similar, suggesting that maybe only few gene products control variation in survival (Krebs et al., 1996).

In contrast to adults, variation in larval resistance is very different (Krebs and Loeschcke, 1995b). Excerpting from that work, a plot of resistance in seven populations indicates that individuals in two are least resistant both as adults and as larvae, but for five others, relatively high resistance at one stage is not correlated to relative resistance of the other (Fig. 2). Relation-

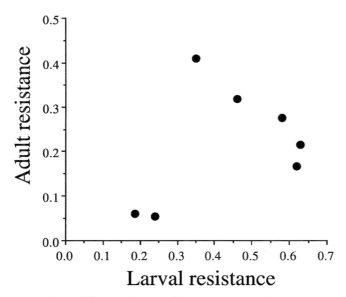

Figure 2. A comparison of thermotolerance of larvae and adults of seven *D. buzzatii* populations. Data are for adults treated by 75 min at 38°C, rested 4 h at 25°C and exposed to 41.5°C for 110 min in an incubator (Krebs and Loeschcke, 1995a). To stress larvae, they were first transferred to medium in vials, and placed in an incubator at 39°C for 6 h (Krebs and Loeschcke, 1995b).

ships between other stages also were weak or absent. This result prompted a test of whether selection to alter the thermotolerance gives a general response across development. Both adults and larvae respond to selection on thermotolerance but fail to show a correlated response on the alternate stage (Loeschcke and Krebs, 1996).

## Mechanisms of resistance

Exposure to high temperatures induces expression of heat shock proteins in most organisms, and these proteins are important for protecting cells against high temperature and other stresses (Lindquist and Craig, 1988; Parsell and Lindquist, 1994; Morimoto et al., 1994). In cell lines of *D. melanogaster*, expression of Hsp70, the primary inducible heat shock protein in this species, is most rapid at 36°C (DiDomenico et al., 1982), and the concentration of Hsp70 is elevated for some time after exposure to stress. In lines varying in *hsp70* copy number, higher levels of Hsp70 are related to higher thermotolerance in eggs (Welte et al., 1993), larvae and pupae (Feder et al., 1996). Krebs and Feder (1997a) also found a positive correlation between expression of Hsp70 and high-temperature toler-ance in larvae among lines from an orchard population. J. Dahlgaard, V. Loeschcke, P. Michalak and J. Justesen (submitted) found that Hsp70 expression in adult *D. melanogaster* was positively associated with survival. Alone, however, the level of Hsp70 could not explain the curve describing survival to heat stress at different times elapsed after acclimation. Hsp70 expression is highest immediately or shortly after pretreatment (Fig. 3), depending on the treatment temperature and duration, and decays there-after until it reaches the level of non-treated individuals in about 24 h. The increase in survival following acclimation, however, persisted much longer, and optimal survival rates were reached after those at which maximum Hsp70 expression occurred. Similarly, larval thermotolerance effects after pretreatment also outlast increases in Hsp70 concentration (Krebs and Feder, 1997b). These results suggest that factors other than or in addition to Hsp70 concentration are responsible for variation in heat resistance in *Drosophila*.

Specific genes or alternative mechanisms for altering thermotolerance, however, have largely eluded identification, and only elements related to the heat shock response have received strong support for a general role in thermotolerance. In addition to *hsp70*, variation at two loci, *hsr-omega* and *hsp68*, responded indirectly to selection for increased resistance to knock-down by heat in adult *D. melanogaster* (McColl et al., 1996). Increased osmolite concentrations increase resistance to thermal stresses in yeast (Parsell and Lindquist, 1994), but analogous changes at high temperatures are not known in *Drosophila*.

An alternative possibility discussed by Hoffmann and Parsons (1991) is that another general stress response, a reduction in metabolic activity, could

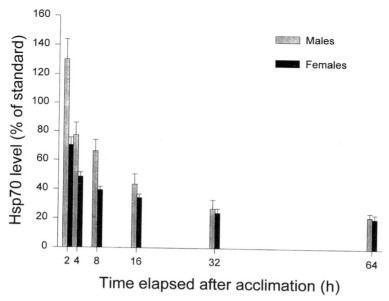

Figure 3. Hsp70 expression as a function of time elapsed after acclimation to 37°C for 55 min in males and females of *D. melanogaster*. Data are relative to a standard, a homogenate of adult flies that were exposed to 38°C for 45 min and allowed to recover for 5 h before freezing. (After J. Dahlgaard, V. Loeschcke, P. Michalak and J. Justesen, *submitted*.)

act as a mechanism to explain the cross-resistance found between several stress factors, usually chemical in nature, but including responses to desiccation. However, larvae from thermotolerant strains appear to develop as fast or faster than those from less-resistant strains. (Krebs and Loeschcke, 1995b; Loeschcke and Krebs, 1996). To test metabolic rate effects on heat resistance, we compared the basal adult metabolic rate of young female adults, larval and adult heat resistance, developmental time, longevity and early female fecundity in 100 isofemale lines of *D. buzzatii*. Metabolic rate was correlated significantly only with fecundity ($r = 0.28$, $p < 0.01$). (V. Loeschcke, R. A. Krebs and J. Dahlgaard, unpublished data).

Energetic effects may constrain stress responses (Coleman et al., 1995), and if energetically costly, fecundity is one trait that is likely to become affected as individuals acclimate to stress. Although few studies have addressed trade-offs with stress tolerance, their occurrence spans over a variety of organisms and stress regimes (Hoffmann, 1995). Pretreatment increases heat resistance in *D. melanogaster*, and a cost in decreased fecundity of acclimated flies occurs for a duration similar to that for the gain in thermotolerance. This effect is relatively stronger when coupled with nutritional stress (Krebs and Loeschcke, 1994a). Sex specificity in effects on traits that may be involved in trade-offs with increasing thermotolerance could also constrain evolutionary responses (Krebs and Loeschcke, 1996).

## Stress resistance and inbreeding depression

Inbreeding depression is potentially environmentally dependent (Pray et al., 1994). By consequence, estimates of inbreeding depression may conflict if experiments are performed in different environments, even if individuals originate from exactly the same population. Environmental dependency of inbreeding depression may be expressed as inbreeding by environment interactions, i.e. the phenomenon that inbred individuals respond differently to environmental change in contrast to outbred individuals. A number of fitness-related traits have been studied in a variety of species, considering more than one environment. Generally, fitness of inbred individuals relative to outbred ones is further reduced as the environment becomes less favourable (Parsons, 1959; Griffing and Langridge, 1963; Pederson, 1968; Ruban et al., 1988; Dudash, 1990; Hauser and Loeschcke, 1996; Dahlgaard and Hoffmann, 1997; Bijlsma et al., this volume).

In most natural environments, the occasional episodes of stress are inevitable, and therefore stress resistance is likely to be important for fitness. In *Drosophila*, inbreeding with few exceptions often reduces stress resistance (Dahlgaard et al., 1995; Dahlgaard and Loeschcke, 1997; Dahlgaard and Hoffmann, 1997; Bijlsma et al., this volume), and this may in part explain why inbreeding depression in harsh environments often is increased. Stress, besides its obvious direct effect on the environmental component of variance, also can affect the level of expressed genetic variation (Hoffmann and Parsons, 1991; Imasheva et al., 1997; Brakefield, this volume). The redistribution of genetic variation occurring during inbreeding therefore may have different quantitative effects on certain traits depending on how stressful the environment is perceived.

In adult *Drosophila*, resistance to heat (Fig. 4), acetone and desiccation is reduced with inbreeding; however, the variance of these traits coordinately is increased (Dahlgaard et al., 1995; Dahlgaard and Hoffmann, 1997). Therefore, despite a decrease in mean resistance after inbreeding some inbred lineages often are at least as resistant as the best outbred lineages (Hauser et al., 1994; Dahlgaard and Hoffmann, 1997). The effect of inbreeding on stress resistance consequently may seem to be only of minor importance, as the possibility still exists for natural selection to operate in stressful environments as long as resistance genes have not disappeared from the population by drift.

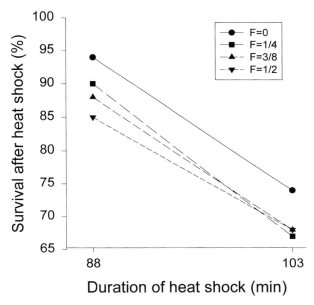

Figure 4. Survival after heat shock of adult *D. buzzatii* males and females of four inbreeding levels. Flies were acclimated at 36.5°C for 75 min before being exposed 24 h later to a potentially lethal temperature of 40.7°C for 88 min or 103 min. Flies were allowed to recover for 22 h before survival was scored. (After Dahlgaard et al., 1995.)

## Multiple stresses

Beardmore (1983) and Bijlsma et al. (this volume), however, have emphasized the potential difficulties that inbred lines or populations may have in adapting to future environmental change. Rephrased and in a context of stress resistance, the question is, How do inbred lines that are highly resistant to one specific stress cope when exposed to a different but still stressful environment? To test the question one could experimentally breed a large number of inbred and outbred lines and test their ability to resist a number of different stresses. Using *D. melanogaster* as a model organism, Dahlgaard and Hoffmann (1997) did not find evidence for correlations between three different stress-resistance traits nor between any of those traits and female progeny production in a stressful and a non-stressful environment. Thus, as perhaps may be expected, inbred lines resistant to a specific stress are not necessarily resistant to other stresses nor can they be characterized by having increased fitness in either stressful nor non-stressful environments.

Considering how inbreeding affects stress resistance, it is not surprising that inbreeding depression often increases in harsh environments. Harsh environments, as opposed to benign environments, may be described as

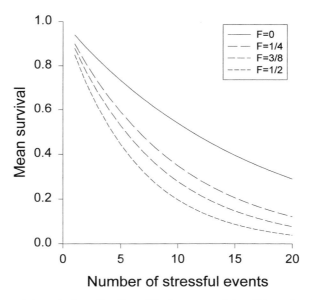

Figure 5. Expected survival as a function of the frequency of stressful events or the number of stresses in the environment. Survival parameters, i.e. resistance, for one stressful event of the different inbreeding groups were chosen from the low stress end (88 min) of Figure 4 with $P_{F=0} = 0.94$; $P_{F=1/4} = 0.90$; $P_{F=3/8} = 0.88$; $P_{F=1/2} = 0.85$. We assumed the same fitness reduction for every stress event (proportional to the inbreeding coefficient). Resistance against the different stresses is assumed to be genetically uncorrelated, and thus survival rates (fitness values) were estimated as being multiplicative.

environments in which stress occurs frequently. Multiple stressful events that each are resisted best by outbred individuals rapidly will enlarge differences in survival between inbreeding groups, particularly if resistance against the different stresses is genetically uncorrelated. A multiplicative model shows expected survival (Fig. 5) as a function of the frequency of stressful events or the number of stresses in the environment. Note that if inbreeding depression is estimated in two different environments, a benign (no stress) and a harsh environment (one or more stresses), inbreeding depression by interaction with the environment may occur even in the absence of interactions between stresses, if only stress resistance is dependent on the level of inbreeding. Inbreeding depression is therefore markedly augmented as the stressfulness of the environment increases. Potential interactions between two or more stresses may contribute further to inbreeding depression and can be revealed as deviations from expected values if resistance to each separate stress is known. Deviations from expected values, however, may also arise due either to negative or positive genetic correlations between traits, and knowledge of potential genetic correlations is thus required before interactions between stresses can be inferred.

For conservation biology these findings imply that with inbreeding in stressful environments, selection and drift simultaneously will operate potentially to facilitate survival of populations. Due to increased variance in stress resistance with inbreeding, the probability of survival may be reasonable if the environment is stressful with respect to only a single factor, but may be much reduced if the environment is stressful in several aspects. However, if a population survives inbreeding in a stressful environment, founders of the next generations may be genetically homogeneous, and the population may consequently be susceptible and unable to adapt to future environmental stress.

## Conclusion

We have reviewed data concerning the ecology and genetics of thermal resistance which may serve as a background to predict whether natural populations of *Drosophila* are able to adapt to increased temperatures, a process that includes more than the ability to tolerate high temperatures. Adaptations enabling the maintenance of normal physiological and reproductive processes are also required as well as the ability to respond to a number of biotic and abiotic changes that may accompany a temperature increase. Evolutionary predictions are by their nature complex, though that does not mean that at least parts of these questions can be answered.

The ability to tolerate increased temperatures for a limited period of time is a critical trait. Assuming a slight increase in the global temperature, an increase in the frequency of short temperature extremes may follow. Evidence favours the likelihood that *Drosophila*, and possibly other insects too, can adapt to short-term high temperature stress. Some descrepancies exist in the order of magnitude of heritability estimates revealed from different experiments, for different species and the actual resistance trait being measured. However, considering the time scale at which increasing temperatures may be expected, even low heritabilities may be sufficient for resistance to evolve. That many genes are involved in resistance to high temperatures is supported by a gradual and continued response to selection in various experiments. Still, however, genes with a major effect on resistance have been identified, and evidence suggests that natural variation is present for the expression of the heat shock protein Hsp70 (Krebs and Feder, 1997a).

It is unclear, however, to what extent an evolutionary route for increased thermal resistance will include variants or increasing copy numbers of major genes as, for instance, *hsp70*. *Hsp70* certainly functions to increase resistance of most life stages in *D. melanogaster* (Welte et al., 1993; Feder et al., 1996; J. Dahlgaard, V. Loeschcke, P. Michalak and J. Justesen, *submitted*). However, not all developmental stages, late embryos for instance, experience increased thermal resistance with increasing quantities of

Hsp70 (Welte et al., 1993). Hsp70 also has been found to be detrimental to growth at normal temperatures (Feder et al., 1992) and may even cause mortality when present in high concentrations in first-instar larvae (Krebs and Feder, 1997b). Adult *D. melanogaster* also experiences substantial costs in terms of reduced fecundity following an activation of the heat shock response (Krebs and Loeschcke, 1994a). Natural selection therefore is likely to favour variants that are not only able to survive high temperatures but which are also able to maintain normal physiological and reproductive processes as the temperature increases. At increasing temperatures the additive genetic variance of traits (de Jong, 1990; Noach et al., 1996) and coefficients of variation (Imasheva et al., 1997) of several morphological characters are significantly increased. Increased genetic variation in quantitative traits under stress (Hoffmann and Parsons, 1991) will thus potentially be abailable for the evolutionary process of increasing thermal resistance.

*Acknowledgments*
We are grateful to the Carlsberg Foundation (No. 93-0280-30 and 96-0359-40) and the Danish Natural Science Research Council (No. 94-01-631) for financial support and the Tempus Programme for supporting the stay of P.M. at the University of Aarhus. R.K.'s contribution to this paper was written while he was at the University of Chicago supported by National Science Foundation grant IBN94-08216 to Dr. Martin Feder. The monoclonal antibody, FFB, was kindly provided by Dr. S. Lindquist.

# References

Beardmore, J.A. (1983) Extinction, survival and genetic variation. *In*. C.M. Schonewald-Cox, S.M. Chambers, B. MacBryde and W.L. Thomas (eds): *Genetics and Conservation*. Benjamin/Cummings, London, pp. 125–151.

Cavicchi, S., Guerra, D., Natali, V., Pezzoli, C. and Giorgi, G. (1989) Temperature-related divergence in experimental populations of *Drosophila melanogaster*. II. Correlation between fitness and body dimensions. *J. Evol. Biol.* 2:235–251.

Cavicchi, S., Guerra, D., La Torre, v. and Huey, R.B. (1995) Chromosomal analysis of heat-shock tolerance in *Drosophila melanogaster* evolving at different temperatures in the laboratory. *Evolution* 49:676–684.

Coleman, J.S., Heckathorn, S.A. and Hallberg, R.L. (1995) Heat-shock proteins and thermotolerance: Linking ecological and molecular perspectives. *Trends Ecol. Evol.* 10:305–306.

Coyne, J.A., Bundgaard, J. and Prout, T. (1987) Geographic variation of tolerance to environmental stress in *Drosophila pseudoobscura*. *Am. Nat.* 122:474–488.

Crill, W.D., Huey, R.B. and Gilchrist, G.W. (1996) Within- and between-generational effects of temperature on the morphology and physiology of *Drosophila melanogaster*. *Evolution* 50:1205–1218.

Dahlgaard, J. and Hoffmann, A.A. (1977) Reduced stress resistance in *Drosophila melanogaster* after inbreeding: General versus specific genes. *In*: J. Dahlgaard, *Inbreeding, acclimation and stress resistance in Drosophila*. Ph.D. dissertation, Institute of Biological Sciences, University of Aarhus.

Dahlgaard, J. and Loeschcke, V. (1997) Effects of inbreeding in three life stages of *Drosophila buzzatii* after embryos were exposed to a high-temperature stress. *Heredity* 78:410–416.

Dahlgaard, J., Krebs, R.A. and Loeschcke, V. (1995) Heat-shock tolerance and inbreeding in *Drosophila buzzatii*. *Heredity* 74:157–163.

de Jong, G. (1990) Quantitative genetics of reactions norms. *J. Evol. Biol* 3:447–468.

DiDomenico, B.J., Bugaisky, G.E. and Lindquist, S. (1982) The heat shock response is self-regulated at both the transcriptional and posttranscriptional levels. *Cell* 31:593–603.

Dobzhansky, T. (1937) *Genetics and the Origin of Species*. Chicago University Press, Chicago.

Dudash, M.R. (1990) Relative fitness of selfed and outcrossed progeny in a self-compatible, protandrous species, *Sabatia angularis* L. (Gentianaceae): A comparison in three environments. *Evolution* 44:1129–1139.

Feder, M.E. (1996) Ecological and evolutionary physiology of stress proteins and the stress response: The *Drosophila melanogaster* model. *In*: I.A. Johnston and A.F. Bennett, (eds): *Phenotypic and Evolutionary Adaptation to Temperature*. Cambridge University Press, Cambridge, pp. 79–102.

Feder, J.H., Rossi, J.M., Solomon, J., Solomon, N. and Lindquist, S. (1992) The consequences of expressing hsp70 in *Drosophila* cells at normal temperatures. *Genes & Devel.* 6:1402–1413.

Feder, M.E., Cartaño, N.V., Milos, L., Krebs, R.A. and Lindquist, S.L. (1996) Effect of engineering *Hsp70* copy number on Hsp70 expression and tolerance of ecologically relevant heat shock in larvae and pupae of *Drosophila melanogaster*. *J. Exp. Biol.* 199:1837–1844.

Feder, M.E., Blair, N. and Figueras, H. (1997) Natural thermal stress and heat-shock protein expression in *Drosophila* larvae and pupae. *Funct. Ecol.* 11:90–100.

Griffing, B. and Langridge, J. (1963) Phenotypic stability of growth in the self-fertilized species, *Arabidopsis thaliana*. *In*: W.D. Hanson and H.F. Robinson (eds): *Symposium on Statistical Genetics and Plant Breeding*. Publ. 982. National Academy of Sciences, National Research Council, Washington, DC, pp. 386–394.

Hauser, T.P. and Loeschcke, V. (1996) Drought stress and inbreeding depression in *Lychnis flos-cuculi* (Caryophyllaceae). *Evolution* 50:1119–1126.

Hauser, T.P., Damgaard, C. and Loeschcke, V. (1994) Effects of inbreeding in small plant populations: Expectations and implications for conservation. *In*: V. Loeschcke, J. Tomiuk and S.K. Jain (eds): *Conservation Genetics*. Birkhäuser, Basel, pp. 115–130.

Hoffmann, A.A. (1995) Acclimation: Increasing survival at a cost. *Trends Ecol. Evol.* 10:1–2.

Hoffmann, A.A. and Parsons, P.A. (1991) *Evolutionary Genetics and Environmental Stress*. Oxford University Press, New York.

Huey, R.B. and Bennett, A.F. (1990) Physiological adjustments to fluctuating thermal environments: An ecological and evolutionary perspective. *In*: R.I. Morimoto, A. Tissières and C. Georgopoulos (eds): *Stress Proteins in Biology and Medicine*. Cold Spring Harbor Laboratory Press, Cold Spring Harbor, NY, pp. 37–59.

Huey, R.B. and Kingsolver, J.G. (1989) Evolution of sensitivity of ectotherm performance. *Trends Ecol. Evol.* 4:131–135.

Huey, R.B., Partridge, L. and Fowler, K. (1991) Thermal sensitivity of *Drosophila melanogaster* responds rapidly to laboratory natural selection. *Evolution* 45:751–756.

Huey, R.B., Crill, W.D., Kingsolver, J.G. and Weber, K.E. (1992) A method for rapid measurement of heat or cold resistance of small insects. *Funct. Ecol.* 6:489–494.

Imasheva, A.G., Loeschcke, V., Lazebny, O.E. and Zhivotovsky, L.A. (1997) Effects of extreme temperatures on quantitative variation and developmental stability in *Drosophila melanogaster* and *Drosophila buzzatii*. *Biol. J. Linn. Soc.* 61, *in press*.

Kilias, G. and Alahiotis, S.N. (1985) Indirect thermal selection in *Drosophila melanogaster* and adaptive consequences. *Theor. Appl. Genet.* 69:645–650.

Krebs, R.A. and Feder, M.E. (1997a) Natural variation in the expression of the heat-shock protein Hsp70 in a population of *Drosophila melanogaster*, and its correlation with tolerance of ecologically relevant thermal stress. *Evolution* 51:173–179.

Krebs, R.A. and Feder, M.E. (1997b) Negative consequences of Hsp70 overexpression in *Drosophila melanogaster* larvae. *Cell, Stress and Chaperones* 2:60–71.

Krebs, R.A. and Loeschcke, V. (1994a) Costs and benefits of activation of the heat-shock response in *Drosophila melanogaster*. *Funct. Ecol.* 8:730–737.

Krebs, R.A. and Loeschcke, V. (1994b) Effects of exposure to short-term heat stress on fitness components in *Drosophila melanogaster*. *J. Evol. Biol.* 7:39–49.

Krebs, R.A. and Loeschcke, V. (1994c) Response to environmental change: Genetic variation and fitness in *Drosophila buzzatii* following temperature stress. *In*: V. Loeschcke, J. Tomiuk and S.K. Jain (eds): *Conservation Genetics*. Birkhäuser, Basel, pp. 309–321.

Krebs, R.A. and Loeschcke, V. (1995a) Resistance to thermal stress in adult *Drosophila buzzatii*: Acclimation and variation among populations. *Biol. J. Linn. Soc.* 56:505–515.

Krebs, R.A. and Loeschcke, V. (1995b) Resistance to thermal stress in preadult *Drosophila buzzatii*: Variation among populations and changes in relative resistance across life stages. *Biol. J. Linn. Soc.* 56:517–531.

Krebs, R.A. and Loeschcke, V. (1996) Acclimation and selection for increased resistance to thermal stress in *Drosophila buzzatii*. *Genetics* 142:471–479.

Krebs, R.A., La Torre, V., Loeschcke, V. and Cavicchi, S. (1996) Heat-shock resistance in *Drosophila* populations: Analysis of variation in reciprocal cross progeny. *Hereditas* 124: 47–55.

Lindquist, S. and Craig, E.A. (1988) The heat-shock proteins. *Annu Rev Genet* 22:631–677.

Lints, F.A. and Bourgois, M. (1987) Phenotypic and genotypic differentiation in cage populations of *Drosophila melanogaster*. *Genet. Sel. Evol.* 19:155–170.

Loeschcke, V. and Krebs, R.A. (1996) Selection for heat-shock resistance in larval and adult *Drosophila buzzatii*: Comparing direct and indirect responses on viability and development. *Evolution* 50:2354–2359.

Loeschcke, V., Krebs, R.A. and Barker, J.S.F. (1994) Genetic variation and acclimation to high temperature stress in *Drosophila buzzatii*. *Biol. J. Linn. Soc.* 52:83–92.

McColl, G., Hoffmann, A.A. and McKechnie, S.W. (1996) Response of two heat shock genes to selection of knockdown resistance in *Drosophila melanogaster*. *Genetics* 143:1615–1627.

Morimoto, R.I., Tissières, A. and Georgopoulos, C. (eds) (1994) *Heat Shock Proteins: Structure, Function and Regulation*. Cold Spring Harbor Laboratory Press, Cold Spring Harbor, NY.

Morrison, W.W. and Milkman, R. (1978) Modification of heat resistance in *Drosophila* by selection. *Nature* 273:49–50.

Mourad, A.E. (1965) Genetic divergence in M. Vetukhov's experimental populations of *Drosophila pseudoobscura*. 2. Longevity. *Genet. Res.* 6:139–146.

Noach, E.J.K., de Jong, G. and Scharloo, W. (1996) Phenotypic plasticity in morphological traits in two populations of *Drosophila melanogaster*. *In*: E.J.K. Noach, *Phenotypic plasticity in* Drosophila melanogaster: *Body size and physiological characters*. Ph.D. dissertation, Faculty of Biological Sciences, University of Utrecht.

Parsell, D.A. and Lindquist, S. (1994) Heat shock proteins and stress tolerance. *In*: R.J. Morimoto, A. Tissières and C. Georgopoulos (eds): *The Biology of Heat Shock Proteins and Molecular Chaperones*. Cold Spring Harbor Laboratory Press, New York, pp. 457–494.

Parsons, P.A. (1959) Genotypic-environmental interactions for various temperatures in *Drosophila melanogaster*. *Genetics* 44:1325–1333.

Parsons, P. (1978) Boundary conditions for *Drosophila* resource utilization in temperate regions, expecially at low temperatures. *Am. Nat.* 112:1063–1074.

Parsons, P.A. (1980) Isofemale strains and evolutionary strategies in natural populations. *Evol. Biol.* 13:175–217.

Pederson, D.G. (1968) Environmental stress, heterozygote advantage and genotype-environment interaction in *Arabidopsis*. *Heredity* 23:127–138.

Pray, L.A., Schwartz, J.M., Goodnight, C.J. and Stevens, L. (1994) Environmental dependency of inbreeding depression: Implications for conservation biology. *Conserv. Biol.* 8:562–568.

Quintana, A. and Prevosti, A. (1990) Genetic and environmental factors in the resistance of *Drosophila subobscura* adults to high temperature shock. 2. Modification of heat resistance by indirect selection. *Theor. Appl. Genet.* 80:847–851.

Ruban, P.S., Cunningham, E.P. and Sharp, P.M. (1988) Heterosis × nutrition interacton in *Drosophila melanogaster*. *Theor. Appl. Genet.* 76:136–142.

Stephanou, G., Alahiotis, S.N., Christodoulou, C. and Marmaras, V.J. (1983) Adaptation of *Drosophila melanogaster* to temperature; Heat-shock proteins and survival in *Drosophila melanogaster*. *Dev. Genet.* 3:299–308.

Ulmasov, K.A., Shammakov, S., Karaev, K. and Evgenev, M.B. (1992) Heat shock proteins and thermoresistance in lizards. *Proc. Natl. Acad. Sci. USA* 89:1666–1670.

Welte, M.A., Tetrault, J.M., Dellavalle, R.P. and Lindquist, S. (1993) A new method for manipulating transgenes: Engineering heat tolerance in a complex, multicellular organism. *Curr. Biol.* 3:842–853.

White, E.B., Debach, P. and Garber, J. (1970) Artificial selection for genetic adaptation to temperature extremes in *Aphytes lingnanansis* (Hymenoptera: Aphelinidae). *Hilgardia* 40:61–192.

# Stress, selection and extinction

Environmental Stress, Adaptation and Evolution
ed. by R. Bijlsma and V. Loeschcke
© 1997 Birkhäuser Verlag Basel/Switzerland

# Genetic and environmental stress, and the persistence of populations

R. Bijlsma[1], Jørgen Bundgaard[2], Anneke C. Boerema[1]
and Welam F. Van Putten[1]

[1] *Department of Genetics, University of Groningen, Kerklaan 30, NL-9751 NN Haren, The Netherlands.*
[2] *Department of Ecology and Genetics, University of Aarhus, Ny Munkegade, Building 540, DK-8000 Aarhus C, Denmark.*

*Summary.* Many populations of endangered species have to cope both with stressful and deteriorating environmental conditions (mostly the primary cause of the endangerment) and with an increase in homozygosity due to genetic drift and/or inbreeding in small isolated populations. The latter will result in genetic stress often accompanied by a decrease in fitness (inbreeding depression). We have studied the consequences of genetic stress, under optimal as well as stressful environmental conditions, for the fitness and persistence of small populations using *Drosophila melanogaster* as a model system. The results show that, already under optimal environmental conditions, an increase in homozygosity or inbreeding both impairs fitness and increases the extinction risk of populations significantly. Under environmental stress, however, these effects become greatly enhanced. More important, the results show that the impact of environmental stress becomes significantly greater for higher inbreeding levels. This explicitly demonstrates that genetic and environmental stress are not independent but can act synergistically. This apparent interaction may have important consequences for the conservation of endangered species.

## Introduction

There are many different definitions of stress (for a discussion see Hoffmann and Parsons, 1991; Forbes and Calow, this volume). Generally, this term is applied to situations where the fitness of individuals or populations is reduced because of changed conditions. Most often, the stress is considered to be environmentally induced, either by physical features of the environment, such as climatic factors and pollutants, or by the biotic environment, such as parasites, competitors and predators. However, stress may also be intrinsic: an increase in homozygosity due to genetic drift and/or inbreeding, in conjunction with natural selection, will cause genetic stress that also often decreases fitness. Several mechanisms are thought to cause such a decrease in fitness, among which (1) an increase in homozygosity for recessive deleterious alleles (the partial dominance hypothesis), and (2) a decrease in the level of heterozygotes for overdominant loci (the overdominance hypothesis) are the most prominent (Charlesworth and Charlesworth, 1987). But other mechanisms might also be important, such as a decrease in the level of genomic coadaptation or an increase in genomic instability (Parsons, 1992; Clark, 1993).

   As many endangered species both have to cope with stressful environmental conditions and are often subject to genetic stress due to small population size, which causes drift and inbreeding (Bijlsma et al., 1994), the
study of these processes and especially their interaction has been targeted
as an area of priority research in conservation biology (Miller and Hedrick,
1993; Miller, 1994). In this chapter we will address some of the questions
in this research field.


*Stress and conservation biology*

In recent years, many plants and animal species have become endangered.
This has primarily been due to severe reductions in suitable habitats and the
concurrent deterioration of the remaining habitats by human activities such
as overexploitation, introduction of exotic species and pollution (Frankel
and Soulé, 1981). Thus, many species have had to cope with various environmental stresses that have caused, and still do cause, a decline in numbers. Importantly, suitable habitats have not only become deteriorated but
also greatly fragmented. Consequently, many endangered species find
themselves confined to small and often isolated habitat patches that can
sustain only a limited number of individuals. Populations that decline in
size become increasingly affected by stochastic events, and as a consequence the risk of populations to go extinct in the near future increases
significantly (see Soulé, 1987). These stochastic processes that threaten
small populations with extinction are often subdivided in four types
(Shaffer, 1981, 1987):

(1)  Demographic stochasticity: random effects that cause variation in
     survival and reproduction of individuals.
(2)  Environmental stochasticity: random fluctuations in the physical and
     biotic environment of a population (e.g. nutrient supply, weather,
     diseases, predators, competitors etc.).
(3)  Natural catastrophes: an extreme form of environmental stochasticity
     such as droughts, floods, fires, storms etc.
(4)  Genetic stochasticity: random genetic drift, inbreeding and founder
     events that cause changes in genetic composition of small populations
     and change survival and reproduction of individuals.

The consequences of genetic stochasticity have become a major issue in
conservation biology. Population genetics theory predicts that small and
isolated populations will be subject to genetic drift and inbreeding, resulting in loss of genetic variation, an increase in homozygosity and often
accompanied by a decrease in fitness (inbreeding depression). This loss of
genetic variation has in fact been observed in small natural populations of
different species (Bijlsma et al., 1991, 1994; Van Treuren et al., 1991; Karron, 1987; Kappe et al., 1995, 1997). Loss of variation may significantly

decrease the adaptability of populations to contemporary environmental conditions and to changing environmental conditions (see chapter by Bürger and Lynch, this volume, and references therein). Inbreeding and concurrent inbreeding depression may significantly decrease the fitness of individuals and populations, and thus may directly affect population persistence. This will evidently cause populations to become more prone to other stress factors in the long term, e.g. with respect to disease resistance and so on. Although the genetic basis of inbreeding depression is still debated (Charlesworth and Charlesworth, 1987), there is little doubt that it can have a significant impact on many animal and plant species. However, only a very few studies up to now have documented the occurrence of inbreeding depression in natural populations (Van Noordwijk and Scharloo, 1981; Hedrick et al., 1996).

As empirical data are scarce and often conflicting, the importance of genetics for the persistence of populations has become increasingly questioned (Caro and Laurenson, 1994; Caughley, 1994). Among the many arguments that have been used to reason that the genetics problems of small populations might be far less than expected, there are two main questions that will be addressed in this chapter.

First, it has been observed that some populations that have lost most of their genetic diversity after going through (a) severe bottleneck(s) nevertheless prosper currently (Ellegren et al., 1993; Hoelzel et al., 1993). This suggests that populations are able to purge their genetic load due to recessive deleterious alleles during the early stages of the inbreeding process and thereby preserve a relatively unaffected long-term fitness (Hedrick, 1994). This may, however, only be valid when environmental conditions are relatively constant and optimal, and it has been suggested that genetic load becomes increasingly expressed under more stressful environmental conditions (Dudash, 1990; Koelewijn, 1993; Dahlgaard et al., 1995; Dahlgaard and Loeschcke, 1997). Second, it has been suggested that in small populations genetic problems may be secondary to ecological problems due to environmental risks. Consequently, extinction is thought to occur before genetic problems become evident (Caughley, 1994; Lande, 1988; Lande and Barrowclough, 1987; Menges, 1991). This does not, however, take into account the possible interaction between genetic and ecological effects: an increase in homozygosity might enhance the susceptibility of populations to environmental stochasticity, especially under stressful conditions. In other words, genetic stress and environmental stress are not necessarily independent, and it is perhaps more realistic to assume that they can be synergistic (Parsons, 1992). As a better understanding of the interaction between ecological and genetic factors is therefore fundamental (Lande, 1988; Menges, 1991), the possible relationship between homozygosity and inbreeding, on the one hand, and fitness in optimal and stressful conditions, on the other hand, was studied in a number of experiments with *Drosophila melanogaster*.

Drosophila *as a model organism*

If we are interested in the behaviour of natural populations of endangered species, why study *Drosophila*? First of all, we want to be able to manipulate the study organism, to experimentally change population size or to manipulate environmental conditions. And it may be clear that this is not advisable or permissible for endangered species, and nature managers, for good reasons, are often very reluctant to permit experimentation with protected species and in protected sites. Second, we need a good understanding of the genetic structure of the organism under study to critically evaluate the results of the experiments. Not knowing the genetic (and ecological) history of populations is often a great obstacle for interpretation of observations in natural populations (Bijlsma et al., 1994). Most important, many of the processes involved, including the genetic component, are stochastic processes. As a result, there is no single deterministic outcome of an experiment, but an array of possible outcomes (Hauser et al., 1994). We thus can only try to estimate the probability of a certain event. This means we need a sufficient amount of replication of each experiment, which clearly is impracticable in the case of natural populations.

Thus *D. melanogaster* seems an ideal organism to fill the gap between theory and nature. The species is not endangered, it is easy to breed in large numbers and it has a short generation time. Moreover, it is genetically one of the most well known organisms, and many sophisticated techniques are available to manipulate the genetic composition and the level of homozygosity. For this and other reasons, *Drosophila* has become a model organism for evaluating conservation problems (Frankham, 1995).

## Homozygosity, fitness and stress

The consequences of an increase in homozygosity for individual fitness were studied using one of the several techniques available for *Drosophila*: the chromosome balancer procedure (Sved and Ayala, 1970; Wallace, 1981; Miller and Hedrick, 1993). Using the second-chromosome balancer *CyO*, which effectively suppresses crossing-over in the entire second chromosome, a large number of independent second chromosomes ($n = 453$) were extracted from a *D. melanogaster* population caught in the wild. This procedure (schematically shown in Fig. 1) results in flies that are either homozygous for one entire second chromosome (almost 45% of the total genome of *D. melanogaster*) or in flies that are heterozygous for the same chromosome over the standard *CyO* balancer chromosome. In the end the viability of the homozygotes can be estimated relative to their accompanying heterozygotes based on the progeny numbers as outlined in the legend of Figure 1. When made homozygous, recessive or partly recessive detrimental alleles become expressed, decreasing the relative viability

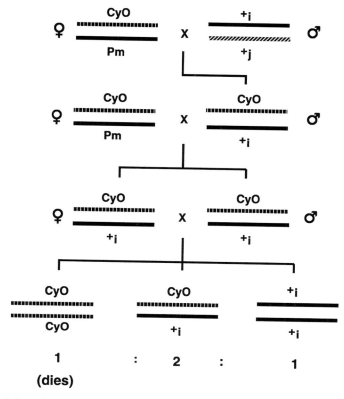

Figure 1. Schematic representation of the principle of the chromosome extraction procedure. By using single males in the first two crosses, all emerging wild-type homozygotes ($+_i/+_i$) in the last cross are isogenic for one entire second chromosome ($+_i$) from the wild population. Because the *CyO* balancer chromosome carries a recessive lethal, *CyO/CyO* homozygotes die in the embryonic stage. Assuming a normal Mendelian segregation, heterozygous (*CyO*/$+_i$) and homozygous ($+_i/+_i$) adults are expected in a ratio of 2:1. The viability of the homozygotes can then be estimated relative to that of the accompanying heterozygotes as two times the number of homozygotes divided by the number of heterozygotes. This ratio is expected to be 1 if both genotypes have the same viability, smaller than 1 if the homozygotes have a lower viability and greater than 1 if they have a higher viability than the heterozygotes.

(RV) of their homozygous carriers. As expected, many chromosomes carried detrimental alleles (Bijlsma et al., 1997a). About 15% of the 453 extracted chromosomes carried lethals or near-lethals (RV < 0.10); another 35% of the chromosomes carried detrimentals (0.10 < RV < 0.80) (Bijlsma et al., 1997a). The remainder, apart from 10% of supervitals (RV > 1.20), showed a close to normal (0.80 < RV < 1.20) viability compared with the heterozygotes. The median value over all extracted chromosomes was 0.78. The picture emerging was not unlike that generally observed for natural populations (Sved and Ayala, 1970; Wallace, 1981; Simmons and Crow,

1977). The observation that some 50% of all second chromosomes significantly decreased viability when made homozygous indicates that the
inbreeding load can be quite substantial in natural populations of *Droso-*
*phila*. An increase in homozygosity due to genetic drift and inbreeding thus
may cause considerable genetic stress in small populations.

However, when the detrimentals become expressed in homozygotes, they
become subject to natural selection. Consequently, if the increase in homozygosity proceeds relatively slowly, these detrimentals will be selected
against and ultimately can become purged from the population (Barrett and
Charlesworth, 1991; Hedrick, 1994; Lande, 1988). This could lead to small
populations that show no inbreeding load at all (Templeton and Read,
1984). We have simulated this process of purging by natural selection by
selecting only those chromosome lines that showed a relative fitness of
0.80–1.20 (quasi-normals) when being homozygous. The viability of these
quasi-normals was again tested under both optimal control conditions and
under high-temperature stress (30°C). The results are shown in Figure 2.
The effects of artificial purging are evident: all lethals and most detrimentals have been purged, and the relative viabilities show a peaked distribution with a median of 0.88. The high-temperature stress (that decreased
total egg-to-adult survival, thus irrespective of the genotype, by almost

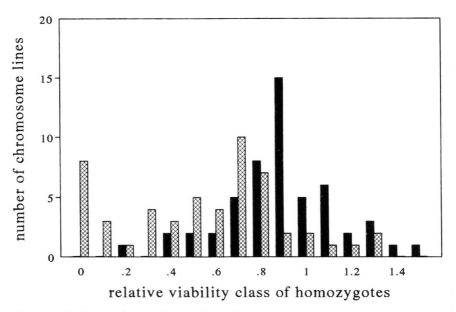

Figure 2. Distribution of the relative viabilities of homozygotes for 53 quasinormal lines determined both under unstressed control conditions (filled bars) and under stressed high-temperature conditions (cross-hatched bars). For calculation of the relative viability see legend to
Figure 1. (Data after Bijlsma et al., 1997a.)

50%) significantly affected the RV of chromosome homozygotes compared with control conditions (Fig. 2). Most chromosome homozygotes showed a strongly decreased relative viability, and 20% of the chromosomes became lethal or near lethal when homozygous at 30°C. The median RV at 30°C was 0.59 and thus significantly lower than in the control. This demonstrates that, although we had purged most detrimental chromosomes under control conditions, new deleterious alleles became expressed under the altered environmental stress conditions. These (conditional) detrimentals, obviously, could not have been purged from the population, because they were either not expressed under optimal control conditions or even beneficial under control conditions. Although the strength of this effect may depend on the nature of the stress environment (see Bijlsma et al., 1997a), the results suggest that purging is environment specific (see also Zhivotovsky, this volume).

These experiments showed the effects of homozygosity and purging only for the viability component of fitness. Other fitness components, however, can easily be estimated based on the fact that in general homozygotes have a lower fitness than the accompanying heterozygotes. If homozygotes and heterozygotes are allowed to compete for several generations in a population, the fitness of the homozygotes can be estimated from the equilibrium frequencies that eventually will be reached in the population as outlined in Table 1. This was done for 50 quasi-normals, and the results are shown in Table 1 for a number of different stress environments. Two facts emerge from the data. First, chromosome homozygotes have, based on the equilibrium frequency, a significantly lower fitness than heterozygotes, which agrees well with previous results (Sved and Ayala,

Table 1. Mean frequency of homozygotes and the estimated mean selection coefficients averaged over all extant quasinormal lines after nine (for ethanol eight) generations for optimal control conditions and four stress environments

|  | Number of extant lines | Mean frequency of homozygotes | Estimated mean SE selection coefficient of the homozygotes |
|---|---|---|---|
| Control | 50 | $0.634^a$ (0.029) | $0.237^a$ (0.023) |
| Ethanol | 46 | $0.637^{a,b}$ (0.036) | $0.249^{a,b}$ (0.028) |
| Crowding | 50 | $0.603^{a,b}$ (0.029) | $0.269^{a,b}$ (0.024) |
| DDT | 50 | $0.555^{b,c}$ (0.042) | $0.335^{b,c}$ (0.038) |
| High temp. | 42 | $0.525^c$ (0.033) | $0.336^c$ (0.029) |

Standard errors are shown in brackets. Means denoted by different letters are significantly different ($p < 0.05$). Selection coefficients were estimated by assuming that the fequency of the wild-type chromosome had reached the equilibrium value for those lines still segregating for both the wild-type and balancer chromosome. As $CyO/CyO$ homozygotes are lethal, a heterotic equilibrium is expected with the frequeny of the wild-type chromosome $p_{eq.} = 1/(1+s)$, $s$ being the selection coefficient against the wild-type homozygotes. For a few lines that became fixed for the wild-type chromosomes, $s$ was assumed to be zero. (Data from Bijlsma et al., 1997a.)

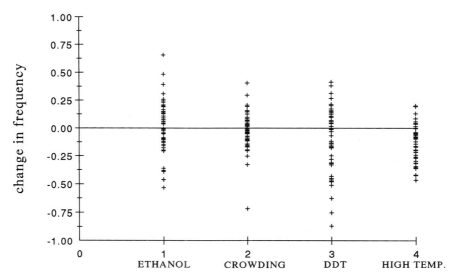

Figure 3. The change in frequency of the wild-type chromosome due to stress for each of the individual chromosome lines compared with the frequency of this chromosome of the same line reached under control conditons. The change is calculated as the frequency observed for stress minus the frequency observed for control conditions. (Modified after Bijlsma et al., 1997a.)

1970; Miller and Hedrick, 1993). On average the homozygotes show a reduction in fitness of about 25%. Second, relative to the heterozygote, the fitness of the homozygotes is further decreased under DDT (dichloro-diphenyltrichloroethane) and high-temperature (HT) stress, compared with control conditions. This again proves that environmental stress can enhance the inbreeding load, most probably because additional recessive detrimentals become expressed under these stress conditions. The fact that some stress environments do not affect the equilibrium frequencies significantly, indicates that different environmental stresses might differ greatly in their effects. These observations agree well with the results of Miller (1994).

That different stresses mediate their effects differently is shown in Figure 3. This figure shows the effect of stress on the equilibrium frequency for each individual chromosome line separately. Especially the difference between the effects of HT and DDT is noteworthy. For HT stress most lines (31 out of 42) show a decrease in equilibrium frequency compared with the values reached under control conditions, but the decrease is of the same order of magnitude for most of these lines. This suggests that HT stress affects most of the lines in a similar way. For DDT, which showed on average the same decrease in mean equilibrium frequency as observed for HT stress (see Tab. 1), only half of the lines (24 out of 50) showed a decrease in equilibrium frequency. It indicates that in many lines DDT stress did not affect or only minimally affected the relative fitness of the homozygotes, while it had a strong effect in others. The difference in the way HT and DDT stresses mediate the decrease in mean equilibrium frequency is corroborated by the significant increase in the standard error observed for DDT (Tab. 1). This contrast in reaction to the different stresses possibly reflects differences in the genetic determination of the sensitivity/resistance to stress. Sensitivity/resistance to HT is most probably a polygenic character (Krebs and Loeschcke, 1996), and it is conceivable that most extracted chromosomes, although for different loci, carry a comparable number of sensitive alleles. In contrast, it is known that, depending on the intensity of selection, DDT resistance can be monogenic (Roush and McKenzie, 1987), and in *Drosophila* can be governed by one or a few major genes that are located at the second chromosome (R. Bijlsma, unpublished data). Consequently, in some chromosome lines the homozygotes carry the resistant allele(s) and may be relatively insensitive to the stress, while others, carrying the non-resistant allele(s), are sensitive.

In summary, *D. melanogaster* populations, like many other organisms, carry a substantial level of recessive deleterious alleles that can become fixed in small populations due to genetic drift and inbreeding. Although purging of this inbreeding load can be effective, especially when the increase in homozygosity proceeds slowly (Hedrick, 1994), our data show that purging is environment-specific and is not necessarily effective under changed environmental conditions. Thus, increased homozygosity caused by bottlenecks, as has been observed for a number of species (Ellegren et al., 1993; Hoelzel et al., 1993), may, due to purging, have no immediate fitness consequences, but this is clearly no guarantee for survival in changing and deteriorating environments. Populations may have become homozygous for alleles that act neutral or near-neutral under given conditions but could become deleterious under other environmental conditions.

This observation has particular significance for managing captive populations of endangered species and reintroduction programs. Though it might seem that relatively optimal captive conditions have bred a healthy population, a reintroduction program may have a high chance of failure: under the new, often stressful and variable, natural conditions, "silent" detrimentals become important. This suggests that breeding programs not only should avoid inbreeding but also that breeding should preferably be done under conditions that mimic future natural situations as closely as possible.

## Inbreeding, stress and population extinction

In the previous section we showed that an increase in homozygosity can significantly impair individual fitness, especially in changing and deteriorating environments. As many endangered species find themselves in this situation, the next question we want to address is, does an increase in homozygosity affect the extinction probability of small populations?

The alleged negative impact of genetic erosion has been widely accepted since the 1980s (Frankel and Soulé, 1981; Soulé, 1987; Loeschke et al., 1994). However, as mentioned before, this assumption has recently come under increasing criticism (Caro and Laurenson, 1994; Caughley, 1994; Merola, 1994), as is clear from this quote from Caro and Laurenson: "*Similarly, although loss of heterozygosity has detrimental effects on individual fitness, no population has gone extinct as a result.*" To get some insight in this problem, we have studied the consequences of different levels of inbreeding for the extinction probability of small *Drosophila* populations under both optimal and stressful environmental conditions (Bijlsma et al., 1997b).

By means of either full sib mating or small population size during a number of generations, lines were obtained that differed in the level of inbreeding, from F = 0 to F = 0.785 (see Tab. 2). To determine if the decrease in individual fitness, as observed in the previous section, could be related to extinction, we estimated the extinction probability of populations that differed in inbreeding coefficient in a so-called extinction experiment. Per inbreeding level, 50 vial populations were established for each of three environmental conditions. Each generation the following procedure was then applied: parents were removed after a few days of egg laying (length depending on the environment) to prevent crowding. All offspring were transferred to new vials with fresh food the next generation, without mixing between the vials. This was repeated for eight generations, the time of transfer varying between environments because of differences in developmental time, but within environments all vials were treated synchronously. Extinction then was defined as the proportion of vials producing

Table 2. Cumulative extinction percentages of small-vial populations for different levels of inbreeding under unstressed (control) and stressed (high temperature and ethanol) environmental conditions after four and eight generations

| Inbreeding level ($F$): | Control | | High temperature | | Ethanol | |
|---|---|---|---|---|---|---|
| | G4 | G8 | G4 | G8 | G4 | G8 |
| 0 | 0 | 24 | 14 | 48 | 6 | 12 |
| 0.10 | 0 | 46 | 12 | 44 | 18 | 46 |
| 0.20 | 0 | 50 | 24 | 54 | 24 | 48 |
| 0.25 | 0 | 26 | 76 | 84 | 40 | 50 |
| 0.35 | 0 | 48 | 74 | 74 | 14 | 56 |
| 0.38 | 10 | 46 | 90 | 96 | 18 | 66 |
| 0.50 | 6 | 50 | 96 | 100 | 38 | 68 |
| 0.56 | 6 | 36 | 98 | 98 | 22 | 70 |
| 0.67 | 2 | 34 | 96 | 98 | 38 | 84 |
| 0.79 | 6 | 50 | 98 | 98 | 52 | 88 |

Data from Bijlsma et al., 1997b.

no offspring in a given generation. This was done for an optimal control environment and two stress environments: HT (28.5°C) and ethanol (food supplemented with 10–17.5% ethanol). The extinction rate in such an experiment has previously been shown to reflect the genetic composition and fitness well (Bijlsma-Meeles and Van Delden, 1974; Bijlsma and Van Delden, 1977).

Based on the assumption that inbreeding decreases the fitness of populations, a correlation between the degree of inbreeding and the proportion of extinction is expected. Table 2 shows that the results are in agreement with this theory. Under control conditions extinction after eight generations is still low, but already significantly higher for the inbred than for the non-inbred populations in most cases. This indicates that inbreeding, even for the low level of F = 0.10, significantly increases the extinction risk of populations. Further, and consistent with the previous purging experiment, when stressed, the effects on extinction become much greater. This is most clear when comparing cumulative extinction percentages in generation 4 for the three environments: for HT, except for the lower levels (F = 0–0.20), extinction is already approaching 100% for most levels of inbreeding, while for ethanol extinction reaches less extreme values, but still considerably higher than for the control. Nevertheless, for both there is still a distinct increase in extinction the higher the inbreeding coefficient of the populations, both for G4 and G8.

Much more important, however, is the observation that the impact of stress becomes significantly enhanced for higher inbreeding coefficients (Fig. 4). This figure shows the relationship between the initial inbreeding level of populations and their relative sensitivity to stress. This stress sensitivity is defined as the additional extinction caused by the stress applied relative to the extinction observed under unstressed control conditions. This sensitivity measure is thus corrected for the fitness differences already observed under control conditions (for details see Bijlsma et al., 1997b). The fitted linear regression shows for both stresses a significant positive relationship (p < 0.001), even though in the case of HT a non-linear model would probably fit the data better.

Thus not only are the effects of inbreeding more manifest under stress, but they are also strongly intensified for higher inbreeding levels. This clearly shows that inbreeding and environmental stress are not independent but act synergistically. This indisputably demonstrates that more inbred populations will be more prone to environmental stochasticity. Hence, deterioration of the environmental conditions increases the extinction probability of inbred populations more than it does non-inbred populations. This agrees well with the recent findings that for white-footed mice and song sparrows inbreeding depression is only observed under stressful natural conditions (Keller et al., 1994; Jiménez et al., 1994), although it is not clear if synergism was involved in these examples.

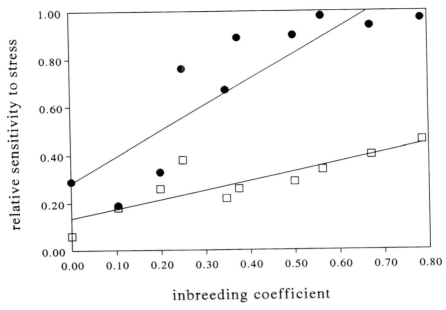

Figure 4. Relationship between the level of inbreeding ($F$), and the sensitivity to high-temperature stress (closed circles) and ethanol stress (open squares). (Modified after Bijlsma et al., 1997b.)


## Conclusions

The data presented in this chapter show that genetic stress resulting from an increase in homozygosity should be considered an important force in conservation biology. It impairs fitness and increases the extinction risk of small populations significantly. Although these negative effects can be neutralized, at least partly, by purging of the inbreeding load, this will only be effective in relatively constant environments, because purging is evidently, at least partly, environment-specific. Changing and deteriorating conditions will constantly evoke the expression of new genetic load. The results also explicitly show that genetic effects should not be considered independently from environmental and demographic effects: inbreeding in combination with environmental stress strongly increases the extinction risk of small populations, and together they can act synergistically. The impact of environmental stress may be less if the environment changes much more gradually (Lynch and Lande, 1993; Bürger and Lynch, this volume), but many endangered species actually have to cope with human-induced catastrophic and rapid changes in their environment. By choosing quite specific and extreme environmental conditions, we may, however, have aggravated the impact of inbreeding, and the balance between genetic and non-genetic causes may well be different under different conditions. Nevertheless, many endangered species/populations, that have lost much of their genetic variability, run a high risk of becoming extinct in the (near) future because of this synergistic interaction between genetic erosion and environmental deterioration, notwithstanding the fact that some of these species/populations presently look quite healthy. This view is supported by the recent data from natural case studies (see also Hedrick et al., 1996). Therefore, understanding the interaction between genetic and non-genetic processes should be the emphasis of further research rather than emphasis on the importance of one over the other.

*Acknowledgments*
We thank E. Bijlsma-Meeles for critically reading the manuscript, and P.W. Hedrick, R. Lande, V. Loeschcke and F.J. Weissing for stimulating discussions on earlier drafts. R.B. was in part supported by a grant from the University of Aarhus.

# References

Barrett, S.C.H. and Charlesworth, D. (1991) Effects of a change in the level of inbreeding on the genetic load. *Nature* 352:522–524.

Bijlsma, R. and Van Delden, W. (1977) Polymorphism at the G6pd and 6Pgd loci in *Drosophila melanogaster*. I. Evidence for selection in experimental populations. *Genet. Res.* 30: 221–236.

Bijlsma, R., Ouborg, N.J. and Van Treuren, R. (1991) Genetic and phenotypic variation in relation to population size in two plant species: *Salvia pratensis* and *Scabiosa columbaria*. *In*: A. Seitz and V. Loeschcke (eds): *Species conservation: a population-biological approach*. Birkhäuser, Basel, pp. 89–101.

Bijlsma, R., Ouborg, N.J. and Van Treuren, R. (1994) On genetic erosion and population extinction in plants: A case study in *Scabiosa columbaria* and *Salvia pratensis*. *In:* V. Loeschcke, J. Tomiuk and S. Jain (eds): *Conservation Genetics*. Birkhäuser, Basel, pp. 255–272.

Bijlsma-Meeles, E. and Van Delden, W. (1974) Intra- and interpopulation selection concerning the alcohol dehydrogenase locus in *Drosophila melanogaster*. *Nature* 247:369–371.

Caro, T.M. and Laurenson, M.K. (1994) Ecological and genetic factors in conservation: A cautionary tail. *Science* 263:485–486.

Caughley, G. (1994) Directions in conservation biology. *J. Anim. Ecol.* 63:215–244.

Charlesworth, D. and Charlesworth, B. (1987) Inbreeding depression and its evolutionary consequences. *Ann. Rev. Ecol. Syst.* 18:237–268.

Clark, G.M. (1993) The genetic basis of developmental stability. I. Relationship between stability, heterozygosity and genomic coadaptation. *Genetica* 89:15–23.

Dahlgaard, J. and Loeschcke, V. (1997) Effects of inbreeding in three life stages of *Drosophila buzzatii* after embryos were exposed to high-temperature stress. *Heredity* 78:410–416.

Dahlgaard, J., Krebs, R.A. and Loeschcke, V. (1995) Heat-shock tolerance and inbreeding in *Drosophila buzzatii*. *Heredity* 74:157–163.

Dudash, M.R. (1990) Relative fitness of selfed and outcrossed progeny in a self-compatible, protandrous species, *Sabatia angularis* L. (*Gentianaceae*): A comparison in three environments. *Evolution* 44:1129–1139.

Ellegren, H., Hartman, G., Johansson, M. and Andersson, L. (1993) Major histocompatibility complex monomorphism and low levels of DNA fingerprinting variability in a reintroduced and rapidly expanding population of beavers. *Proc. Natl. Acad. Sci. USA* 90: 8150–8153.

Frankel, O.H. and Soulé, M.E. (1981) *Conservation and Evolution*. Cambridge University Press, Cambridge.

Frankham, R. (1995) Conservation genetics. *Ann. Rev. Genetics* 29:305–327.

Hauser, T.P., Damgaard, C. and Loeschcke, V. (1994) Effects of inbreeding in small plant populations: Expectations and implications for conservation. *In*: V. Loeschcke, J. Tomiuk and S. Jain (eds): *Conservation Genetics*. Birkhäuser, Basel, pp. 115–129.

Hedrick, P.W. (1994) Purging inbreeding depression and the probability of extinction: Full-sib mating. *Heredity* 73:363–372.

Hedrick, P.W., Lacy, R.C., Allendorf, F.W. and Soulé, M.E. (1996) Directions in conservation biology: Comments on Caughley, *Conserv. Biol.* 10:1312–1320.

Hoelzel, A.R., Halley, J., O'Brien, S.J., Campagna, C., Arnbom, T., Le Boeuf, B., Ralls, K. and Dover, G.A. (1993) Elephant seal genetic variation and the use of simulation models to investigate historical population bottlenecks. *J. Hered.* 84:443–449.

Hoffmann, A.A. and Parsons, P.A. (1991) *Evolutionary Genetics and Environmental Stress*. Oxford University Press, Oxford.

Jiménez, J.A., Hughes, K.A., Alaks, G., Graham, L. and Lacy, R.C. (1994) An experimental study of inbreeding depression in a natural habitat. *Science* 265:271–274.

Kappe, A.L., Van De Zande, L., Vedder, E.J., Bijlsma, R. and Van Delden, W. (1995) Genetic variation in the harbour seal (*Phoca vitulina*) revealed by DNA fingerprinting and RAPD's. *Heredity* 74:647–653.

Kappe, A.L., Bijlsma, R. Osterhaus, A.D.M.E., Van Delden, W. and Van De Zande, L. (1997) Structure and amount of genetic variation at minisatellite loci within the subspecies complex of *Phoca vitulina* (the harbour seal) *Heredity* 78:457–463.

Karron, J.D. (1987) A comparison of the levels of genetic polymorphism and self compatibility in geographically restricted and widespread plant congeners. *Evol. Ecol.* 1:47–58.

Keller, L.F., Arcese, P., Smith, J.N.M., Hochachka, M. and Stearns, S.C. (1994) Selection against inbred song sparrows during a natural population bottleneck. *Nature* 372: 356–357.

Koelewijn, H.P. (1993) On the genetics and ecology of sexual reproduction in *Plantago coronopus*. Ph.D. dissertation, University of Utrecht, the Netherlands.

Krebs, R.A. and Loeschcke, V. (1996) Acclimation and selection for increased resistance to thermal stress in *Drosophila buzzatii*. *Heredity* 142:471–479.

Lande, R. (1988) Genetics and demography in biological conservation. *Science* 241:1455–1450.

Lande, R. and Barrowclough, G.F. (1987) Effective population size, genetic variation and their use in population management. *In*. M.E. Soulé (ed.): *Viable Populations for Conservation*. Cambridge University Press, Cambridge, pp. 87–124.

Loeschcke, V., Tomiuk, J. and Jain, S. (eds) (1994) *Conservation Genetics*. Birkhäuser, Basel.

Lynch, M. and Lande, R. (1993) Evolution and extinction in response to environmental change. *In*: P.M. Kareiva, J.G. Kingolver and R.B. Huey (eds): *Biotic Interactions and Global Change*. Sinauer, Sunderland, pp. 234–250.

Menges, E.S. (1991) The application of minimum viable population theory to plants. *In*: D.A. Falk and K.E. Holsinger (eds): *Genetics and Conservation of Rare Plants*. Oxford University Press, New York, pp. 45–61.

Merola, M. (1994) A reassessment of homozygosity and the case for inbreeding depression in the cheetah, *Acinonyx jubatus*: Implications for conservation. *Conserv. Biol.* 8:961–971.

Miller, P.S. (1994) Is inbreeding depression more severe in a stressful environment? *Zoo Biology* 13:195–208.

Miller, P.S. and Hedrick, P.W. (1993) Inbreeding and fitness in captive populations: Lessons from *Drosophila*. *Zoo Biol.* 12:333–351.

Parsons, P.A. (1992) Fluctuating asymmetry: A biological monitor of environmental and genomic stress. *Heredity* 68:361–364.

Roush, R.T. and McKenzie, J.A. (1987) Ecological genetics of insecticide acaricide resistance. *Ann. Rev. Entomol.* 32:361–380.

Shaffer, M.L. (1981) Minimum population sizes for species conservation. *Bioscience* 31:131–134.

Shaffer, M.L. (1987) Minimum viable populations: Coping with uncertainty. *In*. M.E. Soulé (ed.): *Viable Populations for Conservation*. Cambridge University Press, Cambridge, pp. 69–86.

Simmons, M.J. and Crow, J.F. (1977) Mutations affecting fitness in *Drosophila* populations. *Ann. Rev. Genetics* 11:49–78.

Soulé, M.E. (ed.) (1987) *Viable Populations for Conservation*. Cambridge University Press, Cambridge.

Sved, J.A. and Ayala, F.J. (1970) A population cage test for heterosis in *Drosophila melanogaster*. *Genetics* 66:97–113.

Templeton, A.R. and Read, B. (1984) Factors eliminating inbreeding depression in a captive herd of Speke's gazelle (*Gazella spekei*). *Zoo Biology* 3:177–199.

Van Noordwijk, A.J. and Scharloo, W. (1981) Inbreeding in an island population of the great tit. *Evolution* 35:674–688.

Van Treuren, R., Bijlsma, R., Van Delden, W. and Ouborg, N.J. (1991) The significance of genetic erosion in the process of extinction. 1. Genetic differentiation in *Salvia pratensis* and *Scabiosa columbaria* in relation to population size. *Heredity* 66:181–189.

Wallace, B. (1981) *Basic Population Genetics*. Columbia University Press, New York.

Environmental Stress, Adaptation and Evolution
ed. by R. Bijlsma and V. Loeschcke
© 1997 Birkhäuser Verlag Basel/Switzerland

# Adaptation and extinction in changing environments

Reinhard Bürger [1] and Michael Lynch [2]

[1] *Institute of Mathematics, University of Vienna, Strudlhofgasse 4, A-1090 Wien, Austria*
[2] *Department of Biology, University of Oregon, Eugene, Oregon 97403, USA*

*Summary.* The extinction risk of a population is determined by its demographic properties, the environmental conditions to which it is exposed, and its genetic potential to cope with and adapt to its environment. All these factors may have stochastic as well as directional components. The present chapter reviews several types of models concerned with the vulnerability of small populations to demographic stochasticity and to random and directional changes of the environment. In particular, the influence of mutation and genetic variability on the persistence time of a population is explored, critical rates for environmental change are estimated beyond which extinction on time scales of tens to a few thousand generations is virtually certain, and the extinction risks caused by the above mentioned factors are compared.

## Introduction

Many ecological and demographic factors have been identified that pose a risk for extinction of small and moderately large populations. Some of these may be labeled as random, like demographic and environmental stochasticity or catastrophes; others are of an inherently directional nature, though with random components, like destruction and fragmentation of habitat, reduction of resources, harvesting, introduction of new predators or parasites, or changing climatic conditions. All such factors impose stress on a population because they result in demographic costs, which in turn can generate negative genetic consequences. If these costs are sufficiently large, the likelihood of extinction will be high.

In recent years a central question in conservation biology has been whether, and under what conditions, genetic responses to a changing environment can decrease a population's extinction risk (Lande, 1988; Frankham, 1995a; Lynch, 1996). The evaluation of the possible benefits of genetic variability for individual fitness requires the investigation of mathematical models that incorporate the basic features of the genetics, demography and ecology of a population. Models that incorporate all three of these factors are notoriously difficult to handle analytically, and typically require extensive computer simulation to gain quantitative insights. Analytical treatments are necessarily based on simplifying assumptions and yield either only qualitative insights or approximations for a limited range of parameters. It is the purpose of the present chapter to review

recent theoretical progress in understanding the impact of demographic
and environmental factors with random and directional components on the
extinction risk of genetically monomorphic and polymorphic populations.

We first consider models without genetics that combine density-dependent population regulation with demographic stochasticity, such as variation in family size or sex ratio, and random environmental fluctuations.
Persistence times in such models are highly sensitive to assumptions about
the specific form of population regulation, and synergistic effects between
stochastic factors lead to a significant increase of the extinction risk. For
models of this kind a considerable body of mathematical theory has been
developed, in particular in terms of diffusion approximations. However,
these approximations give only the correct scaling of the extinction time in
terms of parameters like carrying capacity, and tend to underestimate the
extinction risk substantially unless population growth rates are small or
negative. Since understanding such models forms the basis for understanding more realistic models that include genetics, a further development of
this mathematical theory would be highly desirable.

The level of genetic polymorphism in a population is determined by a
stochastic balance between mutation, selection, random genetic drift, and
migration. Since so many factors are involved, the consequences of genetic
variability for the persistence time of a population are manifold. In small
populations, genetic drift is a powerful force because it reduces the standing genetic variance by a factor $1/(2N_e)$ each generation (where $N_e$ is the
effective population size). Also, in small populations recessive deleterious
mutations become fixed through genetic drift, leading to inbreeding
depression. Mutation clearly is the ulitmate source of genetic variation.
However, it has been demonstrated that the majority of new mutations is
nearly neutral or has slightly deleterious effects on overall fitness (e.g.
Mukai, 1964; Houle et al., 1992; Lynch et al., 1995b; Schultz et al., 1997;
Kobota and Lynch, 1996). Such mutations readily accumulate in most
asexual populations, as well as in small to moderately large sexual populations through random genetic drift, thus imposing a steady decay of population mean fitness unless a certain fraction of mutations is advantageous
(Lynch and Gabriel, 1990; Gabriel et al., 1993; Gabriel and Bürger, 1994;
Lynch et al., 1993; Lande, 1994, 1995; Lynch et al., 1995a, b). Once the
deleterious mutation load is sufficiently large that individuals can no
longer replace themselves, the decline in population size precipitates a progressive increase in the rate of accumulation of future deleterious mutations. This self-accelerating process has been called the mutational meltdown. It presents a non-negligible extinction risk to sexual populations of
sizes up to several thousand individuals, as we shall discuss further below.
Thus, under certain demographic conditions, genetic variability may
magnify, rather than reduce, the vulnerability of a population to extinction.

For well-adapted populations, where the prevailing type is optimally
adapted, variability represents a load that decreases mean fitness. This is

also true when the environmental optimum is subject to white noise-like fluctuations but not for autocorrelated environments (cf. Charlesworth, 1993a,b; Bürger and Lynch, 1995; Lande and Shannon, 1996). However, in a directionally changing environment, adaptive genetic variation is essential for long-term population survival, with the rate of adaptive evolution being proportional to the additive genetic variance in the population. Since adaptation is not an instantaneous process but requires time, the mean phenotype will deviate from its optimum value in a changing environment, to a degree depending on the rate of environmental change. Whether a population will survive a selective challenge depends on the magnitude of the reduction in mean fitness induced by the selective load and its impact on population size during the phase of adaptation, as well as on the rate at which natural selection returns the population to a demographically secure situation (Lynch et al., 1991; Huey and Kingsolver, 1993; Lynch and Lande, 1993; Bürger and Lynch, 1995; Gomulkiewicz and Holt, 1995; Lande and Shannon, 1996).

A quantitative-genetic model for evaluating the maximum rate of environmental change that can be tolerated by a population was introduced by Lynch et al. (1991) and Lynch and Lande (1993), and further developed by Bürger and Lynch (1995). This model, as outlined below, takes into account directional and stochastic environmental changes, density-dependent population regulation and demographic stochasticity, as well as genetic stochasticity resulting from the processes of mutation, recombination and random genetic drift. It demonstrates that there is a sharp boundary in the relationship between the rate of environmental change and mean extinction time, such that below a critical rate of environmental change the population can track the environmental trend sufficiently closely to guarantee long-term survival, while above this critical rate extinction occurs rapidly, even in large populations.

## Stochastic demographic factors and extinction

Stochastic factors are usually classified as demographic stochasticity, environmental stochasticity, and catastrophes (Shaffer 1981, 1987). Demographic stochasticity, caused by random fluctuations in birth and death processes, is manifest in variation in family size, sex ratio and other individual properties. Environmental stochasticity arises from frequent perturbations of the environment that affect birth and death rates of all individuals in a similar way. Catastrophes are rare and major perturbations leading to sudden and high mortality. While environmental fluctuations and catastrophes affect small and large populations, the effects of demographic stochasticity alone, basically being sampling effects, tend to average out in large populations. Synergistic interactions between these factors, however, amplify the effects they exert separately.

Various models have been developed to analyze the extinction risk caused by these factors, e.g. MacArthur and Wilson (1967), Richter-Dyn and Goel (1972), Leigh (1981), Goodman (1987a,b), Gabriel and Bürger (1992), Lande (1993), Foley (1994), Ludwig (1976, 1996). The object of primary interest is the distribution of extinction times and its dependence on the parameters of the model, i.e. on the intrinsic growth rate, carrying capacity, kind and magnitude of stochasticity, and mode of population regulation. Here we shall concentrate on populations governed by density-dependent dynamics. The theory of density-independent dynamics (linear birth and death processes) is relatively simple and well understood (Ludwig, 1974; Karlin and Taylor, 1975).

*Population Regulation*

Many different deterministic models for density-dependent population growth have been suggested (cf. MacArthur and Wilson, 1967; May and Oster, 1976; May, 1981). For the basic density-dependent relation

$$N(t+1) = F(N(t)), \tag{1}$$

designed for a finite population of size $N(t)$ with discrete generations, $t = 0, 1, 2,...$, and no age structure or genetics, we consider four possibilities. The first is for a population with an average intrinsic growth rate (or Malthusian parameter) $r$ and a carrying capacity $K$ that functions as a ceiling for the population size (e.g. because there is a limited number of nesting places):

$$F(N) = \begin{cases} Ne^r & \text{for } 0 \le e^r N < K \\ K & \text{otherwise.} \end{cases} \tag{2}$$

The next is the classical Verhulst model defined by

$$F(N) = N \frac{e^r K}{K + (e^r - 1) N}, \tag{3}$$

which leads to monotonic convergence to the carrying capacity $K$. The familiar logistic model is

$$F(N) = \begin{cases} N[1 + \rho(1 - N/K)] & \text{for } N \le K(\rho+1)/\rho \\ 0 & \text{otherwise,} \end{cases} \tag{4}$$

where, in the limit $N \to 0$, $F(N)/N = 1 + \rho = e^r$. As is well known (cf. May, 1981), this may lead to chaotic behavior for moderately large $\rho$ and to rapid extinction for large $\rho$. A "smooth" version of the logistic model is given by

$$F(N) = N \exp[r(1 - N/K)], \tag{5}$$

whose deterministic dynamics are similar to the logistic dynamics. Equation (5) is sometimes called the Ricker stock-recruit relationship. An array of models, including Eqs. (2) and (5), was investigated Gabriel and Bürger (1992).


## Demographic Stochasticity

Random events influencing individual reproduction and mortality lead to variation in family size. It is natural to assume that the actual number of offspring, $N(t+1)$, follows a Poisson distribution with mean $F(N(t))$ given by the deterministic density-dependent relation. This leads to a Markov chain model that is not a branching process. The mean extinction time depends only weakly on the initial population size (cf. Gabriel and Bürger, 1992; Grass, 1996). Let $T(K)$ denote the mean time to extinction (in generations) for a population initially at its carrying capacity. If we assume that extinction occurs at $N = 1$, which is certainly appropriate for sexually reproducing species, the mean extinction time increases approximately exponentially with the carrying capacity $K$. Indeed, when population density is regulated according to Eqs. (2) or (3) and in the limit of large $r$, the exact result is $T(K) = e^K/(K+1)$ and the coefficient of variation (ratio of standard deviation to the mean extinction time) is $\sqrt{1 - e^{-K}}\,(K+1)$ (in Gabriel and Bürger, 1992, extinction was assumed to occur at $N = 0$, and then $T(K) = e^K$).

For small $r$, a diffusion approximation can be derived by treating the per capita growth rate as a random variable with mean $r$, variance $V_i/N$ per unit time, and no autocorrelation. The parameter $V_i$ is the variance in individual fitness per unit time, and $V_i = 1$ if sampling is performed as in the above discrete-time model (Keiding, 1975; Leigh, 1981; Lande, 1993). Employing the population growth model (2) leads to analytically simple formulas (Lande, 1993). With the infinitesimal mean and variance of $F(N)$ being $\mu(N) = RN$ and $\sigma^2(N) = V_i N$, respectively, the mean time to extinction is

$$T(K) = \frac{1}{r}\, e^{-2r/V_i} \left[ \mathrm{Ei}\left(\frac{2r}{V_i}K\right) - \mathrm{Ei}\left(\frac{2r}{V_i}\right) \right] - \frac{\ln K}{r} \tag{6a}$$

where $\mathrm{Ei}(x)$ is the exponential integral. If $2rK/V_i \gg 1$, the previous expression is closely approximated by

$$T(K) \approx \frac{V_i\, e^{2r(K-1)/V_i}}{2r^2 K} \left(1 + \frac{V_i}{2rK}\right) \tag{6b}$$

(cf. Lande, 1993). This nearly exponential scaling is qualitatively consistent with previous results (MacArthur and Wilson, 1967; Richter-Dyn and Goel, 1972; Leigh, 1981; Goodman, 1987b; Gabriel and Bürger, 1992). For $r = 0$, the diffusion approximation for the mean extinction time is $T(K) = 2(K-1- \ln K)/V_i$, while for $r < 0$ it is, to a very close approximation,

$$T(K) \approx - \frac{1}{r} \left[ \ln K + e^{-2r/V_i} \, \text{Ei} \left( \frac{2r}{V_i} \right) \right]. \tag{7}$$

Figure 1(A) displays the mean time to extinction, $T(K)$, as a function of the growth rate $r$ for the population regulation modes (2)–(5) allowing for variation in family size but constant even sex ratio. It is evident that the mean extinction time is highly sensitive to assumptions about the mode of density-dependent population regulation. The diffusion approximation provides an accurate approximation to the mean extinction time in the discrete-time model only if the growth rate $r$ is close to zero (Fig. 1(A)) or negative (results not shown). The decline of the mean extinction time with large $r$ under the two logistic population growth models (4) and (5) results from overshooting of the carrying capacity, and suggests that the logistic population growth models are of rather limited relevance for high-fecundity populations. The slight irregularities of curve (4) are a consequence of the chaotic dynamics of the underlying deterministic density-dependent relation.

If there are two sexes in the population, there will also be variation in sex ratio. It is most natural to assume that the number of males and females are binomially distributed with both sexes having equal probabilities. We now suppose that the expected population size is $F(N\phi_{N,N_f})$, where $N_f$ is the number of females and $\phi_{N,N_f}$ is a factor describing productivity of females as a function of population size and sex ratio. Here we shall assume that $\phi_{N,N_f} = 0$ if $N_f = 0$ or $N_f = N$ (no males), and $\phi_{N,N_f} = 2N_f/N$ otherwise, so that for an even sex ratio, with $N_f = N/2$, $F(N)$ offspring are produced on average. The actual number of offspring is again supposed to obey a Poisson distribution whose mean, then, is $F(N\phi_{N,N_f})$.

Although variation in sex ratio alone (the population size being $K$ unless there are no males or females) poses a risk to extinction only for extremely small populations (the expected extinction time is $T(K) = 2^{K-1}$; cf. Gabriel and Bürger, 1992), it drastically magnifies the extinction risk when combined with variation in family size (Fig. 2). The mean extinction time for large $r$ is somewhat less than $\frac{1}{2}e^{K/2}$ (compare also Gabriel and Bürger, 1992). For $r = 0$ the mean time to extinction under model (2) is approximately $T(K) \approx K$. Nevertheless, since the expected time to extinction scales exponentially with $K$ for both stochastic factors in isolation (if $r > 0$), the expected extinction time for the model with variation in both sex ratio and family size will also scale exponentially with $K$, though with a much smaller exponent (effectively influencing the extinction dynamics as though the carrying capacity were reduced by more than a factor of 1/2).

**(A)**

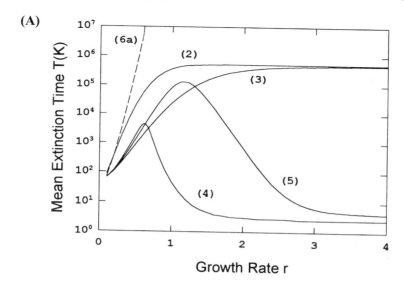

**(B)**

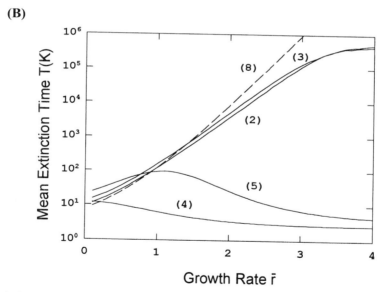

Figure 1. Mean time to extinction as a function of the intrinsic growth rate $r$, for the density-dependent growth models (2)–(5) under demographic and environmental stochasticity. The carrying capacity is $K = 16$. (A) The results in the upper panel are for variation in family size as the sole source of stochasticity. They are obtained from numerical solution of a system of linear equations defined by the transition matrix (cf. Gabriel and Bürger, 1992). The dashed line represents the diffusion approximation (6a). (B) The lower panel displays the mean extinction time resulting from variation in family size and environmental stochasticity with $V_e = 1$. The results are from Monte-Carlo simulation. The dashed line is the diffusion approximation (8).

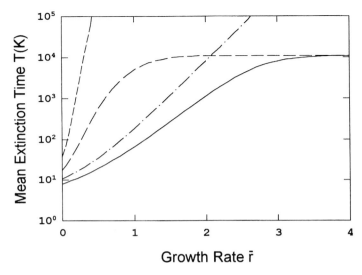

Figure 2. Mean time to extinction as a function of the intrinsic growth rate $\bar{r}$ for the density-dependent relation (2) for various levels of stochasticity. The short-dashed line is the mean extinction time resulting from variation in family size alone, the dash-dotted line is the mean extinction time under variation in family size and environmental stochasticity with $V_e = 1$, the long-dashed line is from variation in family size and sex ratio, and the solid line displays the mean extinction time under the combined action of all three stochastic factors. The carrying capacity is $K = 20$, and the data points were obtained from Monte Carlo simulation.

Therefore, as noted by several authors (cited above), demographic stochasticity will contribute only little to the extinction risk of large populations, unless their growth rate is very small.

*Environmental and Demographic Stochasticity*

Stochastic environmental fluctuations are modeled by assuming that the intrinsic growth rate fluctuates with time and can be described by a random variable with mean $\bar{r}$, variance $V_e$, and no autocorrelation. A discrete-time discrete-state Markov chain model is obtained by drawing $r$ from some probability distribution (in the present study a Gaussian) with mean $\bar{r}$ and variance $V_e$, calculating the expected population size $\bar{N}(t+1)$ under one of the deterministic population regulation models (2)–(5), and finally sampling the population size $N(t+1)$ from a (Poisson) distribution with mean $\bar{N}(t+1)$.

For small $\bar{r}$, such a Markov chain can be approximated by a diffusion process with infinitesimal mean and variance $\mu(N) = (\bar{r} + \frac{1}{2}V_e)N$ and $\sigma^2(N) = V_iN + V_eN^2$, respectively. Thus, with environmental stochasticity, the average growth rate, $\bar{r} + \frac{1}{2}V_e$, of the approximating continuous-time diffusion process is larger than that ($\bar{r}$) of the original discrete-time model, while for demographic stochasticity alone these growth rates coincide. Solving the stochastic differential equation correspondig to the diffusion process with the Ito calculus shows that the logarithm of the population size changes according to $E[\ln N(t)] = E[\ln N(0)] + \bar{r}t$. Therefore, the average intrinsic growth rate $\bar{r}$ of the discrete-time model coincides with the so-called long-run growth rate of the continuous-time stochastic model (cf. Turelli, 1977; Lande, 1993; Engen and Bakke, 1996; Grass, 1996).[1]

For the population growth model (2), the diffusion approximation for the mean extinction time is

$$T(K) = 2 \int\limits_1^K (zV_e + V_i)^{-c-1} \int\limits_z^K \frac{(yV_e + V_i)^c}{y} \, dy \, dz, \qquad (8)$$

where $c = 2\bar{r}/V_e$. Here $K$ acts as a reflecting boundary and, since extinction is assumed to occur at $N = 1$, we have $T(1) = 0$. The integrals in (8) can be evaluated explicitly only for integer $c$. For negative growth rates $\bar{r} < 0$, (8) can be approximated by

$$T(K) = -\frac{1}{\bar{r}} \ln \left( \frac{KV_e + V_i}{V_e + V_i} \right) + O(1), \quad \text{as } K \to \infty. \qquad (9)$$

If $c > 0$ and $KV_e \gg 1$ the asymptotic equality

$$T(K) \simeq \frac{2V_e^{c-1}}{c^2(V_i + V_e)^c} K^c \qquad (10)$$

is obtained as $K \to \infty$. For the logistic population growth model (4), Ludwig (1976) and Leigh (1981) obtained formulas with the same asymptotic scaling.

For the case of no demographic stochasticity ($V_i = 0$), Lande (1993) obtained the exact expression

$$T(K) = \frac{1}{\bar{r}} \left( \frac{K^c - 1}{c} - \ln K \right) \qquad (11)$$

To leading order in $K$, this is larger than (10) by a vactor $V_e^c/(V_i + V_e)^c$.

An important point to notice from (10) and (11) is that with environmental stochasticity the mean time to extinction increases like a power

---

[1] If, instead of assuming that $r$ is a random variable with mean $\bar{r}$ and variance $V_e$, it is assumed that $e^r$ is a random variable with mean $\lambda$ and variance $v_e$, i.e. $E[\Delta N|N] = \lambda N$ and $\text{Var}[\Delta N|N] = V_i N + v_e N^2$ (the $V_i$ again is due to sampling the population size), one obtains a diffusion with infinitesimal mean $\lambda N$ and variance $V_i N + v_e N^2$. Then the long-run growth rate is $\lambda - \frac{1}{2} v_e$. Mathematically, these approaches are equivalent, though the latter one gives the (correct) impression that stochasticity effectively reduces the growth rate.

function in $K$ if the (long-run) growth rate $\bar{r}$ is strictly positive. This is in contrast to purely demographic stochasticity, for which the mean extinction time is approximately an exponential function of $K$.

Figure 1(B) displays the mean time to extinction as a function of the growth rate $\bar{r}$ for the population regulation modes (2)–(5) under environmental stochasticity and variation in family size but constant even sex ratio. It demonstrates, as does Figure 1(A), how sensitive the mean time to extinction is with regard to the mode of population regulation. It also shows that in this case the diffusion approximation is qualitatively accurate unless $\bar{r}$ is too large. Comparison with Figure 1(A) shows that for large $\bar{r}$, environmental stochasticity only slightly enhances the extinction risk caused by variation in family size. Figure 2 compares the mean extinction times from three different combinations of variation in family size, variation in sex ratio and environmental stochasticity with pure variation in family size. For small populations and large $r$, sex-ratio variation combined with variation in family size leads to a higher extinction risk than the present level of environmental stochasticity together with variation in family size. Additional simulations (not shown) indicate that for larger populations or higher $V_e$, the long-dashed and the dash-dotted lines intersect at higher $\bar{r}$-values, while the opposite is true for smaller population sizes and lower $V_e$.

Little is known about the distribution of extinction times in these models except that typically the distribution has a high positive skewness and a coefficient of variation near 1 (results not shown). Some numerical and analytical results suggest that, except for a lag reflecting the fact that no or few populations die out in the initial phase, this distribution can be approximated by an exponential distribution (cf. Nobile et al., 1985; Goodman, 1987a; Gabriel and Bürger, 1992).

Most of the results in this section are based on the simple density-dependent relation (2). However, as Figure 1 shows, this assumption about the mode of density dependence leads to much longer average persistence times than population growth models that allow overshooting of the carrying capacity, unless the intrinsic growth rate is very small. In addition, in the present kind of model, diffusion approximations are often inaccurate predictors of the extinction risk.

In the above models we assumed that the actual number of offspring, $N(t+1)$, is sampled from a Poisson distribution with mean $F(N(t))$, thus after population regulation. However, when employing the population growth model (2), in many circumstances it may be more natural to let the population reproduce first, so that the actual number of offspring is a Poisson variate with mean $N(t)e^r$, then possibly including other features of the life cycle, and performing the culling to $K$ individuals only among adults. Thus, on average, under this life cycle there will be more offspring. Indeed, numerical simulations yield results that are qualitatively similar to the previous model, but extinction times are always somewhat larger, and substantially so for high growth rates $r$ (results not shown). These results, together with the poor performance of the diffusion approximations, under-line the need for further mathematical analyses of the extinction risk of populations subject to random demographic and environment events.

## Random Catastrophes

Since catastrophes, by definition, are rare and large perturbations, they cannot be modeled by diffusion approximations. Lande (1993) modified an approach of Hanson and Tuckwell (1978), derived a differential-difference equation with delay, and proved that the mean time to extinction is again a power function of the carrying capacity if the growth rate between cata-strophes exceeds a compound measure of the catastrophe rate and size. Thus, not unexpectedly, catastrophes pose a higher extinction risk to large populations than demographic stochasticity. However, no general state-ment can be made about the relative importance of catastrophes and random environmental fluctuations, because this depends on the values of the parameters in the model.

## Genetic variability and critical rates of environmental change

Many if not most species that ever have lived were exposed to gradual and prolonged environmental changes during (part of) their evolutionary history. Some have been able to respond evolutionarily and to adapt to the new environmental conditions, while others have become extinct. Currently, evidence is mounting that human activity is causing, and will continue to cause, global changes in the temperature, the chemical composition of the atmosphere and in the food chains of many populations at rates that are likely to be higher than in prehistoric times (e.g. Kareiva et al., 1993). Long-term experiments imposing directional selection on quantitative traits have shown that populations are capable of evolving mean pheno-types that can deviate by at least 10 phenotypic standard deviations from the mean prior to selection (e.g. Jones et al., 1968; Eisen, 1980; Yoo, 1980). Thus, it is clear that populations generally have the evolutionary potential to respond to substantial selective challenges. However, in most experi-ments of this kind, selection plateaus are reached and lines are lost. Al-though, most selection plateaus are caused by the advancement of genes with major effects on the selected trait and negative pleiotropic effects on fitness, the rate of phenotypic evolution of any population will eventually become limited by the availability of newly arising additive genetic variance. Therefore it is natural to consider the maximum rate (or amount) of en-vironmental change that can be tolerated by a species.

Models yielding insight into such questions need to incorporate the features of a changing environment and the demographic and genetic potential of the population. Such quantitative-genetic models were developed by Lynch et al. (1991) for large asexual populations and by Lynch and Lande (1993) for small sexual populations. They considered a continuously changing environment such that the optimum phenotype is, for example, gradually increasing. In this case, the mean phenotype evolves in the same direction as the moving optimum, but lagging behind it. This lag causes a reduction in mean fitness of the population. If the rate of environmental change is sufficiently low and if there is enough genetic variation, the lag will be small and the mean fitness will remain high enough so that the population can maintain a positive growth rate. If, however, the rate of environmental change is too high, the selective load (reduced viability and fecundity) on the population will induce a negative growth rate, and the population will decline in size until extinction occurs. Thus, as shown by Lynch and Lande (1993), there is a critical rate of environmental change, defined by the demographic and genetic features of the population, beyond which extinction is certain. Their theory, however, was essentially a deterministic one, which assumed a constant genetic variance, and a Gaussian phenotypic distribution of the character under selection, and it neglected various sources of stochasticity.

Explicit genetics, as well as population regulation and a weak form of demographic stochasticity, was introduced into the original model by Bürger and Lynch (1995). Its basic features are as follows. Consider a randomly mating, finite population with discrete generations. Individual fitness is determined by a single quantitative character under Gaussian stabilizing selection on viability, with the optimum phenotype $\theta_t$ exhibiting temporal change. The viability of an individual with phenotypic value $z$ is assumed to be

$$W_{z,t} = \exp\left[ -\frac{(z-\theta_t)^2}{2\omega^2} \right], \tag{12}$$

where $\omega^2$ is inversely proportional to the strength of stabilizing selection.

The intention is to investigate the response to environmental change for a population that has been at mutation-selection-drift balance and is well adapted to its environment. The environmental change may be either directional, or stochastic, or a combination of both. A simple model for this is a phenotypic optimum that moves at a constant rate $k$ per generation, fluctuating randomly about its expected position.

$$\theta_t = kt + \varepsilon_\theta, \tag{13}$$

where $\varepsilon_\theta$ represents white noise with variance $\sigma_\theta^2$, mean zero, and no autocorrelation (see also Charlesworth, 1993b). Under this model, the population experiences a mixture of directional and stabilizing selection. The width $\omega$ of the fitness function is assumed to be constant.

The quantitative character under consideration is assumed to be determined by $n$ freely recombining, mutationally equivalent loci. The allelic effects are additive within and between loci, i.e. there is no dominance or epistasis. Mutation occurs with the genic mutation rate $\mu$, and the effects of mutants are drawn from a probability distribution (e.g. a Gaussian) with mean zero and variance $\alpha^2$. In all our simulations the genomic mutation rate is $2n\mu = 0.02$ and the input of mutational variance per generation is $V_m = 2n\mu\alpha^2 = 10^{-3}\sigma_e^2$, as approximately observed in empirical studies (Lande, 1976; Lynch, 1988). The phenotypic value of an individual is the sum of a genetic contribution and a normally distributed environmental effect with mean zero and variance $\sigma_e^2 = 1$. Therefore, the phenotypic mean equals the mean of the additive genetic values, $\bar{g}_t$, and the phenotypic variance is $\sigma_{p,t}^2 = \sigma_{g,t}^2 + \sigma_e^2$, with $\sigma_{g,t}^2$ denoting the additive genetic variance in generation $t$. (The subscript $t$ will be omitted to signify equilibrium values attained for large $t$.) the multiplicative growth rate of the population is $R_t = B\bar{W}_t$, where $\bar{W}_t$ denotes the mean viability.

The population is assumed to be subject to density-dependent population regulation according to the simple model (2), but culling is performed on adults. Thus, the maximum number of breeding adults is $K$, the carrying capacity, which may be interpreted as the maximum number of available nesting places. The $N_t$ ($\leq K$) breeding parents in generation $t$ produce $BN_t$ offspring, an expected $R_tN_t$ of which will survive viability selection. Thus, demographic stochasticity is only induced through viability selection. If the actual number of surviving offspring exeeds $K$, then $K$ individuals are chosen randomly to constitute the next generation of parents. Otherwise, all surviving offspring serve as parents for the next generation. The sex ratio is always $\frac{1}{2}$. With this type of density-dependent population regulation, the number of reproducing adults remains roughly constant at $K$, until $\bar{W}_t < 1/B$, at which point the population cannot replace itself.

The evolutionary dynamics of the selected character is the key to understanding the extinction process. As noted above, in the face of directional environmental change, the mean phenotype $\bar{g}_t$ evolves but lags behind the optimum. Once the mean phenotype lags sufficiently far behind $\theta_t$ that $\bar{W}_t < 1/B$, the population size starts to decline. With a smaller population size, genetic drift reduces the genetic variance, which leads to an even larger lag of the mean phenotype, and a further decrease of mean fitness. Once this synergistic process begins, extinction soon follows. The complexity of the model requires a number of simplifying assumptions for obtaining analytical approximations. Two sets of assumptions in particular facilitate analytical progress.

First, we assume that the rate of environmental change, $k$, and the variance of fluctuations of the fitness function, $\sigma_\theta^2$, are small enough that mean fitness $\bar{W}_t$ remains sufficiently high to ensure that the effective population size $N_e$ is approximately constant. Assuming a normal distribution of phenotypes, the expected dynamics of the mean phenotype can then be calculated (cf. Lynch and Lande, 1993; Bürger and Lynch, 1995). Eventually, the mean phenotype evolves with an expected rate equal to the rate of environmental change $k$, but lags behind the optimum by an average amount of $k\sigma_g^2/(\sigma_p^2 + \omega^2)$. The mean phenotype of any particular population fluctuates around this value with variance $V[\bar{g}] \approx (2N_e)^{-1} (\omega^2 + \sigma_e^2) + \frac{1}{2} \sigma_g^2 \sigma_\theta^2 (\omega^2 + \sigma_e^2)^{-1}$. Using these results, the expected multiplicative growth rate $E(R_t) = BE(\bar{W}_t)$ can be calculated.

The critical rate of environmental change, $k_c$, is defined to be the value of $k$ such that the population can just replace itself, that is, such that $E(R_t) = 1$. It is computed to be

$$k_c = \frac{\sigma_g^2}{\omega^2 + \sigma_p^2} \sqrt{2V \left[ r_{\max} - \frac{1}{2} \ln \frac{V}{\omega^2} \right]}, \tag{14}$$

where $r_{\max} = \ln B$ is the intrinsic growth rate in a constant environment with the population being fully adapted (and monomorphic), and $V = \omega^2 + \sigma_p^2 + \sigma_g^2 + V[\bar{g}]$. Assuming that stabilizing selection is weak and environmental fluctuations small, i.e. $\omega^2 \gg \sigma_p^2 + \sigma_\theta^2$, the rate of critical change can be approximated by an upper bound,

$$k_c < h^2 \frac{\sigma_p}{\omega} \sqrt{2r_{\max} - \frac{1}{2N_e}}, \tag{15}$$

where $h^2$ denotes the heritability of the trait (cf. Lynch and Lande, 1993, and Bürger and Lynch, 1995). Formulas (14) and (15) are deceptively simple because the genetic and phenotypic variance, $\sigma_g^2$ and $\sigma_p^2$, are actually functions of the width of the fitness function, the number of loci, the effective population size, the mutation parameters and the rate of environmental change $k$.

What do the expressions (14) and (15) imply about the maximum sustainable rate of evolution? Given that most morphological and fitness-related traits have heritabilities less than 0.5 (Mousseau and Roff, 1987; Roff and Mousseau, 1987), that $\omega^2$ is typically between 5 and 100 (Turelli, 1984; Endler, 1986), and that on time scales of generations $r_{\max}$ is usually less than 1.0, it is unlikely that even large populations can sustain rates of environmental change exceeding 10% of a phenotypic standard deviation $\sigma_p$ per generation. The situation is more stringent in small populations or if the directional environmental trend is accompanied by environmental fluctuations (cf. Lynch and Lande, 1993; Bürger and Lynch, 1995; and Fig. 4(B)).

The second situation, in which analytical progress can be achieved assumes that the actual rate of environmental change, $k$, is higher than the critical rate $k_c$. In this case, two phases of population dynamics need to be considered. During the first phase, the expected multiplicative growth rate

$E(R_t)$ declines steadily until $E(R_t) = 1$, with the number of reproducing adults remaining constant at $K$ (we assume that initially there are $K$ individuals). During the second phase, as $E(R_t)$ decreases further, the population size progressively declines. We define extinction as having occurred when the population size has been reduced to a single individual (this is clearly justified in dioecious species). Denoting the average lengths of the two phases by $T_1$ and $T_2$, the mean time to extinction is $T_e = T_1 + T_2$. The length of the first phase can be shown to be approximately

$$T_1 \approx -\frac{\omega^2 + \sigma_p^2}{\sigma_g^2} \ln\left(1 - \frac{k_c}{k}\right), \quad \text{for } k > k_c, \tag{16}$$

which, using (15), reduces to

$$T_1 \approx \frac{1}{k}\frac{\omega^2}{\sigma_p}\sqrt{2r_{\max} - \frac{1}{2N_e}}, \tag{17}$$

provided $k \gg k_c$ and $\omega^2 \gg \sigma_p^2 + \sigma_\theta^2$. The length of phase 1 turns out to be rather short, even if the rate of environmental change is only slightly larger than the critical rate. Consider, for instance, the case where $k = \frac{10}{9}k_c$, $\omega^2 = 20$, and $h^2 = \frac{1}{4}$ (corresponding to $\sigma_g^2 \approx \frac{1}{3}$). Then $T_1 \approx 140$ generations. In most cases, the second phase is much shorter than the first.

Because of the complexity of the model, comprehensive computer simulations were performed, adapting the multilocus simulation model of Bürger et al. (1989); see Bürger and Lynch (1995) for a description. For a slowly moving optimum, $k \approx k_c$, these simulations showed that the evolution of the average mean phenotype, $E(\bar{g}_t)$, is well described by the Gaussian theory outlined above. However, the variance of fluctuations of a particular population mean phenotype around the expected value is often an order of magnitude higher than predicted by the theory. In addition, our simulations demonstrated that in populations with an effective size of more than a few dozen individuals, the genetic variance is elevated above the equilibrium value expected at mutation-stabilizing selection-drift balance with a stable optimum, the so-called stochastic-house-of-cards approximation, $2N_e V_m/[1 + \alpha^2 N_e/(\omega^2 + \sigma_e^2)]$ (cf. Bürger et al., 1989; Bürger and Lande, 1994). However, the genetic variance remains well below the neutral mutation-drift equilibrium value of $2N_e V_m$ (Lynch and Hill, 1986).

It has long been well known that in populations with effective sizes of several hundreds or fewer individuals, random genetic drift causes the actual genetic variance in a population to wander around its expected value from generation to generation and that random decreases or increases of genetic variance may persist for many generations (e.g. Keightley and Hill, 1989; Bürger et al., 1989; Zeng and Cockerham, 1991). For a population in a constant environment, Bürger and Lande (1994) showed that the expected

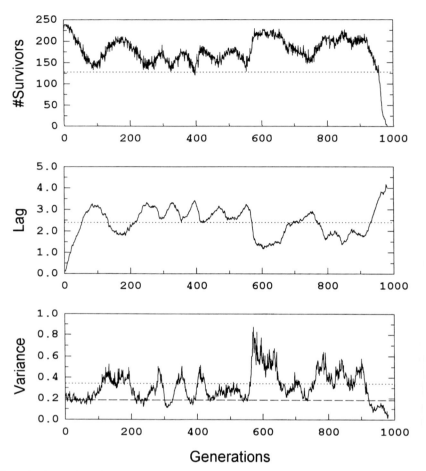

Figure 3. Dynamics of the number of surviving offspring (upper panel), the lag of the mean phenotype (middle panel) and the genetic variance (lower panel) of a selected character from a moving optimum (from Bürger and Lynch, 1995). Initially, the population mean phenotype matches the environmental optimum. The subsequent rate of environmental change ($k = 0.075$) is slightly less than the critical rate ($k_c \approx 0.08$) defined by Eq. (14), yet the population still goes extinct because of a temporary bottleneck in genetic variance caused by random genetic drift. The carrying capacity is $K = 128$, the number of offspring produced by each individual is $B = 2$ and the strength of stabilizing selection is $\omega^2 = 9$. In the lower panel, the dotted and dashed lines denote the equilibrium levels of genetic variance expected under the neutral model and under the stochastic house-of-cards approximation, respectively; in the middle panel, the dotted line is the expected lag of the mean phenotype (see text). Note that the lag increases and the number of surviving offspring decreases during transient periods of low genetic variance. During a final period of high stochastic loss of genetic variation, the lag becomes so great that the per capita rate of replacement is less than 1, and the population goes extinct rapidly.

correlation of the genetic variance in the same population at times separated by $t$ generations is approximately $e^{-t/(2N_e)}$, which declines only to 0.5 after $t = 1.4\, N_e$ generations. For a gradually changing environment, these fluctuations are most pronounced if the rate of environmental change is slightly below the critical rate of environmental change. Figure 3 displays a particular run where the genetic variance gets several boosts because of stochastic mutation events, allowing the population to persist much longer than the average. Once the genetic variance becomes low, however, the population lag increases. This leads to stronger selection, reduced population size, stronger genetic drift and further reduction of genetic variance, and the resultant synergistic interactions finally drive the population to rapid extinction. The figure clearly demonstrates the high autocorrelation of the genetic variance and the central role of genetic variance in enhancing the rate of adaptation or, equivalently, in reducing the lag.

These results emphasize that even in the absence of a bottleneck, populations frequently suffer from temporary losses of genetic variation caused by chance events (see the figures in Bürger et al., 1989). Sometimes, such events deprive small populations of all their genetic variation, and because of its high autocorrelation, the genetic variance may then remain low for a long time. Such excursions of genetic variance to low levels can jeopardize the survival of populations inhabiting changing environments.

All finite populations subject to density-dependent population regulation and exposed to stochastic events are doomed to eventual extinction. Nevertheless, our simulation study demonstrates that there is a sharp boundary in the relationship between the rate of environmental change and mean extinction time, such that below a critical rate of change, the population can track the environmental trend sufficiently closely to guarantee long-term survival. In general, the theoretically predicted critical rate $k_c$ (14) agrees reasonably well with that found from simulations, although there are two opposing forces leading to deviations from that prediction. In small populations or with appreciable environmental fluctuations, the observed critical rate is smaller than predicted, because our theory ignores several sources of stochasticity. In larger populations ($K \geq 500$), the observed critical rate is sometimes larger than predicted because the genetic variance increases after the onset of directional selection, thus elevating the maximum sustainable rate of adaptation. The approximation $T_e$ for the mean extinction time is highly accurate as long as the optimum changes rapidly and mean extinction times are on the order of a few dozens to a few hundreds generations.

The precise role of environmental, demographic and genetic parameters in the extinction risk in a changing environment is relatively complex and discussed in detail in Bürger and Lynch (1995). Here we only summarize the main findings. In the face of a changing environment, larger population size reduces the risk of extinction by mitigating genetic and demographic stochasticity and by increasing the level of genetic variation. However,

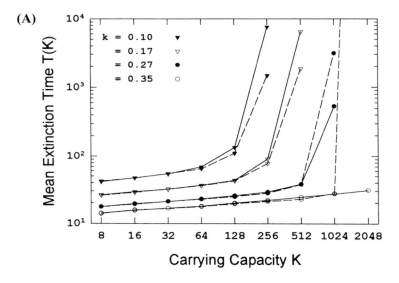

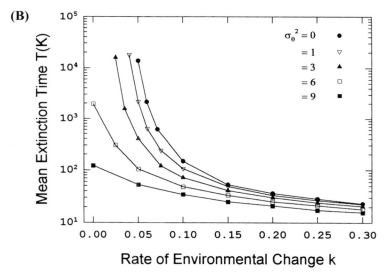

Figure 4. Mean extinction times in a directionally changing environment. (A) Mean time to extinction as a function of the carrying capacity $K$, for different rates of environmental change $k$ and for two different distributions of mutational effects. The solid lines are for a Gaussian distribution, the dashed lines for a reflected gamma distribution with a kurtosis of 11.7 (The latter is defined by $\lambda^\beta |x|^{\beta-1} \exp(-\lambda|x|)/(2\Gamma(\beta))$, where the constants $\lambda$ and $\beta$ can be chosen such that this distribution has a predefined variance and kurtosis. For $\beta = 0.5$, the kurtosis becomes 11.7; see Keightley and Hill, 1989). For all data $B = 2$, $\omega^2 = 9$, $\alpha^2 = 0.05$ and $\sigma_\theta^2 = 0$. (B) Mean time to extinction as a function of the rate of environmental change, for five levels of temporal fluctuations in $\theta$. The other parameters are $B = 2$, $K = 128$ and $\omega^2 = 9$ (after Bürger and Lynch, 1995).

these effects are not necessarily large. For a rapidly changing environment, so that $E(R_t) < 1$, the extinction risk of large populations is only slightly below that of small ones because in this case the mean extinction time $T_e$ increases more slowly than $\ln K$. Actually, since the genetic variance does not increase indefinitely as $K$ increases, the critical rate $k_c$ approaches a constant value as $K$ tends to infinity. This suggests that for sufficiently large $k$, any population, however large, will become extinct rapidly, whereas for smaller $k$, there is a sharp boundary between the extinction risk of small and large populations (cf. Fig. 4(A)). Similarly, a higher birth rate $B$ reduces the extinction risk significantly only if the actual rate of change, $k$, is near the critical rate $k_c$. If $k \gg k_c$, the mean extinction time depends approximately logarithmically on the birth rate $B$.

An interesting effect is observed when considering the dependence of the mean extinction time on the strength of stabilizing selection. For high $k$, the extinction time increases slowly but monotonically with the width $\omega$ of the fitness function. For lower $k$ values (near $k_c$), a maximum is observed at intermediate $\omega$ values, the reason being the following. For weak stabilizing selection ($\omega$ large), the fitness function is flat, the expected lag is large and random departures are only weakly selected against, but the genetic variance becomes almost independent of $\omega$ because it is close to the neutral prediction. For very strong stabilizing selection, the genetic variance becomes very low, and small excursions of the lag drastically reduce population mean fitness. Since the response to selection is approximately proportional to the genetic variance $\sigma_g^2$, the mean extinction time $T_e$ will be maximized at an intermediate $\omega$. A similar prediction, though with a different argument, was proposed by Huey and Kingsolver (1993), who applied the model of Lynch and Lande (1993) to the evolution of thermal sensitivity. This finding suggests that in a slowly, but steadily, changing environment, broad generalists (which are insensitive to moderate perturbations of the environment) and narrow specialists (highly sensitive to environmental changes) will be most vulnerable to extinction.

The preceding results apply to the situation in which the environmental optimum is changing in a deterministic fashion, but we can generally expect the optimum phenotype to vary randomly about its expected value (e.g. Boag and Grant, 1981; Weis et al., 1992). In the present model, the variance of these (white-noise) fluctuations is denoted by $\sigma_\theta^2$. As Figure 4(B) shows, such random perturbations, even if large ($\sigma_\theta^2 = \omega^2$), add little to the extinction risk if the rate of directional environmental change is large. For a slowly changing environment, however, a small amount of environmental stochasticity may decrease the persistence time of a population substantially. It should be emphasized that the present moving-optimum model is based on the simple mode (2) of population regulation and that culling is performed among adults. Thus, under other forms of density-dependent population regulation, the contribution of environmental stochasticity to the extinction risk is likely to be more pronounced.

For a constant or a fluctuating optimum without a directional component of environmental change ($k = 0$), monomorphic populations with phenotypes coinciding with the average optimum phenotype survive longer than genetically variable populations, the effect becoming weaker as population size increases (cf. Bürger and Lynch, 1995). This occurs because variable populations respond to fluctuations of the optimum, but with a time lag of one generation. On the average, this reduces their fitness in the subsequent generation, compared with a monomorphic population at $\bar{g} = 0$, because the perturbations are uncorrelated. In addition, phenotypic variance induces a load by reducing the expected growth rate under stabilizing selection. This finding agrees with Charlesworth (1993a), who found that genetic variability is advantageous only if environmental fluctuations are very large, i.e. $\sigma_\theta^2 > 2(\omega^2 + \sigma_e^2)$, and with Lande and Shannon (1996), who analyzed a continuous-time model. Lande and Shannon calculated the load caused by stabilizing selection and the "evolutionary load" caused by deviation of the population mean phenotype from the optimum phenotype for various forms of deterministic and stochastic changes in the optimum phenotype. They showed that for predictable environmental change, e.g. a moving optimum, a cyclically changing optimum with large amplitude and long-period oscillations, or a stochastically fluctuating optimum with a sufficiently high autocorrelation, additive genetic variance diminishes the total genetic load (the sum of the evolutionary load and that due to stabilizing selection) and, thus, enhances population persistence in such changing environments. For cyclic or autocorrelated environments similar results were obtained by Slatkin and Lande (1976) and Charlesworth (1993a,b).

The assumptions about mutational effects (many mutationally equivalent loci, no direct pleiotropic effects on fitness) and gene action (complete additivity) in the preceding models are rather simplistic and are likely to produce overly optimistic results concerning maximum sustainable rates of evolution and population survival. For example, recent experiments with *Drosophila* indicate that a substantial fraction of mutants, in particular those with large effects, are highly detrimental. The effects of individual mutations tend to have a leptokurtic (L-shaped) distribution, and they behave in a nonadditive fashion, those with large effects tending to be nearly recessive, whereas those with smaller effects may have highly variable levels of dominance (Mackay et al., 1992a; López and López-Fanjul, 1993a, b; Caballero and Keightley, 1994; Fry et al., 1995; Lyman et al., 1996). Thus, the rate of production of quasi-neutral mutational variance per generation, which is likely responsible for much of the standing variation in quantitative traits, may be an order of magnitude lower than the $V_m$ used above (Lande, 1995). Also, a substantial fraction of the standing variation for quantitative traits may often be attributable to a few loci. For example, most of the genetic variance for bristle numbers in natural populations is a function of allelic variation at four or five loci (Frankham, 1988; Long et al., 1995), though in mice more loci seem to contribute to genetic variation in body weight (Keightley et al., 1996).

If the distribution of mutational effects is leptokurtic, less genetic variation is maintained in a population in mutation-selection balance than in a comparable population in which the distribution of mutational effects is Gaussian. This reduction in variance is negligible for small populations or weak stabilizing selection but increases with population size and strength of stabilizing selection (Bürger, 1989; Bürger and Lande, 1994; Bürger and Hofbauer, 1994). For instance with $\omega^2 = 9$ and $K \geq 128$, the equilibrium genetic variance with a reflected gamma distribution of mutational effects with variance $\alpha^2 = 0.05$ and kurtosis of 11.7 is reduced by approximately 40%, compared with a Gaussian mutation distribution. Therefore, the response to directional selection and the mean time to extinction caused by environmental change may be lower than with a Gaussian distribution of mutational effects. This is visible in Figure 4(A) for intermediate population sizes, while for small population sizes and a rapidly changing environment there is no measurable effect. In contrast, for large populations and a rapidly moving optimum, a highly leptokurtic distribution of mutational effects may greatly increase the persistence time (Fig. 4(A)). The reason is that in sufficiently large populations such a distribution generates rare mutants of large and beneficial effect. These mutants may substantially increase the genetic variance within the population, thus enhancing the rate of evolution sufficiently to prevent the population from extinction. This, however, is a stochastic phenomenon occurring only in some populations. For instance, with $K = 2048$ and $k = 0.35$, 5 of 55 replicate populations survived for $10^5$ generations (when the simulation was stopped), their average genetic variance being elevated above the (initial) equilibrium value of $\sigma_g^2 = 0.224$ by a factor of 9. The mean extinction time of the other 50 populations was 35.1 generations with a maximum of 107, and thus only 50% above the mean extinction time of the corresponding populations with a Gaussian distribution of mutational effects. Their variance was elevated only by a factor of 2.

Most bristle-number mutations have negative pleiotropic effects on fitness, independent of their direct effects on the trait (Mackay et al., 1992a, b, 1995; Keightley, 1994; Nuzhdin et al., 1995). Although several quantitative-genetic models involving pleiotropic effects on fitness have been proposed (Turelli, 1985; Wagner, 1989; Barton, 1990; Keightley and Hill, 1990; Kondrashov and Turelli, 1992; Caballero and Keightley, 1994), they have not yet been applied to ecological or conservation problems, presumably because no archetypical model has yet been shaped and because much simpler models have only recently been investigated. In addition, it is likely that in reality more than one character will be under directional selection caused by environmental change, while other pleiotropically connected traits may remain under stabilizing selection. In such a case, population-genetic constraints, caused by the pattern of variation and covariation, may severely restrict the adaptive potential and reduce the rate of evolutionary response (e.g. Bürger, 1986; Wagner, 1988). Work investigating the effects of environmental change on more than one character is in progress.

All of these analyses and results underline the central role of additive genetic variance in providing the potential for adaptive evolution and long-term population survival in predictably, but slowly, changing environments. However, they also show that neither large population size nor high amounts of genetic variation can prevent population extinction in rapidly ($k \geq \sigma_p/10$) changing environments.

## Accumulation of deleterious mutations

An assumption of the mutation model used above is that mutations are conditionally beneficial or deleterious, depending only on whether they move the phenotype towards or away from the current optimum. However, mutation-accumulation experiments as well as indirect estimates of the selection coefficient against segregating alleles for polygenic characters in several species have shown that the vast majority of new mutations are nearly neutral or deleterious, causing on average a $1-2.5\%$ reduction in viability in the heterozygous state (Mukai, 1964, 1979; Crow and Simmons, 1983; Houle et al., 1996; Kibota and Lynch, 1996; Schultz et al., 1997). Here we outline the consequence of genetic deterioration caused by unconditionally deleterious mutations in a constant environment. In contrast to the quantitative-genetic mutation model underlying the analysis of critical rates of environmental change, we now assume that all new mutations are unconditionally deleterious. Although this assumption cannot be strictly valid, the results cited above clearly indicate that the majority of new mutations are indeed slightly deleterious. The rate at which such mutations arise is quite high – on the order of 1.0 and 0.1 per individual per generation in *Drosophila* (Mukai, 1979; Crow and Simmons, 1983; Lynch et al., 1995 b) and *Arabidopsis* (Schultz et al., 1997), respectively.

As before, we consider a randomly mating monoecious population growing in discrete generations, with the simple form of density-dependence (2) following juvenile production, mutation and viability selection each generation. Letting $B$ be the reproductive rate per surviving adult and $N_t$ be the number of reproducing adults at time $t$, the number of progeny produced prior to selection is $BN_t$. The gametes required to produce these progeny are assumed to be drawn randomly, with free recombination and allowing random selfing, from the $N_t$ parents. Newborns are assumed to incur new mutations following a Poisson distribution with an expected genomic mutation rate $U$. All new mutations are assumed to arise at loci that are not currently segregating in the population, a reasonable assumption for small populations. Let the fitness of the three genotypes at some locus be 1, $1-2hs$ and $1-2s$. Unless stated otherwise, all mutations are assumed to have the constant properties $s$ and $h$. Almost all data for the parameters $U$, $s$ and $h$ are from *Drosophila*. A survey suggests that, on average, $U \simeq 1.5$, $s \simeq 0.01$ and $h \simeq 0.36$ (Lynch et al., 1995b).

Following mutation, viability selection operate on the progeny, with the probability of survival to maturity being determined by the fitness function $W(n_1, n_2) = (1-2hs)^{n_1} (1-2s)^{n_2}$, where $n_1$ and $n_2$ are the numbers of loci in the individual that are heterozygous and homozygous for deleterious mutations. This fitness function allows for dominance but assumes that the effects of mutations at different loci are independent. If, following selection, the number of potential adults exceeds the carrying capacity of the environment, $K$, the population is assumed to decline to $K$ by genotype-independent culling prior to the next round of reproduction. So long as the mean viability remains greater than $1/B$, the effective population size under this model is very close to $K$ (Lynch et al., 1995a), which we assume below.

In a small isolated population descending from a large ancestral population, deleterious mutations influence the mean fitness in three ways. First, segregating mutations inherited from the ancestral population may either be purged by natural selection or random genetic drift, or they may rise to fixation by random drift. Second, mutations arising in the isolated population lead to the establishment of a new segregational load defined by the balance between the present forces of drift, mutation and selection. Third, a fraction of the mutations entering the population each generation becomes fixed by random genetic drift at some future time. Unlike the first two sources of mutation load, the fixation of recurrent deleterious mutations leads to a progressive loss in fitness.

Consider first the consequences of "ancestral" deleterious mutations. Assuming that the ancestral population was effectively infinite ($N_e > 5/s$), in mutation-selection balance, and that the deleterious alleles are not completely recessive, the mean fitness of founding individuals is approximately $\overline{W}_0 = e^{-U}$ (Haldane, 1937; Kimura et al., 1963; Bürger and Hofbauer, 1994).

Haldane's principle allows a prediction about the expected number, $\bar{n}$, of deleterious genes per individual in the founder population. Assuming that the frequencies of deleterious alleles in the ancestral population are small, so that they occur only in heterozygotes, the mean fitness of a founding individual is approximately $(1-2hs)^n \simeq e^{-2hsn}$. Equating this with $e^{-U}$, the expected number of deleterious genes carried by a founder individual is $\bar{n} \simeq U/(2hs)$. Using the average values for the parameters $U$, $h$ and $s$ in *Drosophila* (see above), one obtains $\bar{n} \simeq 140$. Therefore, since most individuals are expected to carry unique sets of rare deleterious genes, even small founder populations are likely to harbor segregating deleterious genes at several hundred loci.

The expected asymptotic fitness in the isolated population, resulting solely from fixation of additive mutations in the founding individuals, is $\overline{W}_B = (1-2su_F)^{KU/s}$, where $u_F$ is the fixation probability. This expression is always larger than $\overline{W}_0$, the difference being negligible for small populations, $Ks \ll 1$, when genetic drift overwhelms selection. As $Ks$ increases above one, some or all of the ancestral additive deleterious mutation load will be purged and, when $Ks > 5$, essentially all of it will be eliminated (Lynch et al., 1995b). Fixation of partially recessive mutations may result in a decrease of fitness, if the founding population size is very small, on the order of a few dozen individuals or less. Otherwise, the population will experience a net purging of the mutational load in the founder population.

When considering the consequences of deleterious mutations on a longer time frame, the purging or fixation of the ancestral mutation load by no means gives a complete picture of the consequences of segregating mutations. While the deleterious mutations inherited from the ancestral population are going to fixation or being eliminated, new ones are appearing each generation, and the population eventually re-establishes a new segregational load defined by its own population size.

The development of the new load of segregating mutations was investigated by Lynch et al. (1995b). For finite populations, the expected mutation load may be larger than in infinite populations, the maximum load occurring approximately for $K \simeq 1/(2s)$ if mutations are additive. For larger $K$, the load decreases to the infinite expectation $e^{-U}$, which provides a good approximation if $Ks > 5$. Only for very small population size, $K < 1/(4hs)$, is the load from segregating mutations lower than for infinite populations.

The reduced load of segregating mutations in small populations is more than offset by a greater rate of fixation of detrimental mutations. It has been widely acknowledged that the gradual accumulation of mildly deleterious mutations leads to a continuous decrease of mean fitness of asexual populations via a process known as Muller's ratchet (Felsenstein, 1974; Haigh 1978; Maynard Smith, 1978; Charlesworth et al., 1993; Stephan et al., 1993; Higgs, 1994; Kondrashov, 1994; Butcher, 1995), which is an important source of extinction for asexual populations of any size (Lynch and Gabriel, 1990; Melzer and Koeslag, 1991; Gabriel et al., 1993; Lynch et al., 1993; Gabriel and Bürger, 1994). However, it has generally been assumed that this process is of little relevance to sexual species. Recent analyses and simulations have shown that this is not the case (Gabriel and Bürger, 1994; Lande, 1994, 1995; Lynch et al., 1995a, b). Accumulation of deleterious mutations leads to a progressive fitness loss. As soon as the mean fitness has fallen below $1/B$, the population is no longer able to replace itself. The dwindling population size enhances the rate of fixation of deleterious alleles, and this synergistic process, called the mutational meltdown (Lynch and Gabriel, 1990), leads to rapid extinction. Although for constant fitness effects $s$ and $h$, the mean time to extinction scales nearly exponentially with the carrying capacity, the leading factor being $e^{4N_e s}/N_e$, (Lande, 1994; Lynch et al., 1995a), there are strong synergistic interactions of this kind of genetic stochasticity with demographic and environmental stochasticity (Gabriel et al., 1991; Gabriel and Bürger, 1994; Lynch et al., 1995b), so that sexual populations of effective sizes up to approximately 100 experience a high risk of extinction and may have persistence times as low as several tens to a few hundred generations.

Lande (1994) analyzed the influence of variance in selection coefficients among new mutations and proved that in this case the mean time to extinction increases only as a power of population size. If $s$ is exponentially distributed and mutations are additive, the mean extinction time is asymptotically proportional to $N_e^2$. This indicates a much greater risk of extinction than in the case of a constant selection coefficient. Lande concluded that fixation of new, slightly deleterious mutations poses a considerable risk of extinction on intermediate or long time scales for populations with effective sizes as large as a few thousand. The reason why variable selection coefficients lead to a much larger extinction risk in moderately sized or large populations is that mutations with small, but intermediate, deleterious effects cause most cumulative damage to populations (Kimura et al., 1963; Gabriel et al., 1993; Charlesworth et al., 1993). Mutations with very large effects are eliminated by selection and have essentially no chance of fixation, whereas neutral mutations have no influence on individual fitness. Thus, there is an intermediate value of $s$ that minimizes the time to extinction (Gabriel and Bürger, 1994). Lande (1994) calculated this value, using diffusion theory, and showed that it is approximately $s^* \simeq 0.4/N_e$, which is close to the border between neutrality and selection defined by $2N_e s = 1$.

Finally, it should be pointed out that the total fitness in an isolated population declines only slowly, in particular at the beginning of this process, because it is the product of the (generally increasing) fitness from purging old mutations and the decreasing fitness caused by new segregating mutations and fixation of new deleterious mutations (see Fig. 4 in Lynch et al., 1995b). The contribution of the latter becomes relevant only after some $2N_e$ generations, because the fixation time of deleterious mutations is proportional to $N_e$.

## Discussion

The evaluation of the relative importance of genetic, environmental and demographic factors in causing extinction has been considered to be one of the major questions in conservation genetics (e.g. Lande, 1988; Frankham, 1995a; Lynch, 1996). However, the results of recent theoretical studies, reviewed above, indicate the limitations of treating these primary factors as independent determinants of the risk of extinction. Genetic stochasticity generates demographic stochasticity by inducing changes in population size, while environmental stochasticity induces genetic change. Because these and other interactions between demographic, environmental and genetic change have strong negative synergistic effects on the vulnerability to extinction, models that focus on single factors appear to have rather limited generality and in some cases may be misleading.

Models that contain only demographic and environmental stochasticity suggest that high extinction risk arises only in very small populations and in populations with very low or negative growth rates, although in combination, these stochastic factors may greatly decrease the expected persistence time of a population. The mean extinction time is highly sensitive to assumptions about the mode of population regulation and, to a lesser extent, to assumptions about the stage in the life cycle at which population regulation acts. For instance, culling among adults leads to a longer persistence time than culling among juveniles, the effect being much more pronounced for large $\bar{r}$.

Drawing from empirical data on genomic mutation rates of deleterious alleles, our numerical and analytical work on the mutational meltdown, as well as the analyses of Lande (1994), has demonstrated that accumulation of mildly deleterious mutations poses a significant long-term risk of extinction to even moderately large sexual populations (with effective sizes of up to a thousand or so individuals). Since the genetic effective sizes of populations are often as small as a tenth of the actual number of breeding adults (Frankham, 1995b), this suggests that population sizes of a few thousand individuals or larger are required to prevent gradual deterioration due to long-term deleterious mutation accumulation. (National Research Council, 1995; Lynch, 1996).

In a sense, the failure of a population to survive a selective event is equivalent to a failure to adapt adequately. Our recent theoretical work has identified critical rates of environmental change beyond which populations are incapable of rapid enough adaptive evolution to avoid extinction within several dozens of generations. Such critical rates are formally equivalent to the maximum sustainable rate of evolution. If the rate of environmental change is above the critical rate predicted by our models, a high extinction risk will exist even for large populations. The negative population growth rate that arises when the critical rate is exceeded causes the mean time to extinction to scale logarithmically with the population size. The models underlying this theory, which are based on quantitative-genetic theory, assume a continuous input of new mutations, half of which are advantageous, on which selection can act. The identified critical rates of environmental change are less than 10% of a phenotypic standard deviation and often on the order of 1%. With more than one character affected by environmental change, these rates may be even lower due to pleiotropy and correlations between the characters. Such work is in progress. A further reason why the model of a gradually changing environment is more likely to be too optimistic is that it ignores several factors that reduce persistence time, like variation in birth rates and sex ratio. It is also based on the simple form (2) of population regulation, which is an assumption that is very favorable for population persistence.

Although this model assumes that environmental change continues directionally forever, which is not realistic, for a wide range of parameters extinction times are on the order of less than $10^3$, or even $10^2$, generations. Depending on the generation length of the species, in many cases this will be a reasonable time span for the applicability of the moving-optimum model.

As can be seen from the stochastic-house-of-cards approximation (see above), the equilibrium genetic variance is essentially constant with $N_e \geq 5\omega^2/\alpha^2$ (cf. Bürger et al., 1989) which, for typical parameters like $\alpha^2 = 0.05$ and $\omega^2 = 10$, translates to $N_e \geq 1000$. Therefore, at least on time scales of interest to conservation biologists, populations beyond this size can be viewed as being as genetically secure as possible (ignoring the temporal drift in the variance, whose coefficient of variation declines only in proportion to $N_e^{-1/2}$). Thus, from both the standpoint of deleterious-mutation accumulation and of incorporation of beneficial quantitative mutations, quantitative genetic models for the risk of extinction lead to the conclusion that $N_e$ on the order of 1000 is a good approximate benchmark for populations being genetically compromised as opposed to genetically secure. With regard to other genetical aspects, however, this number may still be much too small. For instance, rare conditionally beneficial mutations, e.g. mutations causing disease resistance, may be found only in very large populations.

The existing genetic models of extinction necessarily neglect numerous factors that may increase or decrease a population's extinction risk. For example, little is known about the necessary rate of compensatory mutations to avoid the mutational meltdown. (But see Wagner and Gabriel (1990) and Peck et al. (1997) for approaches to this problem that are quite different from ours.) Recent numerical results of Schultz and Lynch (1997) suggest that unless the beneficial mutation rate is on the order of 10% or more of the deleterious rate, in populations of size $N_e \leq 100$ or so, beneficial mutations can only slow (not prevent) the loss of fitness caused by deleterious mutation accumulation. Only for population size beyond $N_e = 500$ does the critical fraction of beneficial mutations to halt fitness loss decline to less than 0.1% of the deleterious rate. Schultz and Lynch also show numerically that synergistic epistasis must be very strong to reduce the risk of a mutational meltdown. This is in contrast to conclusions based on simpler models (Crow and Kimura, 1979; Kondrashov, 1994).

For mobile species, an efficient strategy to avoid the selective challenges and their risks caused by a gradually changing environment is migration to more suitable habitats (Pease et al., 1989). This possibility, however, is increasingly limited by habitat fragmentation. Genetic models for the risk of extinction for species that are distributed into multiple demes with limited among-deme migration (i.e. in metapopulations) are essentially undeveloped.

*Acknowledgments*
Our work was supported by the Austrian Science Foundation (FWF) grant P10689 MAT to R.B. and by grants from the National Science Foundation (BSR-9410610), National Institutes of Health (R01-GM36827), and the Oregon Department of Fish and Wildlife to M.L. We thank R. Lande, S. Engen and W. Gabriel for comments and discussion.

# References

Barton, N.H. (1990) Pleiotropic models of quantitative variation. *Genetics* 124:773–782.

Boag, P.T. and Grant, P.R. (1981) Intense natural selection in a population of Darwin's finches (Geospizinae). *Science* 214:82–85.

Bürger, R. (1986) Constraints for the evolution of functionally coupled characters: A nonlinear analysis of a phenotypic model. *Evolution* 40:182–193.

Bürger, R. (1989) Linkage and the maintenance of heritable variation by mutation-selection balance. *Genetics* 121:175–184.

Bürger, R. and Hofbauer, J. (1994) Mutation load and mutation-selection-balance in quantitative genetic traits. *J. Math. Biol.* 32:193–218.

Bürger, R. and Lande, R. (1994) On the distribution of the mean and variance of a quantitative trait under mutation-selection-drift balance. *Genetics* 138:901–912.

Bürger, R. and Lynch, M. (1995) Evolution and extinction in a changing environment: A quantitative-genetic analysis. *Evolution* 49:151–163.

Bürger, R., Wagner, G.P. and Stettinger, F. (1989) How much heritable variation can be maintained in finite populations by mutation-selection balance? *Evolution* 43:1748–1766.

Butcher, D. (1995) Muller's ratchet, epistasis and mutation effects. *Genetics* 141:431–437.

Caballero, A. and Keightley, P.D. (1994) A pleiotropic nonadditive model of variation in quantitative traits. *Genetics* 138:883–900.

Charlesworth, B. (1993a) The evolution of sex and recombination in a varying environment. *J. Heredity* 84:345–450.

Charlesworth, B. (1993b) Directional selection and the evolution of sex and recombination. *Genet. Res.* 61:205–224.

Charlesworth, D., Morgan, M.T. and Charlesworth, B. (1993) Mutation accumulation in finite outbreeding and inbreeding populations. *Genet. Res.* 61:39–56.

Crow, J.F. and Kimura, M. (1979) Efficiency of truncation selection. *Proc. Natl. Acad. Sci. USA* 76:396–399.

Crow, J.F. and Simmons, M.J. (1983) The mutation load in *Drosophila. In*: M. Ashburner, H.L. Carson and J.N. Thompson Jr. (eds): *The Genetics and Biology of* Drosophila, Volume 3c. Academic Press, New York, pp. 1–35.

Eisen, E.J. (1980) Conclusions from long-term selection experiments with mice. *Z. Tierzüchtg. Züchtungsbiol.* 97:305–319.

Endler, J.A. (1986) *Natural Selection in the Wild*. Princeton University, Princeton.

Engen, S. and Bakke, O. (1996) Diffusion approximations to discrete population models; *preprint.*

Felsenstein, J. (1974) The evolutionary advantage of recombination. *Genetics* 78:737–756.

Foley, P. (1994) Predicting extinction times from environmental stochasticity and carrying capacity. *Conserv. Biol.* 8:124–137.

Frankham, R. (1988) Exchanges in the rRNA multigene family as a source of genetic variation. *In*: B.S. Weir, E.J. Eisen, M.M. Goodman and G. Namkoong (eds): *Proceedings of the Second International Conference on Quantitative Genetics*. Sinauer, Sunderland, MA; pp. 236–242.

Frankham, R. (1995a) Conservation genetics. *Annu. Rev. Genetics* 29:305–327.

Frankham, R. (1995b) Effective population size/adult population size ratios in wildlife: A review. *Genet. Res.* 66:95–107.

Fry, J.D., deRonde, K.A. and Mackay, T.F.C. (1995) Polygenic mutation in *Drosophila melanogaster*: genetic analysis of selection lines. *Genetics* 139:1293–1307.

Gabriel, W. and Bürger, R. (1992) Survival of small populations and demographic stochasticity. *Theor. Popul. Biol.* 41:44–71.

Gabriel, W. and Bürger, R. (1994) Extinction risk by mutational meltdown: Synergistic effects between population regulation and genetic drift. *In*: V. Loeschcke, J. Tomiuk and S.K. Jain (eds): *Conservation Genetics*. Birkhäuser, Basel, pp. 69–84.

Gabriel, W., Bürger, R. and Lynch, M. (1991) Population extinction by mutational load and demographic stochasticity. *In*. A. Seitz and v. Loeschcke (eds): *Species Conservation: A Population-Biological Approach*. Birkhäuser, Basel, pp. 49–59.

Gabriel, W., Lynch, M. and Bürger, R. (1993) Muller's ratchet and mutational meltdowns. *Evolution* 47:1744–1757.

Gomulkiewicz, R. and Holt, R.D. (1995) When does evolution by natural selection prevent extinction? *Evolution* 49:201–207.

Goodman, D. (1987a) The demography of change extinction. *In*: M.E. Soulé (ed.): *Viable Populations for Conservation*. Cambridge University Press, Cambridge, pp. 11–34.

Goodman, D. (1987b) Consideration of stochastic demography in the design and management of biological reserves. *Natural Resource Modeling* 1:205–234.

Grass, D. (1996) *Aussterbezeiten bei diskreten und stetigen Populationsmodellen im Vergleich.* Master's thesis, University of Vienna.

Haigh, J. (1978) The accumulation of deleterious genes in a population. *Theor. Pop. Biol.* 14:251–267.

Haldane, J.B.S. (1937) The effect of variation on fitness. *Am. Nat.* 71:337–349.

Hanson, F.B. and Tuckwell, H.C. (1978) Persistence times of populations with large random fluctuations. *Theor. Pop. Biol.* 14:46–61.

Higgs, P.G. (1994) Error thresholds and stationary mutant distributions in multi-locus diploid genetics models. *Genet. Res.* 63:63–78.

Houle, D., Hoffmaster, D.K., Assimacopolous, S. and Charlesworth, B. (1992) The genomic mutation rate for fitness in *Drosophila*. *Nature* 359:58–60.

Houle, D., Morikawa, B. and Lynch, M. (1996) Comparing mutational variabilities. *Genetics* 143:1467–1483.

Huey, R.B. and Kingsolver, J.G. (1993) Evolution of resistance to high temperature in ectotherms. *Am. Nat.* 142:S21–S46.

Jones, L.P., Frankham, R. and Barker, J.S.F. (1968) The effects of population size and selection intensity in selection for a quantitative character in *Drosophila*. *Genet. Res.* 12:249–266.

Kareiva, P.M., Kingsolver, J.G. and Huey, R.B. (eds) (1993) *Biotic Interactions and Global Change*. Sinauer, Sunderland, MA.

Karlin, S. and Taylor, H.M. (1975) *A first Course in Stochastic Processes*, Second Edition Academic Press, New York.

Keiding, N. (1975) Extinction and exponential growth in random environments. *Theor. Pop. Biol.* 8:49–63.

Keightley, P.D. (1994) The distribution of mutation effects on Viability in *D. melanogaster*. *Genetics* 138:1315–1322.

Keightley, P.D. and Hill, W.G. (1989) Quantitative genetic variability maintained by mutation-stabilizing selection balance: Sampling variation and response to subsequent directional selection. *Genet. Res.* 54:45–57.

Keightley, P.D. and Hill, W.G. (1990) Variation maintained in quantitative traits with mutation-selection balance: Pleiotropic side-effects on fitness traits. *Proc. R. Soc. Lond. B.* 242:95–100.

Keightley, P.D., Hardge, T., May, L. and Bulfield, G. (1996) A genetic map of quantitative trait loci for body weight in the mouse. *Genetics* 142:227–235.

Kibota, T.T. and Lynch, M. (1996) The deleterious genomic mutation rate for overall fitness in *Escherichia coli*. *Nature* 381:694–696.

Kimura, M., Maruyama, T. and Crow, J.F. (1963) The mutation load in small populations. *Genetics* 48:1303–1312.

Kondrashow, A.S. (1994) Muller's ratchet under epistatic selection. *Genetics* 136:1469–1473.

Kondrashow, A.S. and Turelli, M. (1992) Deleterious mutations, apparent stabilizing selection and the maintenance of quantitative variation. *Genetics* 132:603–618.

Lande, R. (1976) Natural selection and random genetic drift in phenotypic evolution. *Evolution* 30:314–334.

Lande, R. (1988) Genetics and demography in biological conservation. *Science* 241:1455–1460.

Lande, R. (1993) Risks of population extinction from demographic and environmental stochasticity and random catastrophes. *Am. Nat.* 142:911–927.

Lande, R. (1994) Risk of population extinction from new deleterious mutations. *Evolution* 48:1460–1469.

Lande, R. (1995) Mutation and conservation. *Conserv. Biol.* 9:782–791.

Lande, R. and Shannon, S. (1996) The role of genetic variation in adaptation and population persistence in a changing environment. *Evolution* 50:434–437.

Leigh, E.G. (1981) The average lifetime of a population in a varying environment. *J. Theor. Biol.* 90:213–239.

Long, A.D., Mullaney, S.L., Reid, L.A., Fry, J.D., Langley, C.H. and Mackay, T.F.C. (1995) High-resolution mapping of genetics factors affecting abdominal bristle number in *Drosophila melanogaster*. *Genetics* 139:1273–1291.

López, M.A. and López-Fanjul, C. (1993a) Spontaneous mutation for a quantitative trait in *Drosophila melanogaster*. I. Response to artificial selection. *Genet. Res.* 61:107–116.

López, M.A. and López-Fanjul, C. (1993b) Spontaneous mutation for a quantitative trait in *Drosophila melanogaster*. II. Distribution of mutant effects on the trait and fitness. *Genet. Res.* 61:117–126.

Ludwig, D. (1974) *Stochastic Population Theories*. Lecture Notes in Biomathematics 3. Springer-Verlag, Berlin.

Ludwig, D. (1976) A singular perturbation problem in the theory of population extinction. *SIAM-AMS Proceedings of Symposia on Applied Mathematics* 10:87–104.

Ludwig, D. (1996) The distribution of population survival times. *Am. Nat.* 147:506–526.

Lyman, R.F., Lawrence, F., Nuzhdin, S.V. and Mackay, T.F.C. (1996) Effects of single *P* element insertions on bristle number and viability in *Drosophila melanogaster*. *Genetics* 143:277–292.

Lynch, M. (1988) The rate of polygenic mutation. *Genet. Res.* 51:137–148.

Lynch, M. (1996) A quantitative-genetic perspective on conservation issues. *In*: J. Avise and J. Hamrick (ed): *Conservation Genetics: Case Histories from Nature*. New York: Chapman & Hall, New York, pp. 471–501.

Lynch, M. and Gabriel, W. (1990) Mutation load and the survival of small populations. *Evolution* 44:1725–1737.

Lynch, M. and Hill, W.G. (1986) Phenotypic evolution by neutral mutation. *Evolution* 40:915–935.

Lynch, M. and Lande, R. (1993) Evolution and extinction in response to environmental change. *In*: P.M. Kareiva, J.G. Kingsolver and R.B. Huey (eds): *Biotic Interactions and Global Change*. Sinauer, Sunderland, MA, pp. 234–250.

Lynch, M., Gabriel, W. and Wood, A.M. (1991) Adaptive and demographic responses of plankton populations to environmental change. *Limnol. Oceanogr.* 36:1301–1312.

Lynch, M., Bürger, R., Butcher, D. and Gabriel, W. (1993) The mutational meltdown in asexual populations. *J. Heredity* 84:339–344.

Lynch, M., Conery, J. and Bürger, R. (1995a) Mutational meltdowns in sexual populations. *Evolution* 49:1067–1080.

Lynch, M., Conery, J. and Bürger, R. (1995b) Mutation accumulation and the extinction of small populations. *Am. Natur.* 146:489–518.

MacArthur, R.H. and Wilson, E.O. (1967) *The Theory of Island Biogeography*. Princeton University Press, Princeton.

Mackay, T.F.C., Lyman, R. and Jackson, M.S. (1992a) Effects of *P* elements on quantitative traits in *Drosophila melanogaster*. *Genetics* 130:315–332.

Mackay, T.F.C., Lyman, R., Jackson, M.S., Terzian, C. and Hill, W.G. (1992b) Polygenic mutation in *Drosophila melanogaster*: Estimates from divergence among inbred strains. *Evolution* 46:300–316.

Mackay, T.F.C., Lyman, R.F. and Hill, W.G. (1995) Polygenic mutation in *Drosophila melanogaster*: Non-linear divergence among unselected strains. *Genetics* 139:849–859.

May, R.M. (1981) Models for single populations: *In*: R.M. May (ed.) *Theoretical Ecology*. Blackwell, Oxford, pp. 5–29.

May, R.M. and Oster, G.F. (1976) Bifurcations and dynamic complexity in simple ecological models. *Am. Nat.* 110:573–599.

Maynard Smith, J. (1978) *The Evolution of Sex*. Cambridge University Press, Cambridge.

Melzer, A.L. and Koeslag, J.H. (1991) Mutations do not accumulate in asexual isolates capable of growth and extinction – Muller's ratchet reexamined. *Evolution* 45:649–655.

Mousseau, T.A. and Roff, D.A. (1987) Natural selection and the heritability of fitness components. *Heredity* 59:181–197.

Mukai, T. (1965) The genetic structure of natural populations of *Drosophila melanogaster*. I. Spontaneous mutation rate of polygenes controlling viability. *Genetics* 50:1–19.

Mukai, T. (1979) Polygenic mutations. *In*: J.N. Thompson, Jr. and J.M. Thoday (eds): *Quantitative Genetic Variation*. Academic Press, New York, pp. 177–196.

National Research Council (1995) *Science and the Endangered Species Act*. National Academy Press, Washington, DC.

Nobile, A.G., Ricciardi, L.M. and Sacerdote, L. (1985) Exponential trends of first-passage time densities for a class of diffusion processes with steady-state distributions. *J. Applied Prob.* 22:611–618.

Nuzhdin, S.V., Fry, J.D. and Mackay, T.F.C. (1995) Polygenic mutation in *Drosophila melanogaster*: The causal relationship of bristle number to fitness. *Genetics* 139:861–872.

Pease, C.M., Lande, R. and Bull, J.J. (1989) A model of population growth, dispersal and evolution in a changing environment. *Ecology* 70:1657–1664.

Peck, R.R., Barreau, G. and Heath, S.C. (1997) Imperfect genes, Fisherian mutation and the evolution of sex. *Genetics* 145:1171–1199.

Richter-Dyn, N. and Goel, N.S. (1972) On the extinction of a colonizing species. *Theor. Pop. Biol.* 3:406–433.

Roff, D.A. and Mousseau, T.A. (1987) Quantitative genetics and fitness: Lessons from *Drosophila. Heredity* 58:181–197.

Schultz, S.T. and Lynch, M. (1997) Mutation and extinction: The role of variable mutational effects, synergistic epistasis, beneficial mutations, and degree of outcrossing. *Evolution; in press.*

Schultz, S.T., Willis, J. and Lynch, M. (1997) Spontaneous deleterious mutation in *Arabidopsis. Science; in press.*

Shaffer, M.L. (1981) Minimum population sizes for species conservation. *BioScience* 31: 131–134.

Shaffer, M.L. (1987) Minimum viable populations: Coping with uncertainty. *In*: M.E. Soulé (ed.): *Viable Populations for Conservation*. Cambridge University Press, Cambridge, pp. 69–86.

Slatkin, M. and Lande, R. (1976) Niche width in a fluctuating environment-density independent model. *Am. Nat.* 110:31–55.

Stephan, W., Chao, L. and Smale, J.G. (1993) The advance of Muller's ratchet in a haploid asexual population: Approximate solutions based on diffusion theory. *Genet. Res.* 61: 225–231.

Turelli, M. (1977) Random environments and stochastic calculus. *Theor. Pop. Biol.* 12:140–178.

Turelli, M. (1984) Heritable genetic variation via mutation-selection balance: Lerch's zeta meets the abdominal bristle. *Theor. Pop. Biol.* 25:138–193.

Turelli, M. (1985) Effects of pleiotropy concerning mutation-selection balance for polygenic traits. *Genetics* 111:165–195.

Wagner, G.P. (1988) The influence of variation and of developmental constraints on the rate of multivariate phenotypic evolution. *J. Evol. Biol.* 1:45–66.

Wagner, G.P. (1989) Multivariate mutation-selection balance with constrained pleiotropic effects. *Genetics* 122:223–234.

Wagner, G.P. and Gabriel, W. (1990) Quantitative variation in finite parthenogenetic populations: What stops Muller's ratchet in the absence of recombination? *Evolution* 44:715–731.

Weis, A.E., Abrahamson, W.G. and Anderson, M.C. (1992) Variable selection in *Eurosta's* gall size. I. The extent and nature of variation in phenotypic selection. *Evolution* 46:1674–1697.

Yoo, B.H. (1980) Long-term selection for a quantitative character in large replicate populations of *Drosophila melanogaster*. I. Response to selection. *Genet. Res.* 35:1–17. II. Lethals and visible mutants with large effects. Ibid., pp. 19–31.

Zeng, Z.-B. and Cockerham, C.C. (1991) Variance of neutral genetic variation within and between populations for a quantitative character. *Genetics* 129:535–553.

Environmental Stress, Adaptation and Evolution
ed. by R. Bijlsma and V. Loeschcke
© 1997 Birkhäuser Verlag Basel/Switzerland

# Environmental stress and evolution:
# A theoretical study

Lev A. Zhivotovsky

*Vavilov Institute of General Genetics, Russian Academy of Sciences, 3 Gubkin St.,
Moscow 117809, Russia*

*Summary.* A haploid population subject to recurrent deleterious mutations and two environments that provide two different profiles of selection coefficients over loci are modeled. The population is supposed to inhabit one "home" environment where it evolves the corresponding genetic constitution. One generation of the population is then exposed to the second, "foreign" environment. The decline in the mean population fitness is considered as a measure of stress in the population caused by the foreign environment. I define the relative strength of stress as 1 minus the ratio of the mean fitnesses in the foreign and home environments, and give the corresponding analytical expression. The stress strength is composed of three different contributors: the environmental component of stress, which is determined by purely external, non-genetic factors of the foreign environment (its carrying capacity), and two genetic components. The latter consists of the environment × genetic component, caused by the direct influence of the foreign environment on selection coefficients, and of the evolutionary component that is due to the adaptation of the population to the home environment. Among others, it is shown that even if the home and foreign environments were equivalent, so that both the environmental and environmental × genetic components of stress were absent, stress would still occur in the foreign environment due to the evolutionary component. The model also predicts that stressful foreign environments cause an increase in the genotypic variance of fitness. Some other features of population variability in stressful environments are discussed. A general conclusion that can be drawn from this model is that a certain environment may be claimed as stressful only if considered with respect to both a given population and the environment in which the population has evolved.

## Introduction

How adverse environmental and ecological conditions influence populations has recently appeared to be an important evolutionary problem. Hoffmann and Parsons (1991) have summarized literature on stress, and concluded that stress may challenge population processes and thus have significant evolutionary consequences. Nevertheless, there is still little known about the population genetics mechanisms underlying the evolution of stress. Thus it seems useful to develop models for predicting evolutionarily developed responses to stressful environments.

To understand the relationship between the variability of organisms and the variability of the environment in which they live has always been a central goal of evolutionary biology (Schmalhausen, 1938, 1946; Gause, 1947; Waddington, 1959; Robertson, 1960; Bradshaw, 1965; Levins, 1968; and many others). Some mathematical models have been also developed to analzye how environmental heterogeneity affects phenotypic variability (Via and Lande, 1985; Lynch and Gabriel, 1987; Houston and McNamara, 1992; Gomulkiewicz and Kirkpatrick, 1992; Kawecki and Stearns, 1993; Gavrilets and Scheiner, 1993; Van Tienderen, 1994; de Jong, 1995; Zhivotovsky et al., 1996a; and others).

The same models might be applied to include stress as a component of a heterogeneous environment. However, stress involves specific feature that should be incorporated in the corresponding model. Indeed, stress changes individual physiological reactions, which, in turn, may lead to a lower level of adaptive traits, say, to worse survival, reduced fecundity and lesser mating success, as in *Drosophila* (Krebs and Loeschcke, 1994), or to failure to spermiate and ovulate, impairment of spawning, and timing of reproduction, as in migratory fish (Donaldson, 1990). Moreover, the reduction in fitness varies among genotypes. For example, in early work on *Drosophila* it was already shown that different genotypes may have different fitness and change their fitness ranking under environmental changes (Timofeeff-Ressovsky, 1933, 1935; Dobzhansky and Spasky, 1944; Dubinin and Tiniakov, 1945; Dobzhansky et al., 1955). Thus, to study stress in populations, we need to know the behavior of adaptive traits under environmental changes.

It has been recently shown in a model of the fitness evolution under heterogeneous environmental conditions that in rare and poor environments populations show lower mean fitness as compared with more frequent and productive environments (Zhivotovsky et al., 1996c). The purpose of this study is to develop further this idea and consider a population that has evolved in a "home" environment under selection against recurrent deleterious mutations at multiple loci and then been tested in an unusual, "foreign" environment. That harmful mutations may be an important evolutionary force was pointed out by Fisher (1930) in his theory of dominance, and recently analyzed theoretically for the problems of evolution of genetic recombination and sex (Kondrashov, 1984; Charlesworth, 1990). Our study extends this to the problem of evolution of stress and shows that foreign environments may occur to be stressful in terms of the mean population fitnesses.

## The model

### Population in a home environment

Consider a randomly mating haploid population with non-overlapping generations that inhabits a "home" environment $E_h$. Fitness (an adaptive trait) is supposed to be determined by $n$ loci. Each locus $i$ has a frequent "wild" allele $A_i$. The relative fitness of the genotype carrying all allels $A$, which is obtained by averaging individual fitnesses, is supposed to be normalized to 1, the baseline fitness level in the home environment. Let the alleles be subject to mutation with rate $\mu$, so that at each locus $i$ the allele $A_i$ may mutate to a deleterious allele $a_i$, decreasing the fitness by a small positive value $h_i$, the *relative selection coefficient*. Although non-allelic mutations may epistatically interact and show synergism (Mukai, 1969), it is not known whether this synergism is a common phenomenon. Accordingly, we simplify the model and consider a case of independent additive influence of mutations on genotypic fitnesses. For example, if a genotype carries the mutant alleles at the first three loci, with the rest of the loci containing the alleles $A$, its fitness is $1 - \sum_{i=1}^{3} h_i$. If $p_i$ is the frequency of the mutant allele $a_i$ at locus $i$, the mean fitness of the population is $1 - \sum_{i=1}^{n} h_i p_i$. The sequence of the (relative) selection coefficients ordered over loci, $\{h_1, h_2, \ldots, h_n\}$, will be called *selection profile*.

Each $h_i$ may be viewed as a realization of a continuous random value. Very little is known about the distribution of selection coefficients, although the majority of new mutations at structural loci, and polygenic mutations as well, are deleterious (Kimura, 1983; Mukai, 1969). Nevertheless, data obtained in early work on viability mutations in *Drosophila melanogaster* (Timofeeff-Ressovsky, 1935) allow the conclusion that such a distribution should have a maximum at some intermediate values of $h$ and disappear with extremely small and large $h$'s (if we neglect supervital mutations, which are much rarer than deleterious mutations).

To model a selection profile for a home population, we introduce an assumption according to which selection coefficient $h$ follows a gamma distribution over loci with positive mean $\bar{h}$ and variance $\sigma_h^2$; we also require the number of loci to be large (see Appendix A for details). Different mutations may have different influences on fitness. In the interest of simplicity the model does not include differences among allelic mutations but focuses on the variability of selection coefficients over loci (quantitatively, the principal results of this study remain the same if variation in selection coefficients among allelic mutations is also included, though the formal analysis requires an analytical technique dealing with multilocus, multiallele dynamics which is not yet properly developed). Additionally, we assume that the selection coefficients do not vary too widely among the loci. To be exact, we require that the variation coefficient, $\sigma_h/\bar{h}$, be less than 1; denote $C_h \overset{df}{=} (\sigma_h/\bar{h})^2$.

It is natural to assume that mutation rates are much smaller than the corresponding coefficients of selection against mutations (Mukai, 1964, 1969). If so, the frequencies of $a$ alleles do not deviate far from 0. Hence, we may neglect mutations of $a_i$ to $A_i$, and thus consider only mutations of the frequent wild alleles $A_i$ to the rare deleterious alleles $a_i$. Denote by $U = \sum \mu$ the total mutation rate at the loci, i.e. the average number of newly arisen deleterious mutations per genome per generation.

Let the population evolve sufficiently long in a home environment $E_h$. Assuming that genetic linkage between the loci is not tight (large recombination rates relative to small selection coefficients), and thus neglecting linkage disequilibrium, it follows that the frequency of the mutant allele at each locus $i$, $p_i$, ultimately converges to

$$\hat{p}_i \approx \frac{\mu}{h_i(1 + U)} \tag{1}$$

(Zhivotovsky et al., 1996c).

Let $\bar{w}_h$ and $V_h$ be the mean fitness and the variance of genotypic fitnesses in the population attained in the mutation-selection equilibrium. Obviously,

$$\bar{w}_h \overset{df}{=} 1 - \sum_i h_i \hat{p}_i \approx \frac{1}{1 + U}. \tag{2}$$

The variance of genotypic fitnesses, according to the theory of quantitative genetics (Falconer, 1989), is $\sum_i \hat{p}_i(1 - \hat{p}_i)h_i^2$. From (1) and Appendix A,

$$V_h \approx \bar{h}\,\frac{U}{1 + U}. \tag{3}$$

Since the fitness of a certain genotype (and its components such as via- bility, mating success, fecundity etc., as well) is the average taken over individuals with the same genotype, $V_h$ is actually the genotypic variance of a fitness trait in the home environment.

### Population in a foreign environment

Let one generation of the population experience a different, "foreign" en- vironment $E_f$ that appears so rarely that it does not significantly influence the equilibrium allele frequencies or did not appear at all in the population history. A foreign environment may also be viewed as the conditions of a laboratory experiment to which the generation (or a sample from) is exposed. The foreign environment may modify the expression of the mutant alleles and, thus, provide a different profile of the relative selection coefficients over the same set of loci, $\{f_1, f_2, \ldots, f_n\}$, with mean $\bar{f}$ and variance $\sigma_f^2$.

The difference between $f_i$ and $h_i$ may vary over loci, since different mutations may differently react on changing environmental conditions. To incorporate this in the model, I consider a linear relationship between them:

$$f_i = \theta h_i + \xi_i, \tag{4}$$

where $\xi$'s are statistically independent of $h$'s with mean $\bar{\xi}$ and variance $\sigma_\xi^2$. Here no other property of the distribution of $f_i$'s is required. Moreover, unlike the $h_i$'s, the $f_i$'s are allowed to be positive, i.e. in the foreign environment the mutant alleles $a$ may be more favorable than the wild alleles $A$.

The parameter $\theta$ (together with $\bar{\xi}$ and $\sigma_\xi^2$) shows how environments $E_h$ and $E_f$ are close to one another in terms of the similarity of the corresponding selection profiles. In particular, if $\theta = 0$, the foreign environment $E_f$ differs from $E_h$ with no correlation between their selection profiles. In the other extreme, with all $\xi_i$'s zeroes (i.e. if $\bar{\xi} = 0$ and $\sigma_\xi^2 = 0$), the selection profiles are completely correlated and differ from one another only by the factor $\theta$, which in this case indicates whether selection in the foreign environment is stronger ($\theta > 1$) or weaker ($\theta < 1$, if positive) than that in the home environment. One can introduce a usual correlation coefficient, $\rho$, between the selection profiles in the home and foreign environments. From (4),

$$\bar{f} = \theta \bar{h} + \bar{\xi}, \quad \sigma_f^2 = \theta^2 \sigma_h^2 + \sigma_\xi^2, \quad \rho = \theta \frac{\sigma_h}{\sigma_f} . \tag{5}$$

Note that $\rho$ is an arbitrary value from the range $[-1, 1]$.

Besides genetic factors, given by the selection profiles, there may be non-genetic factors that influence the mean population fitnesses in both environments. Say the foreign environment causes an increase in mortality by the same value for each genotype. Such a genotype-independent influence of environmental conditions on absolute fitnesses can be introduced by a factor $\varphi$, which is less than 1 if the environment $E_f$ is less favorable than $E_h$, and vice versa. Hence, in the foreign environment, the fitness of the genotype with all alleles $A$ should be $\varphi$, and hence $\varphi$ plays the role of a scaling factor as a baseline fitness level in the foreign environment. For example, the genotype with alleles $a$ at the first three loci has fitness $\varphi(1 - \sum_{i=1}^3 f_i)$. The factor $\varphi$ may be interpreted as the *relative carrying capacity* of a foreign environment with respect to the carrying capacity of the home environment.

By definition, the mean fitness of the population in the foreign environment is

$$\bar{w}_j \overset{df}{=} \varphi \left( 1 - \sum_{i=1}^n f_i p_i \right) \approx \varphi \left( 1 - \frac{\mu}{1 + U} \sum_{i=1}^n \frac{f_i}{h_i} \right) .$$

Use Eqs (4), (5) and (14) to obtain

$$\bar{w}_f \approx \varphi \bar{w}_h \left( 1 - \frac{U}{1 - C_h} \left[ \left( \frac{\bar{f}}{\bar{h}} - 1 \right) + (1 - \theta) C_h \right] \right) . \tag{6}$$

The genotypic variance of fitnesses is $V_f \overset{df}{=} \varphi^2 \sum_i \hat{p}_i (1 - \hat{p}_i) f_i^2$. After some algebra using Eqs (1), (4), (5) and (14), we get

$$V_f \approx \varphi^2 \frac{V_h}{1 - C_h} \left[ \frac{\bar{f}^2}{\bar{h}^2} (1 + C_f) - 2\theta C_h \frac{\bar{f}}{\bar{h}} \right]. \tag{7}$$

Also, one can easily calculate the covariance between genotypic fitnesses in foreign and home environments, $Cov_{fh} \approx \varphi \bar{f} U/(1 + U)$. Thus, the corresponding coefficient of regression of genotypic fitness in the foreign environments on that in the home environment, $R_{fh} \overset{df}{=} Cov_{fh}/V_h$, is

$$R_{fh}^2 \approx \varphi \frac{\bar{f}}{\bar{h}} . \tag{8}$$

**Note:** If mutation rates vary among loci, $\mu_i$, the results remain the same with some additional minor assumptions and with the difference that each locus $i$ must be taken with the weight $\mu_i/U$, where $U = \sum_{i=1}^{n} \mu_i$. Particularly,

$$\bar{h} = \sum_{i=1}^{n} \frac{\mu_i}{U} h_i, \quad \bar{f} = \sum_{i=1}^{n} \frac{\mu_i}{U} f_i ,$$

$$\sigma_h^2 = \sum_i \frac{\mu_i}{U} h_i^2 - \bar{h}^2, \quad \sigma_f^2 = \sum_i \frac{\mu_i}{U} f_i^2 - \bar{f}^2 ,$$

$$\rho = \sum_i \frac{\mu_i}{U} f_i h_i - \bar{f} \bar{h} ,$$

## Results and discussion

The principal goal of this theoretical study is to understand which environment can be claimed to be stressful. For this, I consider the mean fitness as a basic measure of population success, and therefore compare its values, $\bar{w}_h$ and $\bar{w}_f$, realized in two environments, $E_h$ and $E_f$, respectively.

For a given population, the environment $E_f$ is called stressful with respect to environment $E_h$ if the mean fitness realized by the population in $E_f$ is less than that in $E_h$, $\bar{w}_f < \bar{w}_h$, and vice versa. I introduce the *relative strength of stress*.

$$S \overset{df}{=} 1 - \frac{\bar{w}_f}{\bar{w}_h} , \tag{9}$$

analogously to the definition of the selection barrier between two populations, based on the ratio of the mean fitnesses (Zhivotovsky and

Christiansen, 1995). The absence of stress corresponds to $S = 0$; the maximal value $S = 1$ corresponds to zero fitness in environment $E_f$. If environment $E_f$ is less stressful than $E_h$, $S$ becomes negative with minimal limit $-\infty$. Because of this, it may also be useful to introduce a slightly different, although closely related, measure of stress strength, namely the relative difference $S_r \overset{df}{=} (\bar{w}_h - \bar{w}_f)/(\bar{w}_h + \bar{w}_f) = S/(2-S)$; this measure is symmetric with extremes $-1$ and $1$ and equals $0$ in the absence of stress. Another related measure, with logarithmic transformation, may also be considered, $S_l \overset{df}{=} \ln(\bar{w}_h/\bar{w}_f) = -\ln(1-S)$, which limits are $-\infty$ and $\infty$, and zero stress corresponds to $S_l = 0$. However, for the purposes of this paper, it is not important which of these measures is taken, and I use the simpler index $S$, although the other ones may statistically be more appropriate in applications.

How an environment becomes stressful is modeled here by specifying the home and foreign environments. These environments provide different profiles of the coefficients of selection against deleterious mutations at $n$ loci: $\{h_1, h_2, \ldots, h_n\}$ and $\{f_1, f_2, \ldots, f_n\}$, respectively. The principal difference between the environments is that *the population has evolved in the home environment*, while the foreign environment has not contributed to the evolution of the population. In other words, the population has been evolutionarily adapted to the conditions of environment $E_h$; by contrast, $E_f$ is an environment to which the population previously has not been subjected or has appeared too rarely to be evolutionarily significant. To emphasize that the population has been sufficiently long evolving in the home environment, it is assumed to attain its mutation-drift equilibrium. Since the environment may greatly influence absolute fitnesses, not only relative genotypic fitnesses, I also incorporate in the model a possibility that the environments have different abilities in supporting populations: the carrying capacity of the home environment is supposed to be normalized to 1; that of the foreign environment is denoted by $\varphi$ and may be greater or less than 1.

As follows from Eqs (2) and (6), the stress strength is

$$S \approx 1 - \varphi\left(1 - \frac{U}{1 - C_h}\left[\left(\frac{\bar{f}}{\bar{h}} - 1\right) + \left(1 - \rho\frac{\sigma_f}{\sigma_h}\right)C_h\right]\right). \tag{10}$$

One can see that $S$ is composed of environmental and genetic components. The *environmental component* is due to the difference in the carrying capacities of home and foreign environments (1 and $\varphi$, respectively) that are assumed to be independent of genotype, i.e. caused by, say, non-specific mortality or fecundity that depends on only one environment.

The genetic component consists of two parts. The first is determined by the direct influence of the foreign environment on selection coefficients. We see from Eq. (10) that the larger the average within-locus selection coefficient in the foreign environment compared with that in the home environment, $\bar{f} > \bar{h}$, the greater the stress, and we call this direct influence of the foreign environment on selection coefficients the *environmental × genetic component* of stress.

To emphasize the nature of the second genetic component of stress, let us consider a special case of the equivalence of the foreign and home environments, namely equal carrying capacities ($\varphi = 1$) and the same values of the statistical moments of selection coefficients calculated over loci ($\bar{f} = \bar{h}$, $\sigma_f^2 = \sigma_h^2$ etc.). In other words, in this case both the environmental and environmental × genetic components of stress are absent. The only difference between the equivalent foreign and home environments is that the selection profiles do not coincide, although they can be obtained from one another by permutations (in this case, the correlation between the profiles, $\rho$, is less than 1). It follows from Eq. (10) that the foreign environment turns out to be stressful for the population as compared with the home environment despite their equivalence, with strength

$$ S \approx (1-\rho) \, \frac{UC_h}{1-C_h} \; . \tag{11} $$

This formula shows that the lesser the correlation between the selection proviles and the higher the level of variation in selection coefficients calculated over loci in the home environment, the greater the stress displayed in the foreign environment. Formula (11) demonstrates the presence of the second genetic contributor to stress, here called the *evolutionary component*. (Note that the genetic components and the environmental component interact with each other in a complex way, and here I will not tackle the problem of their estimates. Nevertheless, Eq. (8) shows that there is a possibility based on the regression analysis to separate the first two contributors from the evolutionary component.)

The principal difference between the environmental × genetic and evolutionary components of stress occurring in a foreign environment is that the former one is mainly due to the direct influence of the foreign environment on the penetration and expressivity of harmful mutations, while the latter component is due to the properties of the home environment which the population has evolutionarily adapted to.

Consider, for instance, the following extreme case. Let a foreign environment be more productive and softer for a given population than the home environment, i.e. we assume it has a larger carrying capacity ($\varphi > 1$, which is due to the environmental component) or/and provides less average selection pressure ($\bar{f} > \bar{h}$, which is due to the environmental-genetic component). It is interesting that even in this case the foreign environment may be still stressful for the population compared with the home environment, which may happen if the environments loosely correlate to each other and

$C_h$ is large. If stress occurs, it will be due to the evolutionary component. The condition for this immediately follows from Eq. (10) with $S > 0$.

That the foreign environment $E_f$ gives rise to stress in a population already adapted to the home environment $E_h$ is due to the low frequency of the genotypes that might have higher fitness in the foreign environment but were unfavorable in the home environment. If the population evolved in $E_f$, then the previously unfavorable genotypes would be selected for, and thus, in turn, environment $E_h$ would become stressful for this population, as compared with $E_f$.

So stress in a given population, caused by an unusual environment, can be properly evaluated only if compared with the reaction of the population on the environment in which it has evolved. I conclude that *the evolutionary component of stress occurs as a side-effect whereby certain genotypes selected to allow a population to adapt to home environments perform worse in foreign environments too rare to have been a selective force.*

Another important property following from the model is that the genotypic variance of fitnesses in a stressful foreign environment is higher than that in the home environment. Indeed, in the case of the equivalence of the environments ($\varphi = 1$, $\bar{f} = \bar{h}$, $\sigma_f = \sigma_h$ etc.), it follows from Eqs (3) and (7) that the ratio of the both variances becomes

$$\frac{V_f}{V_h} \approx \frac{1 + C_h(1 - 2\rho)}{1 - C_h} ,$$

(12)

which always exceeds 1. In a general case of different carrying capacities of the environments compared ($\varphi \neq 1$), the mean and the variance of genotypic fitnesses in each environment depend linearly and quadratically, respectively, on the carrying capacity (see Eqs (6) and (7)). Because the variation coefficients, $CV_h = \sqrt{V_h}/\bar{w}_h$ and $CV_f = \sqrt{V_f}/\bar{w}_f$ are free of these unknown parameters, they are expected to reflect the increase in variability of genotypic fitnesses in stressful environments more diretly than the corresponding variances.

It follows from Eqs (10) and (12) that the more a foreign environment deviates from the home environment as measured in the values of the corresponding parameters, the stronger the stress, and we may expect that in outlying inexperienced environments populations will show the stress properties predicted by the model. We have already mentioned that the strength of stress occurring in a foreign environment increases with the increase in variation of selection coefficients in the home environment, $C_h$; the genotypic variance $V_f$ also increases with $C_h$. Therefore, the presence in a population of both slightly deleterious mutations and mutations with significant damage leads to stronger stress in populations if subject to unusual environmental conditions. Also, the increase in the total mutation rate per genome in a home environment, $U$, causes stronger stress in a foreign environment. Note that although formulae (10), (11) and (12) evaluate the foreign environment with respect to the home environment, the latter is actually not known. In applications we may consider one of the environments as a reference environment (presumably, closer to the home environment than the others) and then use Eqs (6), (7) and (8) to compare the environments.

We emphasize that the results of this study are relevant to the genotypic variances, $V_h$ and $V_f$, and do not refer to the random variances which describe phenotypic variation among individuals of the same genotype. In modeling the evolution of phenotypic traits under weak stabilizing selection with Lerner's homeostasis, it was concluded that both genotypic and random variances may evolve to a lower level (Zhivotovsky et al., 1996a), and thus are expected to be larger under inexperienced environmental conditions. However, selection on adaptive traits includes a directional component and thus needs to be analyzed in more detail. Because random variance may increase in a changing environment (Cohen, 1993), how the random variances of adaptive traits behave under stress is still unknown.

We may apply the results of this study to adaptive traits such as viability, fecundity and so on. For example, Bennet and Lenski (1993) studied the response of an *Escherichia coli* population to long-term thermal selection and found that the maximal mean fitness occurred at the temperature at which this population was selected ("home environment" in the terminology of this chapter). Additionally, the population contained a line that deviated significantly from the rest of the lines, and this deviation was smaller at the temperature at which this population was selected. This corresponds well to the results of our model. Note that our theoretical predictions concern the total genotypic fitness and may not be specific about separate fitness components. For example, Gebhardt and Stearns (1988) found that low temperatures negatively influence the developmental rate of *Drosophila melanogaster*, while high temperatures suppress viability; considered together, these traits perform better at an intermediate temperature, which plays a role of the home environment for this species with respect to this particular environmental parameter.

Quantitative traits, other than adaptive, are not fully described by the model used in this study, although some findings remain valid for a trait under stabilizing selection (Zhivotovsky et al., 1996a). Nevertheless, the conclusions may be relevant to some traits directly related to fitness, such as body size in *Drosophila*. Experimental observations are in qualitative agreement with our theoretical predictions. For example, Robertson (1960) found decreased mean and increased variance of body size of *D. melanogaster* on deficient diets compared with those on a standard live yeast medium. Tantawy and Mallah (1961) and Zhivotovsky et al. (1996b) found that the mean and variance of wing size of *D. melanogaster* and *D. simulans* attained their minimum and maximum, respectively, at extremely low and high temperatures, compared with those at intermediate temperature. Other examples can be found in Hoffmann and Parsons (1991). However, these experiments do not say which of the components of phenotypic variation (genetic or random) increased under stress.

Our model assumes that a population evolves in a homogeneous "home" environment and that a foreign environment plays no role in the evolution. As to a heterogeneous home environment represented by several spatially distributed "niches", it follows from Zhivotovsky et al., 1996c) that each niche contributes to the population evolution proportionally to its weight, obtained by multiplying the frequency of the niche and its productivity. Thus, a common and productive environment (niche) is important, and a rare and poor environment would seem relatively unimportant, for evolution of fitness. A similar result was obtained for a model of the quantitative trait evolution in a randomly mating population of diploids (Zhivotovsky et al., 1996a), which agrees well with the conclusion by Houston and McNamara (1992) and Kawecki and Stearns (1993) regarding life history-related traits in clonal organisms. So the results of the model presented here seem to hold for populations that evolve in a heterogeneous environment, namely, that minor niches, with small weights, serve as foreign environments and thus are stressful for populations. They provide a decrease in mean fitness and an increase in genotypic variance of fitnesses as compared with those in the major niches which have large weights and, hence, serve as home environments.

Although the model used in this study is rather simple, it highlights some interesting properties of stress in evolving populations. The approach given can be developed further to involve other evolutionary problems. For example, in modeling evolution in heterogeneous environments, it is important to consider a system of two or more subpopulations adapted to their own environments and connected by migration flows. Since the environment of each subpopulation may evolutionarily become stressful for the rest of the subpopulations, a kind of genetic barrier arises between each pair of subpopulations. This may lead to the appearance of hybrid zones (Barton and Bengtsson, 1986) and provide lower fitness among the hybrids and other descendants of immigrants (Zhivotovsky and Christiansen, 1995). Also, evolution of reaction norms requires modifiyng phenotypic traits by environment via regulatory genes (Schlichting and Pigliucci, 1995), and thus the latter may be important in the context of stress evolution. Additionally, modeling shows that individual behavior may evolve to avoid uncomfortable environments or develop the mechanisms of individual tolerance to them (Zhivotovsky et al., 1996a), and thus the environment component, $\varphi$, if small, may increase during evolution if the environments occur often or even very rarely if extremely damageable; this is also important to analyze theoretically. Alleles with polymorphic frequencies, not only deleterious mutations, may induce additional stress and thus need to be incorporated into a more general model. Other genetic parameters, such as mutation rates and recombination rates, may be subject to direct or indirect selection and increase under stress (Hoffmann and Parsons, 1991), and their evolution may be related to the occurrence of rare environmental and ecological conditions.

*Acknowledgments*
Research supported in part by RFBR grant 95-04-11445 and NIH grant 1 R03 TW00491-01.

## References

Barton, N.H. and Bengtsson, B.O. (1986) The barrier to gene exchange between hybridizing populations. *Heredity* 56:357–376.

Bennett, A.F. and Lenski R.E. (1993) Evolutionary adaptation to temperature. II. Thermal niches of experimental lines of *Escherichia coli*. *Evolution* 47:1–12.

Bradshaw, A.D. (1965) Evolutionary significance of phenotypic plasticity in plants. *Advances in Genetics* 13:115–155.

Charlesworth, B. (1990) Mutation-selection balance and the evolutionary advantage of sex and recombination. *Genet. Res.* 55:199–221.

Cohen, D. (1993) Fitness in random environments. *In*: J. Yoshimura and C.W. Clark (eds): *Adaptation in Stochastic Environments*. Springer-Verlag, New York, pp. 8–25.

de Jong, G. (1995) Phenotypic plasticity as a product of selection in a variable environment. *Am. Nat.* 145:493–512.

Dobzhansky, Th. and Spassky, B. (1994) Genetics of natural populations. XI. Manifestation of genetic variants in *Drosophila pseudoobscura* in different environment. *Genetics* 29:270–290.

Dobzhansky, Th., Pavlovsky, O., Spassky, B. and Spassky, N. (1955) Genetics of natural populations. XXIII. Biological role of deleterious recessives in populations of *Drosophila pseudoobscura*. *Genetics* 40:781–796.

Donaldson, E.M. (1990) Reproductive indices as measures of the effects of environmental stressors in fish. *In*: S.M. Adams (ed.): *Biological Indicators of Stress in Fish American Fisheries Society Symposium*. American Fisheries Society, Bethesda, Md., pp. 109–122.

Dubinin, N.P. and Tiniakov, G.G. (1945) Seasonal cycles and the concentrations of inversions in populations of *Drosophila funebris*. *Am. Nat.* 79:570–572.

Falconer, D.S. (1989) *Introduction to Quantitative Genetics*, Third Edition. Longman, London.

Fisher, R.A. (1930) *The Genetical Theory of Natural Selection*. Oxford University Press, Oxford.

Gause, G.F. (1947) Problems of evolution. *Trans. Conn. Acad. Sci.* 37:17–68.

Gavrilets, S. and Scheiner, S.M. (1993) The genetics of phenotypic plasticity. V. Evolution of reaction norm shape. *J. Evol. Biol.* 1:335–354.

Gebhardt, M.D. and Stearns, S.C. (1988) Reaction norms for development time and weight at eclosion in *Drosophila mercatorum*. *J. Evol. Biol.* 1:335–354.

Gomulkiewicz, R. and Kirkpatrick, M. (1992) Quantitative genetics and the evolution of reaction norms. *Evolution* 46:390–411.

Hoffmann, A.A. and Parsons, P.A. (1991) *Evolutionary Genetics and Environmental Stress*. Oxford University Press, Oxford.

Houston, A.I. and McNamara, J.M. (1992) Phenotypic plasticity as a state dependent life-history decision. *Evol. Ecol.* 6:243–253.

Kawecki, T.J. and Stearns, S.C. (1993) The evolution of life histories in spatially heterogeneous environments: Optimal reaction norms revisited. *Evol. Ecol.* 7:155–174.

Kimura, M. (1983) *The Neutral Theory of Molecular Evolution*. Cambridge University Press, Cambridge.

Kondrashow, A.S. (1984) Deleterious mutations as an evolutionary factor. I. The advantage of recombination. *Genet. Res.* 44:199–217.

Krebs, R.A. and Loeschcke, V. (1994) Effects of exposure to short-term heat stress on fitness components in *Drosophila melanogaster*. *J. Evol. Biol.* 7:39–49.

Levins, R. (1968) *Evolution in Changing Environments*. Princeton University Press, Princeton.

Lynch, M. and Gabriel, W. (1987) Environmental tolerance. *Am. Nat.* 129:283–303.

Mukai, T. (1964) The genetic structure of natural populations of *Drosophila melanogaster*. I. Spontaneous mutation rate of polygenes controlling viability. *Genetics* 50 : 1–19.

Mukai, T. (1969) The genetic structure of natural populations of *Drosophila melanogaster*. VIII. Synergistic interactions of spontaneous mutant polygenes controlling viability. *Genetics* 61 : 149–161.

Robertson, F.W. (1960) The ecological genetics of growth in *Drosophila*. 1. Body size and development time on different diets. *Genet. Res.* 1 : 288–304.

Schlichting, C.D. and Pigliucci, M. (1995) Gene regulation, quantitative genetics and the evolution of reaction norms. *Evol. Ecol.* 9 : 154–168.

Schmalhausen, I.I. (1938) *The Organism as a Whole in Development and Evolution*. Izd. Akad. Nauk SSSR, Moscow (in Russian).

Schmalhausen, I.I. (1946) *Factors of Evolution (The Theory of Stabilizing Selection)*. Izd. Akad. Nauk SSSR. Moscow (in Russian). English translations: 1949, 1986. The University of Chicago Press, Chicago.

Tantawy, A.O. and Mallah, G.S. (1961) Studies on natural populations of *Drosophila*. I. Heat resistance and geographical variation in *Drosophila melanogaster* and *D. simulans*. *Evolution* 15 : 1–14.

Timofeeff-Ressovsky, N.V. (1933) Über die relative Vitalität von *Drosophila melanogaster* Meigen und *Drosophila funebris* Fabricius unter verschiedenen Zuchtbedingungen. *Zusammenhang mit den Verbreitungsarealen dieser Arten. Arch. Naturgesch*, N.F. 2 : 285–290.

Timofeeff-Ressovsky, N.V. (1935) Auslösung von Vitalitätsmutationen durch Röntgenbestrahlung bei *Drosophila melanogaster. Nachrichten von der Gesellschaft der Wissenschaft zu Göttingen. Biologie. Neue Folge* 1 : 163–180.

Van Tienderen, P.H. (1994) Selection on reaction norms, genetic correlations and constraints. *Genet. Res.* 64 : 115–125.

Via, S. and Lande, R. (1985) Genotype-environment interaction and the evolution of phenotypic plasticity. *Evolution* 39 : 505–522.

Waddington, C.H. (1959) Canalisation of development and genetic assimilation of acquired characters. *Nature* 183 : 1654–1655.

Zhivotovsky, L.A. and Christiansen, F.B. (1995) The selection barrier between populations subject to stabilizing selection. *Evolution* 49 : 490–501.

Zhivotovsky, L.A., Feldman, M.W. and Bergman, A. (1996a) On the evolution of phenotypic plasticity in a spatially heterogeneous environment. *Evolution* 50 : 547–558.

Zhivotovsky, L.A., Imasheva, A.G., David, J.R., Lazenby, O.E. and Cariou, M.-L. (1996b) Phenotypic plasticity of wing size and shape in *Drosophila melanogaster* and *D. simulans*. *Russian J. Genet.* 32 : 447–452 (in Russian with English translation).

Zhivotovsky, L.A., Feldman, M.W. and Bergman, A. (1996c) Fitness patterns and phenotypic plasticity in a spatially heterogeneous environment. *Genet. Res.* 68 : 241–248.

## Appendix A:
## The gamma-distribution model for deleterious mutations

1. We assume that the number of loci, $n$, is sufficiently large, and the mutation rate is of the order $1/n$, so that $\mu = U/n + o(1/n)$, with $U = O(1)$. Let the reductions in fitness, $h_i$, caused by the mutant alleles, be a random sample of size $n$ from a continuous distribution with a density probability function $F(h)$, whose expectation $\varepsilon(h)$ and variance $\varepsilon(h^2) - (\varepsilon(h))^2$ are denoted by $\bar{h}$ and $\sigma_h^2$, respectively. Under this assumption, the sum $\sum_{i=1}^{n}(\mu/U)h_i$, which is simply the sample mean, $\sum_{i=1}^{n} h_i/n$, converges to $\varepsilon(h)$ if $n$ infinitely increases.

2. As mentioned earlier, we assume $F(h)$ to be a gamma function,

$$F(h) = \frac{1}{\beta\Gamma(\alpha)}\left(\frac{h}{\beta}\right)^{\alpha-1} e^{-h/\beta}, \quad 0 \le h < \infty, \tag{13}$$

with positive parameters $\alpha = \bar{h}^2/\sigma_h^2$ and $\beta = \sigma_h^2/\bar{h}$. We are interested in the case of $\alpha > 1$ for which the variation coefficient, $\sigma_h/\bar{h}$, is less than 1.

It can be proven by integrating of the product of the density function (13) and $h^{-1}$ that the following property holds:

$$\varepsilon\left(\frac{1}{h}\right) = \frac{1}{\bar{h}(1 - C_h)}, \tag{14}$$

where $C_h = (\sigma_h/\bar{h})^2$ is the squared variation coefficient for random value $h$.

Environmental Stress, Adaptation and Evolution
ed. by R. Bijlsma and V. Loeschcke
© 1997 Birkhäuser Verlag Basel/Switzerland

# Stress, developmental stability and sexual selection

Anders P. Møller

*Laboratoire d'Ecologie, CNRS URA 258, Université Pierre et Marie Curie, Bât. A, 7ème étage, 7 quai St. Bernard, Case 237, F-75252 Paris Cedex 5, France*

*Summary.* Sexual selection may give rise to increases in the general level of stress experienced by individuals, either because intense directional selection reduces the ability of individuals to control the stable development of their phenotype, or because extravagant secondary sexual characters on their own impose stress on their bearers. Sexual selection often acts against individuals with asymmetric or otherwise deviant phenotypes, particularly if such phenotypic deviance occurs in secondary sexual characters. A small number of studies suggests that such characters also are more susceptible to the disruptive effects of deviant environmental conditions than are ordinary morphological characters. Plants often show extensive phenotypic asymmetry, and pollinators avoid asymmetric flowers, either because they are generally less attractive or provide fewer pollinator rewards. Floral symmetry may give rise to sexual selection with direct or indirect fitness benefits, as in animals. Sexual selection in animals may result in selection for relatively larger male body size, an overall increase in body size of a lineage and an increased risk of extinction (Cope's rule). Reduced stress resistance associated with intense sexual selection may contribute to this trend.

## Introduction

Sexual selection may superficially appear not to be associated with stress and evolution. However, stress understood as any energy-dissipating process that may cause temporary or permanent damage to a biological system (Alekseeva et al., 1992; Ozernyuk et al., 1992) may play an important role in sexual selection, and sexual selection itself may increase the stress perceived by organisms owing to the severe costs selection imposes on biological systems. This chapter reviews sexual selection theory and how it relates to development stability, one measure of stress experienced by organisms; it also describes how secondary sexual characters and ordinary traits reflect condition and stress. Plants, unlike animals, are sessile organisms that must cope with local environmental conditions. This may render sexual selection in plants more susceptible to stress than in animals. Finally, sexual selection has been hypothesized to result in evolutionary change to a relatively larger male body size, a reduction in population size and a higher susceptibility to stress. Extinction probabilities are thus hypothesized to increase with more intense sexual selection.

Sexual selection is an important evolutionary process that results from the reproductive advantages that certain individuals have over others caused by male-male competition or female choice (Darwin, 1871; Andersson, 1994). Sexual selection may result in the evolution and mainteanance of extravagant traits that reduce male survival prospects, but these are simultaneously balanced by mating advantages accruing to the most extravagantly ornamented individuals. A number of different models of sexual selection have been suggested, and these are traditionally separated into direct and indirect fitness benefit models. The former models posit that male ornamentation reliably reflects male qualities provided to the female, and that choosy females benefit from their choice of mate in terms of benefits such as parental care. Indirect fitness benefits may also be signalled reliably by the extravagancy of a male trait, and females may benefit from mate choice in the male trait reflects genetically based attractiveness or superior viability. In the first case, choosy females benefit via their attractive sons and choosy daughters, in the second case via their offspring of higher than average viability (review in Andersson, 1994). Common to signals of both direct benefits and viability is that they reliably convey information on the phenotypic or genetic quality of the signaller, which is not the case for arbitrary signals of attractiveness. Condition-dependent expression of a male sex trait in reliable signal models may result in phenotype-dependent expression of the sex trait, and the ability to cope with the costs of the trait will result in a reduction in the total amount of energy available for other activities such as maintenance, growth and survival. The male trait will eventually evolve to ever more extreme expressions and impose a viability cost that is balanced by mating advantages. This cost may even reduce the direct fitness benefits obtained by the female. The population size may decrease due to sexual selection, and the sex ratio may become female-biased as a consequence of the costs of male ornamentation.

Developmental stability reflects the ability of individuals to undergo stable development under given environmental conditions (review in Møller and Swaddle, 1997). The stable development of regular phenotypes is a costly process that can only be achieved successfully at the cost of other activities. Developmental instability can be estimated from fluctuating asymmetry, the frequency of phenodeviants and a number of other measures (reviews in Ludwig, 1932; Palmer and Strobeck, 1986; Parsons, 1990; Graham et al., 1993; Møller and Swaddle, 1997). Developmental instability is caused by a range of different environmental (e.g. temperature, food quality and quantity, pollutants, parasitism) and genetic factors (e.g. inbreeding, hybridization, novel mutants, chromosomal anomalies; reviews in Palmer and Strobeck, 1986; Parsons, 1990; Møller and Swaddle, 1997). All these different factors can be considered to impose stress on biological systems and thereby increase the energetic costs of stress resistance (see Parsons, this volume). Since energy has to be allocated to maintenance, growth, storage, reproduction and survival, there will be trade-offs in allocation among different activities. Any energy that is allocated to stress resistance cannot be used for maintenance of developmental stability during growth. There is intense directional selection against aberrant, asymmetric phenotypes due to both natural and sexual selection (reviews in Møller and Swaddle, 1997; see also Brakefield, this volume), and the development of a regular phenotype therefore has very high priority in the overall allocation of energy among different activities. The magnitude of developmental stability of an individual thus may become a reliable reflection of the phenotypic or genotypic quality of an individual in its environment (Møller, 1990). The degree of developmental instability depends not only on the amount of stress imposed by various environmental and genetic factors but also on the functional importance of a trait and its recent selective history. Functionally important traits generally demonstrate lower levels of asymmetry than do less important traits, and this inverse correlation between functional importance and asymmetry is caused by important traits being developmentally well canalized (Møller and Swaddle, 1997). The recent selective history of a trait similarly affects the level of asymmetry by means of genetic modifiers that control the expression of a trait (Møller and Pomiankowski, 1993a, b; Møller and Swaddle, 1997). Stabilizing selection incorporates genetic modifiers and therefore tends to reduce asymmetry, while directional (and disruptive) selection selects against genetic modifiers and therefore increases asymmetry. Developmental instability has a significant additive genetic component (Møller and Thornhill, 1997a, b), and offspring will therefore resemble their parents with respect to asymmetry and phenodeviance. This heritability and the potential reliable signalling of quality by developmental stability suggests that individual fluctuating asymmetry or phenodeviance may play an important role in sexual selection.

Choosy individuals that mate with symmetric partners may obtain direct or indirect fitness advantages (Møller, 1990; 1993b, 1994; Thornhill and Sauer, 1992; Watson and Thornhill, 1994), or choice of mate may simply reflect a pre-existing bias for symmetrical phenotypes (Møller, 1992b). While the latter explanation may account for the initial evolution of the preference, it cannot account for the maintenance of the preference if females pay a fitness cost for their choice of mate. For example, it is likely that female reproduction is delayed by a preference for symmetrical mates, and that the female preference is therefore costly.

## Developmental instability and sexual selection

The role of developmental instability for sexual selection has recently been reviewed by Møller and Thornhill (1997c). Although descriptive and experimental studies have reached different conclusions, there is a clear overall pattern arising from the meta-analysis of Thornhill and Møller. Meta-analysis is a stringent way of statistically analysing a hypothesis from a number of different studies and identifying variables that independently affect the relationship (Arnquist and Wooster, 1995). This is done by calculating a standardized effect size (in this case a correlation coefficient), determining the overall effect and heterogeneity in the effect among studies, and finding explanatory variables that can account for this heterogeneity. The relationship between fluctuating asymmetry and mating success has been estimated for 114 samples in 52 studies of 36 species of animals. Of these samples, 86 showed a negative relation between asymmetry and mating success (i.e. more asymmetric individuals perform worse), while 27 showed a positive relation, and one was reported as zero, a highly significant pattern (p (one-tailed) < 0.00005). A similar conclusion was reached based on a proper meta-analysis that took sample size into account.

A number of potentially confounding variables may have affected the relationship between mating success and asymmetry, since the effect size (a standardized measure of the overall relationship adjusted for sample size) was highly significantly heterogeneous. The heterogeneity was unaffected by the sex studied, whether the trait was directly involved in movement or the internal validity of the study. The finding that the general utility of the trait with respect to mobility did not predict effect size may at first glance seem odd, since asymmetry in such characters may be assumed to have a strongly negative effect on overall performance. Whether characters become the target of sexual selection for symmetric phenotypes apparently depends more on the ability of males and females to perceive asymmetry in a particular trait. Three factors accounted for some of the heterogeneity in effect size: (1) tests for fluctuating asymmetry, (2) whether an experimental approach was adopted and (3) whether the trait under study was a secondary sexual character. First, studies in which specific tests for fluctuating asymmetry were made and fluctuating asymmetry was found demonstrated a stronger effect than studies without such a test (Fig. 1 (A); Møller and Thornhill, 1996c). Second, studies based on an environmental approach found much stronger negative relationships than did studies based on observations (Fig. 1 (B); Møller and Thornhill, 1997c), probably because experimental studies are able to control the effects of confounding variables and because experiments may have been performed mainly on systems with previous indications of an effect. Finally, the effect size was significantly larger for secondary sexual characters compared with normal traits, perhaps because the former are more susceptible to the causes of developmental instability (Fig. 1 (C); Møller and Thornhill, 1997c).

In conclusion, there is overwhelming evidence for sexual selection being associated with developmental instability, particularly in secondary sexual characters.

Figure 1. Effect size standardized for sample size of the relationship between sexual selection and fluctuating asymmetry with respect to (A) statistical tests for fluctuating asymmetry, (B) whether studies were based on experiments or observations and (C) whether or not the trait investigated was a secondary sexual character. Values are means (SE). P values refer to one-tailed comparisons of transformed effect sizes. (Adapted from Møller and Thornhill, 1997c.)

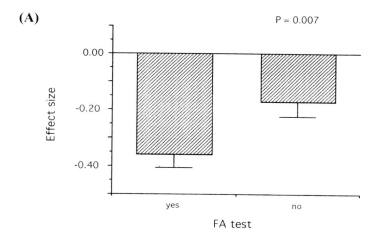

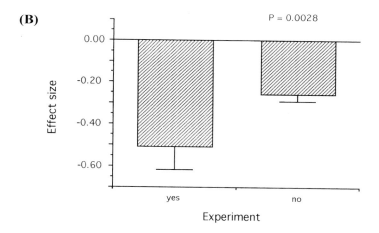

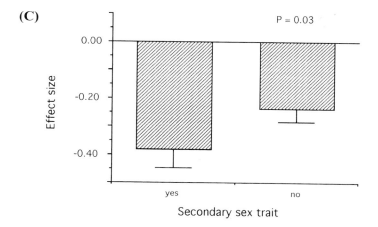

## Comparative studies of developmental instability of secondary sexual characters

A number of studies have addressed the question whether secondary sexual characters demonstrate larger degrees of relative asymmetry than do ordinary morphological characters. The reason such an effect might be expected is that intense directional selection originating from male-male competition or female mate preferences is predicted to result in selection against genetic modifiers that control the stable development of a character (Møller and Pomiankowski, 1993a, b). Laboratory selection experiments imposing directional selection on a character have indeed found an increase in asymmetry, while stabilizing selection gives rise to the opposite effect, a decrease in the level of fluctuating asymmetry (review in Møller and Swaddle, 1997). Some comparative studies of secondary sexual characters and comparable characters in closely related species apparently not subject to intense directional sexual selection have indicated that secondary sexual characters are moare asymmetric than ordinary morpholocial characters (Møller and Höglund, 1991; Møller 1992a; Manning and Chamberlain, 1993b; J. J. Cuervo and A. P. Møller, unpublished data). However, studies based on other species or taxa have failed to find such a relationship (Balmford et al., 1993; Tomkins and Simmons, 1995). The reasons for this discrepancy among studies include several possibilities, but one worth emphasizing here is that the level of fluctuating asymmetry in secondary sexual characters in feather ornaments of birds appears to depend on the intensity of sexual selection as determined indirectly from the skew in male mating success as reflected by the mating system. Ornamented socially monogamous bird species have greater asymmetry than non-ornamented sister taxa in accordance with the prediction (J. J. Cuervo and A. P. Møller, unpublished data). However, polygynous and particularly lekking bird species with the most extreme skew in male mating success have almost as low a degree of asymmetry in their secondary sexual characters as is found in ordinary morphological characters of non-ornamented socially monogamous species (J. J. Cuervo and A. P. Møller, unpublished data). These results suggest that fluctuating asymmetry initially increases when a trait becomes the target of a directional mating preference. If the intensity of sexual selection increases, this high level of asymmetry may decrease considerably.

## Are secondary sexual characters particularly susceptible to stress?

If secondary sexual characters are reliable condition indicators, they might be expected to be more susceptible to the effects of stress than ordinary morphological traits, and the effects of stressors on character expression and asymmetry should therefore be more clear for sexually selected traits (Møller, 1990). Secondary sex traits also have a long history of intense directional selection that may have selected against genetic modifiers that control the expression of a trait (Møller and Pomiankowski, 1993a, b; Møller and Swaddle, 1997). Whether secondary sex traits are more susceptible to stress than other traits has been investigated a number of times. Andersson (1982, 1994) has reviewed the literature on condition-dependent expression of secondary sexual characters, and there is overwhelming evidence for such effects.

Studies of differential susceptibility of secondary sexual characters to the causes of developmental instability can be classified into observational and experimental ones; the former suffer from an inability to infer causality. Two observational studies have addressed whether secondary sexual characters have higher degrees of asymmetry than ordinary morphological traits. Radioactive contamination in Chernobyl in 1986 caused a differential increase in asymmetry in tail length (a secondary sexual character) of male barn swallows, *Hirundo rustica*, but not in three ordinary morphological traits (Møller, 1993a). Similarly, tail asymmetry in male barn swallows was differentially susceptible to annual variation in precipitation (and presumably food availability) in the African winter quarters when compared with ordinary morphological characters (A. P. Møller, unpublished data). Asymmetry of canine teeth of male gorillas, *Gorilla gorilla*, has increased since the last century, apparently because of environmental deterioration (Manning and Chamberlain, 1993a). Canines play an important role in male-male competition in the gorilla. Experimental manipulation of the amount of haematophagous mites in the nests of barn swallows differentially increased the level of asymmetry in the outermost tail feathers, but not in wing length or in the length of central tail feathers, which are not subject to a directional mate preference (Møller, 1992c). Finally, Folstad et al. (1996) demonstrated that asymmetry in antlers, but not in other skull characters, was reduced when reindeer, *Rangifer tarandus*, received anti-helminth treatments.

In conclusion, the studies done so far suggest that secondary sexual characters indeed are more susceptible to stress than are ordinary morphological characters.

## Sexual selection in plants, and developmental instability

Sexual selection in plants differs from that in animals by involving abiotic or biotic mediators of pollen transfer, and any sexual selection process thus relies on the response of such third parties to developmental stability of pollen as well as floral traits used to attract pollinators. In this section I will briefly discuss (1) processes of sexual selection in plants in relation to developmental stability; (2) patterns of developmental stability in flowers; and (3) pollinator preferences for flowers with a regular phenotype.

*Processes of sexual selection in plants in relation to developmental
stability*

Even though plants differ from animals, we may still consider the three different fitness benefits associated with sexual selection in animals. Direct
fitness benefits may take on many forms, such as enhanced fertility and the
absence of directly transmitted diseases. If plants with symmetric flowers
have pollen with superior fertilizing ability, this may result in a direct fitness advantage being associated with floral symmetry.

Indirect fitness benefits include good genes and arbitrary attractiveness
effects. If floral symmetry has a genetic basis, fertilization of ovules by
pollen from symmetric flowers will automaticaly result in plants in the next
generation experiencing an advantage in terms of developmental stability.
Observations on *Epilobium angustifolium* are consistent with this mechanism. Pollen from symmetrical flowers is indeed superior in fertilizing
ovules, and only a small proportion of such ovules are aborted (Møller,
1996). Floral symmetry also reflects the ability of ovules to become ferti-
lized, since a larger proportion of ovules are aborted in plants with asymmetric flowers independent of the symmetry of the pollen donor (Møller,
1996). Floral symmetry in this species has a genetic basis as determined
from resemblance between parents and offspring raised under standardized
greenhouse conditions. These results suggest that similar effects of developmental stability are found at different stages of the life cycle, since floral
symmetry of both pollen donors and recipients was associated with embryo
malformation at the embryonic stage of the following generation. Similarly, experiments on *Lychnis viscaria* have shown that the fertilizing ability
of pollen is inversely related to floral symmetry, even though there is no
effect of floral symmetry on abortion rates (Eriksson, 1996). This suggests
that both pollen-fertilizing ability and embryonic growth may be directly
related to floral symmetry.

Finally, sexual selection may take place as a result of arbitrary attractiveness. Pollinators or abiotic factors are involved in the fertilization process in plants, and a purely attractive feature of flowers without pollinator
rewards is therefore difficult to envisage. However, such a mechanism may
take place after pollen has been deposited by a pollinator, because the genotype of the pollen may interact directly with the style during pollen tube
growth. If pollen from certain fathers experiences an advantage in terms of
pollen tube growth because it is considered attractive by the female function of the flower, this may result in a coevolutionary process between the
female preference and the male attractiveness similar to the Fisherian
process in animals.

Both direct and indirect fitness benefits may rely on pollinator services
caused by nectar or pollen rewards, or flowers mimicking the appearance
of insects. Preliminary results from a number of different plant species
have shown that nectar production, but also a standing crop of nectar, is

inversely related to flower asymmetry (Møller, 1995; Møller and Eriksson, 1995). Symmetric flowers also have nectar with higher sugar content than do asymmetric flowers (J. T. Manning, personal communication). Flower-visiting insects thus will benefit from discriminating between symmetric and asymmetric flowers of the same species, because they maximize their pollen or nectar gain rates.

*Patterns of developmental stability in flowers*

A number of studies have now determined the consistency of floral symmetry during the life of a flower, among flowers on the same plant and among flowers of different plants.

Flowers and leaves are arranged hierarchically on a plant, and growth and therefore resources are differentially allocated to the topmost shoots, leaves and flowers as mediated by plant hormones such as auxins (Sachs, 1994; Sachs et al., 1993). This results potentially in topmost flowers being disproportionally symmetric due to an excess of resources. Such an effect of allocation should be particularly prominent under resource limitation, when lower-rank flowers receive less resources than necessary for stable development. Resource allocation should result in an increased variance in floral morphology and thus in reduced repeatability within plants. If there is a trade-off in stable development between flowers of the same individual, we might even expect to find negative repeatabilities. A number of plant species studied have shown that floral symmetry is significantly repeatable in some of these, while other species clearly demonstrate negative repeatabilities (Møller and Eriksson, 1994). This may indicate that trade-offs exist among flowers. Two studies have examined to which extent flower position affects floral asymmetry. Sherry and Lord (1996) found no effect of position in *Clarkia tembloriensis*, and Eriksson (1996) found a similar lack of position effect in *Lychnis viscaria*.

Floral asymmetry was highly repeatable, in comparisons made among plants flowering simultaneously (Møller and Eriksson, 1994; Møller, 1995), but also among plants flowering at different times of the same season (A. P. Møller, unpublished data) and in different seasons (Eriksson, 1996; A. P. Møller, unpublished data). Floral asymmetries are therefore often consistent at a number of different levels.

Finally, if specific genotypes are better able to cope with adverse environmental conditions during development under a range of environmental conditions, we should expect asymmetries to be positively correlated across environments. This was in fact the case among *Epilobium angustifolium* receiving three different treatments (pure water (a control treatment), saline water and water with artificial fertilizer) (A. P. Møller, unpublished data). There was a clear increase in floral asymmetry caused by the two stress treatments, but floral asymmetries of ramets of the same clone were positively correlated across treatments. Some individuals were therefore consistently better able to cope with the effects of adverse environments.

*Pollinator preferences for flowers with a regular phenotype*

Honey bees preferentially use both bilateral and radial symmetry as cues when learning to discriminate between potential nectar sources (Horridge and Zhang, 1995; Giurfa et al., 1996), and visual symmetry detectors may function as a basis for discrimination between flowers varying in degree of asymmetry. The first study to address the question of whether insects preferentially visited symmetrical flowers was performed by Møller and Eriksson (1995), who compared the level of symmetry in flowers visited by an insect and the nearest neighbouring flower without an insect. A number of species of flowering plants showed less asymmetry in the insect-visited flowers. An experimental study based on symmetric and asymmetric clipping of petals tested whether such preferences were independent of pollinator rewards. Bumblebees, *Bombus* spp., preferred to visit flowers of *Epilobium angustifolium* whose two lower petals were clipped symmetrically rather than asymmetrically, demonstrating that the bumblebees discriminated against asymmetry (Møller, 1995). A subsequent experiment was based on four coloured (yellow or red in different tests) discs of different diameter with the black centre positioned symmetrically (30-, 24-, and 18-mm diameter, respectively) or asymmetrically (24-mm diameter) (Møller and Sorci, 1997). A replicate was terminated when one disc had been visited, and it was replaced with another disc to avoid any effects of pheromones on attractiveness. As in the experiment with *E. angustofolium*, insects (mainly Diptera, but also Hymenoptera and Lepidoptera) preferred to visit larger discs over smaller ones and symmetrical over asymmetrical ones in experiments performed both in Spain and in Denmark. This again demonstrates a preference for floral symmetry, even in the complete absence of any pollinator rewards. Contrarily, Midgley and Johnson (1997) found that daisies with asymmetrically positioned petal spots were not visited less often than individuals with symmetrical petal spots.

In conclusion, a preference for floral symmetry has been demonstrated from observations and experiments for a number of different insect species, but a single study has been unable to replicate this finding.

## Cope's rule, sexual selection, stress and extinction

Which are the consequences of sexual selection for evolution, ability to cope with stress and prospects of survival for species? Sexual selection arises from the effects of male-male competition and female choice, and the outcome is an increase in the size of males relative to that of the female as well as an increase in ornamentation caused by female choice. Such evolutionary change may have several important consequences. First, if a large proportion of available energy is allocated to sexual selection, the ability to cope with stress caused directly by the development and maintenance of secondary sexual characters may be reduced (Parsons, 1995). Secondary sexual characters often evolve well beyond what is optimal with respect to natural selection, and the level of stress imposed by the secondary sexual character itself may therefore be considerable. A resultant smaller amount of energy may then be available for coping with maintenance and costly immune responses to combat virulent parasites. A contributing factor might be the lack of energy arising from the development and maintenance of a secondary sexual character that results in a reduced ability to cope with stress during adverse environmental conditions and periods of overall environmental deterioration. Intense directional selection caused by sexual selection may render secondary sexual characters, but also other traits, more susceptible to the effects of adverse environmental conditions (Møller and Pomiankowski, 1993a, b; Møller and Swaddle, 1997). Species subjected to intense directional sexual selection thus may become more susceptible to environmental perturbations than comparable species not subjected to similarly intense sexual selection.

Second, Cope's rule states that the size of individuals within a taxon increases over evolutionary time, and within a taxon, species characterized by relatively large individuals are more likely to go extinct (Romer, 1966; Eisenberg, 1981). In accordance with this prediction, extinct mammalian taxa were significantly larger than surviving, closely related taxa, and large body size is generally considered to be associated with male-male competition in mammals (McLain, 1993). Extinction may be a direct result of increased body size, lower total population size and thus increased risks due to stochastic processes. Alternatively, extinction may be due to stress by secondary sexual characters reducing the ability of individuals to cope with adverse environmental conditions. Sexual selection may be considered an extreme case of specialization with respect to mating success, and this may reduce the ability to cope with general adverse environmental conditions. Sexual selection may also result in a reduction in additive genetic variance in traits important for viability, and this may have serious consequences for the ability of individuals to cope with extreme environments. This latter explanation was supported by a recent analysis of the survival prospects of birds introduced to Tahiti and Hawaii (McLain et al., 1995). Sexually dichromatic species were more likely to go extinct than sexually monochromatic species. Analysis of the same phenomenon for birds introduced to New Zealand, but this time controlling for the effects of the number of individuals introduced, also provided evidence for a statistically significant increased effect of sexual dichromatism on the probability of extinction (Sorci et al., 1997). This result suggests that particular features of sexually dichromatic species render them more prone to sexual selection. Similarly, a recent analysis of extinct bird species of the world demonstrated that extinct species more often were sexually dichromatic than were surviving species of the same families (A.P. Møller, unpublished data). An analysis of threatened bird species on the red data list of the world also showed a larger frequency of sexually dichromatic species than expected from the appearance of other members of the same families (A.P. Møller, unpublished data).

In conclusion, sexual selection may lead to increased size of males relative to females, increased investment in secondary sexual characters and increased inability to cope with adverse environmental conditions. Intensely sexually selected species of birds are indeed more susceptible to extinction than less sexually selected species. Sexual selection thus may be an important evolutionary force that results in extreme specialization with respect to mating, rendering such taxa particularly prone to high risks of extinction.

## Future prospects

The aspects of sexual selection, developmental stability and stress reviewed in this chapter are mostly based on indirect or relatively limited observational evidence. An essential prerequisite for further progress is that the ideas presented here be subjected to experimental and comparative tests. Such studies will not be easy to perform, because many of the favoured laboratory animals lack conspicuous secondary sexual characters. Laboratory studies also suffer from the potential danger of being conducted under relatively benign conditions that give rise to little or no environmental stress. Experimental manipulation of environmental conditions may alleviate this problem.

Further comparative studies of proneness to extinction in relation to sexual selection may be of both theoretical and practical importance for conservation. Much work on conservation biology has been very focused on stochastic, genetic and demographic parameters without taking the ecology and natural history of the species in question fully into account.

*Acknowledgments*
The chapter benefitted from constructive comments and suggestions from R. Bijlsma and V. Loeschcke. I was supported by grants from the Swedish and Danish Natural Science Research Councils.

## References

Alekseeva, T.A., Zinichev, V.V. and Zotin, A.I. (1992) Energiy criteria of reliability and stability of development. *Acta Zoologici Fennici* 191:159–165.
Andersson, M. (1982) Sexual selection, natural selection and quality advertisement. *Biol. Linn. Soc.* 17:375–393.

Andersson, M. (1994) *Sexual Selection*. Princeton University Press, Princeton.

Arnqvist, G. and Wooster, D. (1995) Meta-analysis: Synthesizing research findings in ecology and evolution. *Trends Ecol. Evol.* 10:236–240.

Balmford, A., Jones, I.L. and Thomas, A.L.R. (1993) On avian asymmetry: Evidence of natural selection for symmetrical tails and wings in birds. *Proc. Roy. Soc. London B* 252: 245–251.

Darwin, C. (1871) *The Descent of Man, and Selection in Relation to Sex*. John Murray, London.

Eisenberg, J.F. (1981) *The Mammalian Radiations*. University of Chicago Press, Chicago.

Eriksson, M. (1996) Consequences for plant reproduction of pollinator preference for symmetric flowers. Ph.D. dissertation, Uppsala University, Sweden.

Folstad, I., Arneberg, P. and Karter, A.J. (1996) Parasites and antler asymmetry. *Oecologia* 105:556–558.

Giurfa, M., Eichmann, B. and Menzel, R. (1996) Symmetry perception in an insect. *Nature* 382:458–461.

Graham, J.H., Freeman, D.C. and Emlen, J.M. (1993) Developmental stability: A sensitive indicator of populations under stress. *In*: W.G. Landis, J.S. Hughes and M.A. Lewis (eds): *Environmental Toxicology and Risk Assessment*. American Society for Testing and Materials, Philadelphia, pp. 136–158.

Horridge, G.A. and Zhang, S.W. (1995) Pattern vision in honeybees (*Apis mellifera*): Flower-like patterns with no predominant orientation. *J. Insect Physiol.* 41:681–688.

Ludwig, W. (1932) *Das Rechts-Links Problem im Tierreich und beim Menschen*. Springer-Verlag, Berlin.

Manning, J.T. and Chamberlain, A.T. (1993a) Fluctuating asymmetry in gorilla canines: A sensitive indicator of environmental stress. *Proc. Roy. Soc. London B* 255:189–193.

Manning, J.T. and Chamberlain, A.T. (1993b) Fluctuating asymmetry, sexual selection and canine teeth in primates. *Proc. Roy. Soc. London B* 251:83–87.

McLain, D.K. (1993) Cope's rule, sexual selection and the loss of ecological plasticity. *Oikos* 68:490–500.

McLain, D.K., Moulton, M.P. and Redfearn, T.P. (1995) Sexual selection and the risk of extinction of introduced birds on oceanic island. *Oikos* 74:27–34.

Midgley, J.J. and Johnson, S.D. (1997) Some pollinators do not prefer symmetrically marked or shaped daisy (Asteraceae) flowers. *Evol. Ecol.; in press.*

Møller, A.P. (1990) Fluctuating asymmetry in male sexual ornaments may reliably reveal male quality. *Anim. Behav.* 40:1185–1187.

Møller, A.P. (1992a) Patterns of fluctuating asymmetry in weapons: Evidence for reliable signalling of quality in beetle horns and bird spurs. *Proc. Roy. Soc. London B* 248:199–206.

Møller, A.P. (1992b) Fluctuating asymmetry and the evolution of signals. *In*: P. Bateson and M. Gomendio (eds): *Workshop on Behavioural Mechanisms in Evolutionary Perspective*. Instituto Juan March de Estudios e Investigaciones, Madrid, pp. 96–98.

Møller, A.P. (1992c) Parasites differentially increase the degree of fluctuating asymmetry in secondary sexual characters. *J. Evol. Biol.* 5:691–699.

Møller, A.P. (1993a) Morphology and sexual selection in the barn swallow *Hirundo rustica* in Chernobyl, Ukraine. *Proc. Roy. Soc. London B* 252:51–57.

Møller, A.P. (1993b) Developmental stability, sexual selection and the evolution of secondary sexual characters. *Etología* 3:199–208.

Møller, A.P. (1994) Symmetrical male sexual ornaments, paternal care and offspring quality. *Behav. Ecol.* 5:188–194.

Møller, A.P. (1995) Bumblebee preference for symmetrical flowers. *Proc. Nat. Acad. Sci. USA* 92:2288–2292.

Møller, A.P. (1996) Floral asymmetry, embryo abortion and developmental selection in plants. *Proc. Roy. Soc. London B* 263:53–56.

Møller, A.P. and Eriksson, M. (1994) Patterns of fluctuating asymmetry in flowers: Implications for honest honest signalling for pollinators. *J. Evol. Biol.* 7:97–113.

Møller, A.P. and Eriksson, M. (1995) Flower asymmetry and sexual selection in plants. *Oikos* 73:15–22.

Møller, A.P. and Höglund, J. (1991) Patterns of fluctuating asymmetry in avian feather ornaments: Implications for models of sexual selection. *Proc. Roy. Soc. London B* 245:1–5.

Møller, A.P. and Pomiankowski, A. (1993a) Fluctuating asymmetry and sexual selection. *Genetica* 89:267–279.

Møller, A.P. and Pomiankowksi, A. (1993b) Punctuated equilibria or gradual evolution: Fluctuating asymmetry and variation in the rate of evolution. *J. Theor. Biol.* 161: 359–367.

Møller, A.P. and Sorci, G. (1997) Insect preference for symmetrical flower models.

Møller, A.P. and Swaddle, J.P. (1997) *Asymmetry, Developmental Stability and Evolution.* Oxford University Press, Oxford.

Møller, A.P. and Thornhill, R. (1997a) A meta-analysis of the heritability of developmental stability. *J. Evol. Biol.* 10: 1–16.

Møller, A.P. and Thornhill, R. (1997b) Developmental stability is heritable. *J. Evol. Biol.* 10: 69–76.

Møller, A.P. and Thornhill, R. (1997c) Developmental stability and sexual selection: A meta-analysis. *Am. Nat.; in press.*

Ozernyuk, N.D., Dyomin, V.I., Prokofyev, E.A. and Androsova, I.M. (1992) Energy homeostasis and developmental stability. *Acta Zoologici Fennici* 191: 167–175.

Palmer, A.R. and Strobeck, C. (1986) Fluctuating asymmetry: Measurement, analysis and patterns. *Ann. Rev. Ecol. Syst.* 17: 391–421.

Parsons, P.A. (1990) Fluctuating asymmetry: An epigenetic measure of stress. *Biol. Rev.* 65: 131–145.

Parsons, P.A. (1995) Stress and limits to adaptation. Sexual selection. *J. Evol. Biol.* 8: 455–461.

Romer, A.S. (1966) *Vertebrate Paleontology.* University of Chicago Press, Chicago.

Sachs, T. (1994) Variable development as a basis for robust pattern formation. *J. Theor. Biol.* 170: 423–425.

Sachs, T., Novoplansky, A. and Cohen, D. (1993) Plants as competing populations of redundant organs. *Plant Cell Environ.* 16: 765–770.

Sherry, R.A. and Lord, E.M. (1996) Developmental stability in flowers of *Clarkia tembloriensis* (Onagraceae). *J. Evol. Biol.* 9: 911–930.

Sorci, G., Møller, A.P. and Clobert, J. (1996) Plumage dichromatism of birds predicts introduction success in New Zealand. *J. Anim. Ecol.; in press.*

Thornhill, R. and Sauer, K.P. (1992) Genetic sire effects on the fighting ability of sons and daughters and mating success of sons in the scorpionfly (*Panorpa vulgaris*). *Anim. Behav.* 43: 255–264.

Tomkins, J.L. and Simmons, L.W. (1995) Patterns of fluctuating asymmetry in earwig forceps: No evidence for reliable signalling. *Proc. Roy. Soc. London B* 259: 89–96.

Watson, P.J. and Thornhill, R. (1994) Fluctuating asymmetry and sexual selection. *Trends Ecol. Evol.* 9: 21–25.

# Evolution and stress

Environmental Stress, Adaptation and Evolution
ed. by R. Bijlsma and V. Loeschcke

# Genetic variability and adaptation to stress

François Taddei[1], Marin Vulić, Miroslav Radman and Ivan Matić

*Laboratoire de Mutagenèse, Institut Jacques Monod, 2 Place Jussieu, F-75251 Paris, France*
[1] *on leave of absence from Ecole Nationale du Génie Rural des Eaux et des Forêts, Paris, France*

*Summary.* Besides an immediate cellular adaptation to stress, organisms can resist such challenges through changes in their genetic material. These changes can be due to mutation or acquisition of pre-evolved functions via horizontal transfer. In this chapter we will review evidence from bacterial genetics that suggests that the frequency of such events can increase in response to stress by activating mutagenic response (e.g. the SOS response) and by inhibiting antimutagenic activities (e.g. mismatch repair system, MRS). Natural selection, by favoring adaptations, can also select for the mechanism(s) that has/have generated the adaptive changes by hitchhiking. These mutator mechanisms can sometimes respond very specifically, though blindly, to the challenge of the environment. Such stress-induced increases in mutation rates enhance genetic polymorphism, which is the structural component of the barrier to genetic exchange. Since SOS and MRS are the enzymatic controls of this barrier, the modulation of these systems can lead to a burst of speciation.

## Stress and adaptation

Adverse environmental conditions – stresses – are at the basis of much evolutionary change. During periods of severe stress, because of the intense selective pressure, fundamental changes such as species extinctions and bursts of change within species are likely to occur. The outcome will depend on the genetic variability present in a given lineage.

We refer to stress as a disturbance of the normal functioning of a biological system (its steady state or homeostasis) by a change in an environmental factor whose amplitude and persistence are such that it causes reduction in fitness or increased mortality. Since small and/or steady departures from steady-state conditions may render the individual more resistant to further change through induction of appropriate responses, they may be viewed as causing a beneficial outcome, and are not referred to as stresses.

Even though mutation is generally recognized as the ultimate source of evolutionary change, its relative importance in the adaptive process has been underestimated. The contribution of mutation to the overall genetic variation present within a species has been thought to be insignificant in comparison to the already existing variation, because the waiting time for a beneficial mutation may be too long, and the probability that the occurrence of such a mutation will coincide with an environmental shift requiring an adaptive response seems infinitesimally small. However, recent evidence suggests that during periods of environmental challenge and subsequent stress of the organism, the rates of many mutational events increase dramatically. Therefore, mutation can be stress-responsive. The implications for adaptation of such a notion are important; at challenging moments in evolutionary history when major adaptive shifts are required, mechanisms exist that increase the probability that appropriate variants will be provided.

To answer questions about the influence of stress on the generation of genetic diversity and adaptation, experimental approaches are desirable, because it is very difficult to determine *a posteriori* the processes that produced some pattern. For example, mutation rates estimated from DNA sequence comparisons are apparently different in different lineages, in the same lineage over time and even at different locations within a single genome (Li, 1993). It is still unclear what the most important factors are that cause the observed rate heterogeneity, and their identification is fundamentally important to our understanding of evolution. The large statistical uncertainty of these estimations further underscores the need for experimental approaches to this and related problems.

It might seem a daunting task because of the complexity of the systems, but there are experimental systems that allow the needed studies to be readily carried out. Model organisms such as bacteria have many advantages, such as short generation time, large population sizes, relatively small haploid genomes, controllable environmental variables, a possibility for direct comparison of initial and derived genotypes, availability of molecular data and substantial knowledge of molecular mechanisms. Moreover, due to the overall conservation of a number of these mechanisms from bacteria to humans, bacterial genetics can serve as a source of new paradigms, piloting research into the links between environment and genetic changes.

This chapter will therefore focus on genetic changes induced by stress, setting aside cellular epigenetic responses to stress (for a review see Morris, 1993; Feige et al., 1996). The links between stress and genetic variability will be reviewed in two sections centered respectively on molecular and evolutionary aspects. The first describes molecular mechanisms which generate genetic variability and gives examples of their regulation that indicate that under stres the level of mutation increases. The second section will try to show that natural selection might favor such an increased mutation rate in the course of adaptation to an environmental challenge. Based on our knowledge of mutagenesis and recombination on the molecular level, the implications for the speciation process of this modulation of evolutionary rates will then be discussed.

## Molecular mechanisms contributing to genetic variability

Experimental investigations with micro-organisms carried out in the past decades have unravelled a number of molecular mechanisms which con-

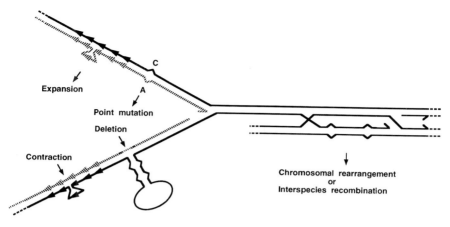

Figure 1. The molecular mechanisms that regulate genetic variability. The MRS controls genetic stability. The *E. coli mut⁻* mutants show large increases : (1) deletion events resulting from excision of transposable elements (Tex phenotype) (Lundblad and Kleckner, 1985); (2) spontaneous base substitution and frameshift mutagenesis (mutator phenotype) (Friedberg et al., 1995); (3) large chromosomal rearrangements resulting from crossovers between diverged DNA repeats (chromosomal instability phenotype) (Petit et al., 1991; Radman and Wagner, 1993; Radman et al., 1993); (4) recombination of genetic markers (hyper-rec phenotype) (Feinstein and Low, 1986); and (5) recombination with related species such as *Salmonella* (disrupted genetic barrier phenotype) (Rayssiguier et al., 1989). SOS induction increases the frequency of all of these genetic modifications, which are suppressed by the activity of the MRS (Friedberg et al., 1995). In black: parental strands; in gray: newly synthesized strand; arrays of arrows indicate repeats of mono-, di- and trinucelotides; mismatched bases are indicated by opposing arrowheads.

tribute, each in its specific way, to the overall generation of genetic variation (Fig. 1). Evolutionary theory has tended to pay less attention to these mechanisms than to the result of mutation, that is, to the variants and how the variants can be fixed in populations. Yet the understanding of mechanisms providing the variation has a direct bearing on the larger questions.

These mechanisms can be grouped as following:

### 1) *Mutagenesis through DNA replication and repair infidelity*
Base alterations arise from the inherent instability of the specific chemical bonds that constitute the normal chemistry of nucleotides under physiological conditions as well as from reaction with a variety of chemical compounds and physical agents present in the environment. The incorporation of such bases during DNA replication results in nucleotide substitutions. If the base alteration happens within a DNA molecule, its mutagenicity will depend on the fidelity of repair mechanisms present in each living cell. Besides misincorporation, mispositioning of template and newly synthesized strands relative to each other also adds to the repertoire of replication errors, resulting in small insertions, small deletions and small duplications.

## 2) *Recombinational DNA rearrangements*

Besides homologous recombination at corresponding allelic regions, which increases genetic variety, recombination can also occur between DNA sequences of sufficient homology located at different sites in the genome. This can give rise to various types of DNA rearrangements, including duplications, deletions, inversions as well as fusion and dissociation of replicons. All these events can be mediated by various mobile genetic elements or can result from other illegitimate processes (Arber, 1995).

## 3) *Acquisition of genetic information from other organisms*

The nature of genetic exchange in prokaryotes makes it a powerful source of genetic variability unique to them (Matic et al., 1996). It can be brought about by transformation or by action of mobile genetic elements such as plasmids, transposons, insertion sequences and phages. The large overlap in host ranges of such genetic exchange vectors allows sequential DNA transmission among almost all bacterial species. Promiscuous genetic exchange gives bacteria a special adaptive advantage: it allows cells to acquire traits developed and refined by organisms from other taxa or even other phyla that confer, for example, resistance to antibiotics, toxins and metals and the ability to degrade new compounds (e.g. xenobiotics). One consequence is a pool of genes that is accessible to virtually every bacterial species.

The majority of events involved in each of these processes produce nonadaptive alterations which are not fixed and therefore of no evolutionary relevance. However, amongst relevant alterations, one can assume that the acquisition of functional genetic information is perhaps the most efficient process, because for the recipient it may represent a functional improvement in a single step. Nucleotide substitutions and other local sequence variations are expected to be more frequent events representing steps in a long-term steady development and improvement of biological functions. Recombinational rearrangements probably represent an evolutionary efficiency between that of gene acquisition and that of point mutations.

Molecular mutagenesis mechanisms that result in qualitatively very different types of mutations all have specific roles in the evolutionary process. Therefore, different mutagenesis processes should not be considered as alternative possibilities. Rather, a parallel action of various mechanisms of mutagenesis is essential for the evolutionary process. Although extensive knowledge about these processes has been collected, it is still quite difficult to determine their relative contributions to any particular case of evolution or, even more, to untangle their roles more generally.

Besides enzymes that are specifically involved in any one of the processes which generate variability mentioned above, there are some enzymatic systems whose actions control or at least influence all of them. We will focus on two such systems: SOS and the mismatch repair system (MRS).

## Mismatch repair system down-regulates genetic variability

Enzymes that recognize and process mismatches were identified first in prokaryotes and then in eukaryotes based on their conservation from bacteria to humans (Radman et al., 1995; Modrich and Lahue, 1996). Their primary function is the control of genetic integrity and stability (for reviews see Modrich, 1991; Radman and Wagner, 1993).

A base-pair mismatch can be defined as any mispaired or unpaired base in a DNA duplex. Such heteroduplex DNA molecules contain non-identical genetic information on the two complementary strands. There are at least three ways in which mismatches can arise in DNA: (1) by physical or chemical damage to the DNA; (2) during DNA replication; and (3) during homologous recombination between non-identical DNA sequences.

Mismatch repair transforms a heteroduplex molecule into a homoduplex which yields pure progeny of a single parental genotype. The MRS thus controls the fidelity of chromosomal replication (by eliminating DNA replication errors such as substitution and frame-shift mutations), of transposon excision and of homologous recombination (preventing chromosomal rearrangements and interspecies recombination).

Bacteria possess multiple mismatch repair pathways that are distinguished on the basis of mismatch specificity and the size of excision repair tracts associated with the correction. We will focus our discussion on the *Escherichia coli* methyl-directed repair (MutSLHU) which is the best-described mismatch repair system (Fig. 2) (Modrich, 1991). This system is characterized by broad mismatch specificity and repair in a strand-specific manner. The strand specificity of repair is governed by secondary signals that may be located far from the mismatch. As the excision-repair tracts associated with this pathway can be a kilobase long or longer, this repair system is referred to as the "long-patch repair system".

Recombination between non-identical DNA sequences gives rise to mismatched heteroduplexes. The density of mispairs within this region reflects the extent of divergence of the two sequences and hence the degree to which imperfect homology is acceptable to the recombination enzymes catalyzing the reaction. The mismatch repair proteins recognize those mismatches and block RecA-catalyzed strand transfer (Worth Jr et al., 1994). The inactivation of mismatch repair genes in *E. coli* results in a 50- to 3000-fold increase in interspecies conjugational recombination between *E. coli* and *Salmonella typhimurium*, whose chromosomes are about 16% divergent at the DNA sequence level (Rayssiguier et al., 1989). Petit et al. (1991) have observed that the frequency of chromosomal duplications resulting from recombination between *E. coli rhsA* and *rhsB* loci (0.9% divergent) increases 10- to 15-fold in *mutL* or *mutS* backgrounds.

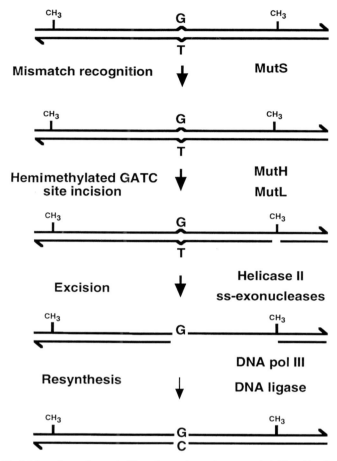

Figure 2. Model for mismatch repair. The mismatch repair process is initiated by the MutS protein, which recognizes and binds to mismatches. The biochemical function of the MutL protein is unknown. It may act as a molecular matchmaker that allows formation of a repair complex containing MutS, MutL and MutH proteins at mismatched base pairs and hemimethylated d(GATC) sites. The state of adenine methylation of GATC sequences determines strand specificity of mismatch repair. The methylation lags behind DNA replication, leaving newly synthesized daughter-strand DNA undermethylated relative to the parental strand. The hemimethylated DNA is corrected on the unmethylated strand, with the modified strand serving as template. The nick introduced by MutH, which is hemimethylated d(GATC)-specific endonuclease, serves to initiate the excision of newly synthesized DNA from the nick to just past the mismatch. The last step of mismatch repair is gap repair by DNA polymerase. Repair directed by a persistent strand break requires neither MutH nor the presence of GATC sites within the heteroduplex, but does depend on the other protein components of the methyl-directed system. For example, the MutS and MutL proteins can edit recombination between similar non-identical sequences even in the absence of the MutH protein, which is not the case for the repair of replication errors.

*Regulation of the MRS*

The expression of the *mutS* and *mutL* mismatch repair genes is not induced by mutagen treatments or in different mutator backgrounds [i.e. when mutation rate is enhanced by some genetic defect (Pang et al., 1985)]. On the contrary it has been shown that the efficiency of the MRS can be decreased in some genetic backgrounds or under particular environmental conditions as a result of either diminution of the amount of mismatch repair proteins or of their titration by high concentration of mismatched DNA. For example, the mismatch repair system is saturated in *E. coli* cells which are treated with DNA-damaging agents introducing DNA lesions recognized by mismatch repair system (Cupples et al., 1990).

Also, MutS protein can be titrated out in cells containing mismatches in retron-encoded multicopy single-stranded DNA. Overexpression of such molecules results in increased mutagenesis and frequency of interspecies recombination, which are phenotypes of *mut⁻* cells (Maas et al., 1994; Maas et al., 1996). These phenomena are suppressed in the presence of a multicopy plasmid carrying the *mutS* gene.

Furthermore, the mismatch repair proteins can be titrated out in *mutD5* mutants, in which there is a proofreading defect in DNA polymerase III, leading to very strong mutator phenotype (Damagnez et al., 1989; Schaaper and Radman, 1989). This mismatch repair defect is observed only in rich medium during the exponential growth phase, and it has been attributed to saturation of the MRS by excess of replication errors due to the proof reading defect. In these cells, it is a limitation in MutL and MutH, rather than in the MutS protein that diminishes mismatch repair capacity. These results suggest that mismatch repair proteins are not present in excess and may be limiting in some cases, even in exponentially growing cells.

There is growing evidence that mismatch-repair capacity decreases in stressed bacterial cells, during a stationary phase or during nutritional deprivation. Schaaper and Radman (1989) observed about five-fold reduction of mismatch repair activity in *E. coli* cells in late log phase compared with cells from early log phase. The expression of *mutS* and *mutL* genes from multicopy plasmids increases the efficiency of error correction in cells during a stationary phase (Foster et al., 1995). The level of MutS protein becomes barely detectable when *E. coli* cells enter the late stationary phase or when they are deprived of a usable carbon source. The transcription of *mutS*, *mutL* and *mutH* also decreases to undetectable levels in late-stationary-phase cells (Feng et al., 1996).

*The SOS response up-regulates genetic variability*

SOS response is the set of physiological responses induced by exposure of *E. coli* cells to a variety of stressful conditions which damage DNA or interfere with DNA replication, like exposure to diverse physical or chemical DNA-damaging agents (for reviews see Oishi, 1988; Friedberg et al., 1995). SOS induction also occurs in bacteria in aging colonies on agar plates, in exhausted liquid media and during phosphate starvation (Dri and Moreau, 1993; Taddei et al., 1995). Starvation may be stressful because under such conditions cells might have to cope with an accumulation of DNA-damaging metabolic byproducts. Alternatively, the starvation might lead to blockage of replication due to depletion of DNA precursors.

SOS induction has also been detected during conjugation (Matic et al., 1995), during restriction of foreign or host DNA (Dharmalingam and Goldberg, 1980), during transposition of Tn*10* (Roberts and Kleckner, 1988) and IS*1* (Lane et al., 1994), during activation of the postsegregational killing system of the F plasmid (Cohen-Fix and Livneh, 1994) and when plasmid or phage DNA persists in single-stranded form (Higashitani et al., 1992; Gigliani et al., 1993). Many Gram-negative and Gram-positive bacteria possess SOS regulatory systems similar to that in *E. coli* with respect to the induced physiological responses and to the specificity of key regulatory interactions.

Regulation of the SOS system is mediated through the *recA* and *lexA* genes (Fig. 3). The LexA protein acts as a repressor for more than 20 genes, including the *recA* and *lexA* genes, while the RecA protein functions as a positive regulatory element in SOS induction. The inferred inducing signal is persistent single-stranded DNA, to which the RecA protein binds and becomes activated as a co-protease (RecA*) that promotes proteolytic cleavage of the LexA repressor, thus derepressing the SOS regulon.

This complex inducible system has two general features that favor survival of a stressed bacterial population: increased capacity for DNA repair and enhanced generation or genetic variation. It can be argued that the genetic variation component is a side effect of the DNA repair activities; however, some SOS functions with the potential to increase genetic diversity have no obvious DNA repair functions. A bacterial population in danger of extinction has little to lose by accelerated generation of genetic modifications in a quest for new genetic potential. Some genes which are directly involved in error-free DNA repair may also be involved in generation of genetic variability. For example, in *E. coli* the SOS induction promotes repair of double-strand beaks and daughter-strand gap recombi-

---

Figure 3. Regulation of the SOS system. SOS responses share the common characteristic of being regulated by the RecA* protein (closed square), and most of them, but not all, are negatively controlled by the LexA repressor (closed circle). The RecA* protein may have an additional role in SOS induction; it might process proteins other than LexA or activate transcription of unknown genes which are not under LexA control. The induction of the SOS system is gradual, and the level of induction is dependent on the degree to which the LexA pool has been decreased in response to a given SOS-inducing treatment. Weak SOS-inducing treatments lead

## NON-INDUCED STATE

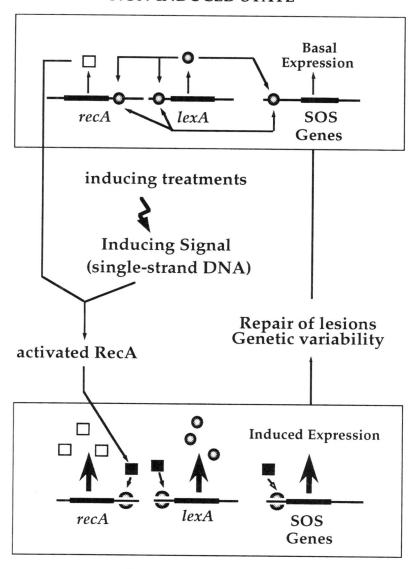

## INDUCED STATE

to increased expression of error-free nucleotide excision and recombinational repair genes, while other SOS responses, such as error-prone translesion synthesis or prophage induction do not occur unless the cell receives a stronger SOS-inducing treatment. As the cell begins to recover from the SOS-inducing treatment, the SOS-inducing signals disappear as a consequence of various DNA repair processes, and RecA molecules return to their non-activated state (open square). Continued synthesis of LexA molecules then leads to an increase in the LexA pool which, in turn, leads to repression of the SOS genes and a return to the uninduced state.

national repair, but it is also required for efficient conjugational and trans-
ductional recombination. SOS induction stimulates recombination between
partially homologous *E. coli rhsA* and *rhsB* sequences (0.9% divergence),
leading to chromosomal rearrangements (Dimpfl and Echols, 1989).

SOS induction-stimulated targeted (to DNA lesions) mutagenesis can be
considered a DNA repair function as well as a function that generates
genetic variation. Activation of the RecA protein, inactivation of the LexA
repressor and subsequent induction of *recA* gene expression are necessary
for the accumulation of "adaptive mutations" in F' *lac* genes during long-
term carbon source starvation (Cairns and Foster, 1991). SOS induction
also increases the transposition frequencies of Tn*5* and Tn*10* (Kuan et al.,
1991; Levy et al., 1993) and induces the expression of the *himA* gene,
whose protein product is required for site-specific recombination.

SOS response is induced in competent *Bacillus subtilis* cells in the
absence of exogenous DNA-damaging treatments (Yasbin et al., 1992). The
initial competence-specific induction of the RecA protein seems to be
independent of the SOS regulon. However, in competent *B. subtilis* cells,
single-stranded gaps are formed in chromosomal DNA, which, together
with transforming single-stranded DNA, provides a substrate for the activation of RecA. Activated RecA then derepresses SOS functions, leading to
a further increase in RecA concentration, which enhances the efficiency of
the transformational recombination.

Treatment of competent *E. coli* cells with SOS-inducing agents increases the efficiency of transformation with plasmid DNA (Vericat et al.,
1988). It is possible that this phenomenon is linked with association of
RecA* with the cell membrane of the SOS-induced *E. coli* cells, which has
been shown to change the levels of major outer membrane proteins. It is
also known that the RecA* protein increases the permeability of the *E. coli*
cell membrane (Swenson and Schenley, 1974; Tessman and Peterson,
1985). Therefore, it may be that modification in structure and permeability
of the membrane could facilitate entry of plasmid DNA.

Numerous SOS functions can be implicated in genetic exchange. During
interspecies conjugation, the DNA sequence divergence between genomes
of different species slows down the RecA-mediated recombination steps,
resulting in activation in the RecA protein co-protease activity. Consequently, the induction of the SOS response enhances interspecies recombination (Matic et al., 1995). Thus, interspecies conjugation acts as an
intracellular stress inducer in the recipient cells. Paradoxically, the DNA
sequence divergence, which is a major component of the interspecies
genetic barrier, helps cells to partially overcome this obstacle by triggering
SOS response.

The RecA* protein also promotes cleavage of repressors of a number of
temperate bacteriophages such as lambda, 434, 21, P22, f80 and coliphage
186. Prophage induction differs from other SOS responses in that it is an
irreversible process that causes cell death, but it can stimulate transduction.

SOS response-dependent restriction alleviation can increase the frequency of transduction and conjugation (Drake et al., 1983; Kowalczykowski, 1994). Furthermore, it has been found that double-strand exonuclease (ExoV) is inhibited in SOS induced cells, which might confer a hyper-recombinogenic phenotype (Thaler et al., 1988; Kannan and Dharmalingam, 1990; Rinken and Wackernagel, 1991).

### SOS and MRS fine-tune genetic variability

By the mechanisms described above, the actions of the SOS and MRS determine the rate of genetic diversification (Fig. 4). In general, the mismatch repair system is responsible for the constant maintenance of genome stability and its faithful transmission from one generation to the next; however, the activity of this down-regulator of genetic variability can be diminished. The SOS system is inducible and serves as an up-regulator of genetic variability under stress. Therefore, an abrupt environmental change which significantly deviates from the previous "stable" state can down-regulate the MRS (as discussed above) and fully induce SOS response (whose activities then become generators of variability), increasing the potential for adaptation either by acquisition of beneficial mutation(s) or by acquisition of functions by horizontal transfer.

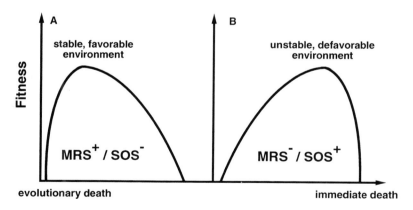

**The rate of accumulation of DNA sequence polymorphism**

Figure 4. Environment and the mutation rate. A mutation rate for an organism that is too low (A) would cause an evolutionary death due to the incapacity for adaptation to the changing environment, and one that is too high would cause an immediate death due to the error catastrophe (B). The optimal mutation rate for an organism (defined as the one maximizing fitness) lies between these extreme values. The rate varies depending on the fit between genotype and environment: the less adapted the population, the higher the mutation rate necessary to create individuals which will be adapted to the new environmental conditions. This would be accomplished through the effects of different states of the MR and SOS systems.

The genetic basis of SOS and MRS is, just like the rest of the genome, a dynamic structure. So it too contains variability, and the extent of this variability is controlled by the action of the very products it encodes, as just described. Another way of seeing this is that mutator genes control their own mutation rate. With an understanding of the mechanisms generating variability outlined above, we can begin to look at the evolutionary outcome of the response of these systems to stress followed by natural selection.

## Evolutionary implications of the fine-tuning of genetic variability

*The speed of the Red Queen*

The fine-tuning of mutation rates by MRS and SOS challenges the classical view of a mutation rate necessarily set at a minimum. Since the vast majority of mutations are neutral or deleterious, Kimura proposed that mutation rates are necessarily as low as possible (Kimura, 1968). This hypothesis is supported by the existence of numerous antimutator mechanisms ("evolutionary brakes" such as the MRS) and by the constancy of mutation rate per genome [1 mutation per 300 genomes replicated for DNA-based microorganisms (Drake, 1991)]. In this view, constraints like speed of replication and more generally cost of mutators prevent a null mutation rate to be reached. Comparative analysis of mutation rates, as well as findings from the fields of molecular biology and population genetics, further supported this minimal mutation rate paradigm.

But this paradigm, as well as the entire neo-Darwinian view, was challenged by work on "directed mutations" in *E. coli*. In the most studied cases of what was then called "adaptive mutagenesis", it was shown that mechanisms specific to the system submitted to selection (transposable element or plasmid) were induced by stress rather than by the selection itself (Shapiro, 1995). The debate then focused on whether mutations accumulated during the period of starvation are harmful but unavoidable given the level of stress [sick watchmaker hypothesis (Lenski and Sniegowski, 1995)] or whether they were due to specific genetic programs [genetic engineer hypothesis (Shapiro, 1995)].

As most experimental systems used in the adaptive mutation controversy have been linked with selfish elements (phage, transposon or plasmid), we can also speculate that selective pressure acting at their level rather than at the cellular level might be responsible for the observed phenomena. As described previously, replication of phages, plasmids and transposons is induced by stressful conditions which trigger SOS response. When their hosts are under stress, in an effort to increase their fitness, these selfish elements might transfer themselves to a new host or enhance the mutation rate of their current one.

If a host could enhance its own mutation rate in response to stress, we could ask whether the optimal mutation rate varies with the environment (Fig. 4). The rate would be minimal when conditions are favorable to growth and would increase in cases of poor adaptation to the environment. In the absence of a neo-Lamarckian mechanism generating only beneficial mutations, one way to maximize the advantage of a mechanism generating genetic variability is to induce it only in bacteria in environments where conditions are not favorable to growth.

Considering the effects of MR and SOS systems, such a scenario is not difficult to imagine at a molecular level. In a favorable environment the mutation rate would be maintained at a low level primarily through the action of the MRS, whereas during environmental stress conditions (radiation, chemical agents, starvation conditions, intra- and/or interspecific competition etc.) when SOS response is induced and the MRS down-regulated, it would be increased.

## Selecting for mutators

Whenever adaptation of organisms is imperfect, intraspecific competition favors mechanisms generating first-adapted genotypes. A mutator allele might help to generate such an adapted genotype. If recombination is low enough, selection for the fittest genotype would also select for the mutator mechanism that has generated it by hitchhiking. In the absence of a system generating only useful mutations, mutator alleles will have a cost due to the deleterious mutations they will cause.

A higher mutation rate is submitted to divergent selective pressure: counterselected at the interindividual level and selected for in an intergroup competition (the same holds true for other mechanisms like sex and migration that increase the distribution of fitness). What sort of evolutionary pattern this creates is being modelled using a population genetics approach. As we await these models, we can already make some general remarks.

The balance between the costs and benefits of a mechanism generating genetic variability will determine if it can be maintained. At least two parameters associated with a given mechanism are important to consider: its localization (either restricted to certain loci or generalized to the whole genome) and its duration (either induced only in periods of stress or permanent). Numerous such mechanisms have survived the test of natural selection. Since space and time are independent parameters, one can find both localized and generalized mechanisms which are permanent or inducible (Tab. 1).

There might even be a synergy among these different mechanisms. A given mutation hot spot might become hotter in mismatch repair-deficient or SOS-induced cells. Under a given selective pressure, such phenomena might make some mutation sites even hotter, whereas others would remain

Table 1. The mechanism that generate genetic variability

| Mutator mechanisms | Inducible | Constitutive |
|---|---|---|
| Localized | site-specific recombination (Arber, 1995) | mutation hot spot, e.g. repeated nucleotide motive $(X)_n$ (Moxon et al., 1994) |
| Generalized | induction of the SOS response, down-regulation of MRS | $MRS^-$ mutants |

cold. This could give the illusion of "directedness" of mutations, but it should rather be viewed as an effect of past selective pressure.

In the case of pathogenic bacteria, which are submitted to strong selective pressure, including that imposed by the immune system, some of these mechanisms which increase the speed of evolution are observed more frequently. Symmetrically, hosts are able to synthesize libraries of antibacterial molecules which are generated by inducible mechanisms analogous to, and often even homologous to, the bacterial ones (e.g. chromosomal rearrangement, gene conversion, hot spots of mutagenesis) (Weill and Reynaud, 1996). By restricting the generation or variability to specialized cells (the B and T cells of the immune system), the hosts diminish the costs associated with the generation of genetic variability, a strategy which seems to be used by bacteria as well (Higgins, 1992).

The importance of non-minimal mutation rates would be heightened if adaptation of organisms to their environment were at best ephemeral because of the impact of the evolution of different organisms on the environment (Ridley, 1993). This hypothesis is called the Red Queen hypothesis, a reference to the Lewis Carroll character who runs without moving forward because the landscape moves too. Under this assumption, mutation rates should vary depending on how fast the environment changes. The fate of a mechanism generating genetic variability would thus be linked with the intensity of the selective pressure (see also Sheldon, this volume).

Experiments performed in chemostat (a liquid, homogeneous and chemically stable environment) tested the possibility of selecting for mutator genes. Bacterial competition in this new environment led to the selection of mutator strains (Chao and Cox, 1983), whereas after 2000 generations in this environment the mutation rate of mutator strains was reduced (Tröbner and Piechocki, 1984).

These observations suggest that $Mut^-$ cells might be over-represented within stressed or recently adapted bacterial populations. However, in a stable and favorable environment, the advantage of a mutator phenotype will progressively deteriorate, leading to overgrowth of $Mut^+$ cells (Tröbner and Piechocki, 1984). The constant oscillation between stable and unstable environments in nature might result in the constant oscillation of the ratio of $Mut^-$ to $Mut^+$ cells in bacterial populations (Fig. 5B).

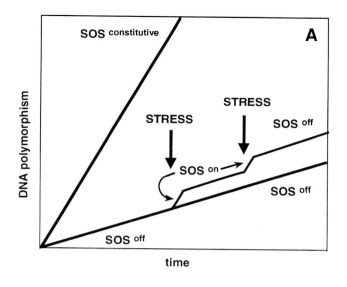

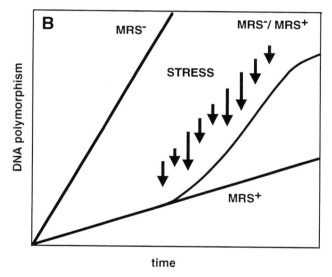

Figure 5. The influence of the environment on the rate of accumulation of DNA sequence poly-morphism. The accumulation of DNA sequence polymorphism depends on the selected rate of mutagenesis. In a stable environment, DNA sequence polymorphism increases slowly. (A) The stressful conditions that induce SOS response transiently increase mutagenesis and con-sequently the rate of accumulation of DNA sequence polymorphism. This inducible system that creates genetic variability only when needed minimizes the cost of deleterious mutations to the organism. (B) The mismatch repair capacity decreases in the stressed bacteria. At the popula-tion level, when bacteria are exposed to a stressful condition for a longer time period, the mutator alleles are selected for, because mutators are more likely to acquire adaptive muta-tion(s) than are wild-type cells. Consequently, the rate of mutagenesis increases on the popula-tion level until adaptation of the bacterial population is achieved. At this point the advantage of mutators will deteriorate, leading to the overgrowth of wild-type.

In nature, mutator strains are constantly generated from wild-type bacteria through mutations that inactivate *mut* genes. Strains defective for MRS are expected to arise at rates of $10^{-5}$ per genomic replication (Ninio, 1991). However, in natural populations of *E. coli*, about 1% of cells show enhanced mutagenesis as measured by antibiotic resistance (Jyssum, 1960; Gross and Siegel, 1981; Tröbner and Piechocki, 1984), of which the majority are expected to be mismatch repair mutants.

One possible explanation for this enrichment of mutators in natural bacterial populations is that under adverse environmental conditions the Mut⁻ cells are more likely to acquire adaptive mutation(s) than the wild-type cells. However, results obtained in chemostat also revealed a paradoxical feature: mutators seemed to be counterselected when rare and favored when abundant (Chao and Cox, 1983). The mechanisms by which mutators become abundant in nature without competing out cells having a lower mutation rate remains obscure.

*Stress, mutators and speciation*

Among constraints on the speed of evolution, the selective pressures of the past seem to play a major role. Not only have they selected for the adaptations of organisms to their environment but also for the mechanisms which generated them. Such mechanisms can sometimes respond very specifically, though blindly, to the stress induced by environmental changes.

How this occurs can now be imagined in molecular terms. The various states of MR and SOS systems, which are dependent on the extent of environmental stress, affect the mutation rate (Fig. 5). Given that mutation causes DNA sequence change, it follows that the rate of DNA diversification is also subject to environmental control. The acceleration of the speed of evolution during adaptive radiation of bacterial species (Woese, 1987) might support this view.

The scenario for a stress-induced burst of evolutionary change would be the following: during a period of intense stress, high mutation rates due to mismatch repair deficiency and/or SOS induction are to be expected, leading to the rapid diversification of an initially homogeneous population. If this situation persists long enough, polymorphism might accumulate to such an extent as to generate a barrier to genetic exchange. At the end of such a period of diversification, adaptation of the organisms to their new niches would favor restoration of a lower mutation rate. Since the same enzymatic systems, SOS and mismatch repair, also directly determine the effectiveness of genetic barriers based on sequence divergence, return to mismatch repair proficiency and repressed SOS would result in an immediate establishment of multiple genetic barriers within a highly polymorphic population, causing the speciation burst which could be seen as a punctuated equilibrium event (Fig. 6).

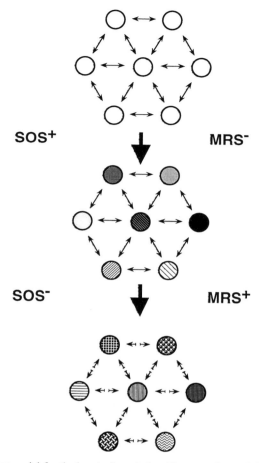

Figure 6. Molecular model for the burst of speciation. The opposing activities of the MR and SOS systems provide a potential for the fine tuning of both the rates of sequence diversification and the extent of genetic isolation, i.e. speciation. In a new habitat and/or when exposed to adverse environmental conditions, a bacterial population is in a period of high mutation rates due to induction of SOS responses, down-regulation of MRS and the selection for mutators at the population level. The adaptation of a bacterial population to the new environmental conditions leads to repression of the SOS system and the return to MR proficiency. Within such a highly polymorphic population, the accumulated DNA sequence polymorphism reduces genetic exchange and leads to establishment of multiple genetic barriers.

*Acknowledgments*
We thank M. Bianchetta for his help in the writing of this manuscript. This work was supported by grants from the Association pour la Recherche contre le Cancer and Actions Concertées Coordonnées-Sciences du Vivant du Ministère de l'Enseignement Supérieur et de la Recherche de France.

# References

Arber, W. (1995) The generation of variation in bacterial genomes. *J. Mol. Evol.* 40:7–12.

Cairns, J. and Foster, P.L. (1991) Adaptive reversion of a frameshift mutation in *Escherichia coli. Genetics* 128:695–701.

Chao, L. and Cox, E.C. (1983) Competition between high and low mutating strains of *Escherichia coli. Evolution* 37:125–134.

Cohen-Fix, O. and Livneh, Z. (1994) *In vitro* UV mutagenesis associated with nucleotide excision-repair gaps in *Escherichia coli. J. Biol. Chem.* 269:4953–4958.

Cupples, C.G., Cabrera, M., Cruz, C. and Miller, J.H. (1990) A set of lacZ mutations in *Escherichia coli* that allow rapid detection of specific frameshift mutations. *Genetics* 125:275–280.

Damagenez, V., Doutriaux, M.P. and Radman, M. (1989) Saturation of mismatch repair in the *mutD5* mutator strain of *Escherichia coli. J. Bacteriol.* 171:4494–4497.

Dharmalingam, K. and Goldberg, E.B. (1980) Restriction *in vivo*. V. Induction of SOS functions in *Escherichia coli* by restricted T4 phage DNA, and alleviation of restriction by SOS functions. *Mol. Gen. Genet.* 178:51–58.

Dimpfl, J. and Echols, H. (1989) Duplication mutation as an SOS response in *Escherichia coli*: Enhanced duplication formation by a constitutively activated RecA. *Genetics* 123:255–260.

Drake, J.W. (1991) A constant rate of spontaneous mutation in DNA-based microbes. *Proc. Natl. Acad. Sci. USA* 88:7160–7164.

Drake, J.W., Glickman, B.W. and Ripley, L.W. (1983) Updating the theory of mutation. *Am. Sci.* 71:621–630.

Dri, A.-M. and Moreau, P.L. (1993) Phosphate starvation and low temperature as well as ultraviolet irradiation transcriptionally induce the *Escherichia coli* LexA-controlled gene *sfiA. Mol. Microbiol.* 8:697–706.

Feige, U., Morimoto, R.I., Yahara, I. and Polla, B.S. (1996) *Stress-Inducible Cellular Responses.* Birkhäuser, Basel.

Feinstein, S.I. and Low, K.B. (1986) Hyper-recombining recipient strains in bacterial conjugation. *Genetics* 113:13–33.

Feng, G., Tsui, H.-C.T. and Winkler, M.E. (1996) Depletion of the cellular amounts of the MutS and MutH methyl-directed mismatch repair proteins in stationary-phase *Escherichia coli* K-12 cells. *J. Bacteriol.* 178:2388–2396.

Foster, P.L., Gudmundson, G., Trimarchi, J.M., Cai, H. and Goodman, M.F. (1995) Proofreading-defective DNA polymerase II increases adaptive mutation in *E. coli. Proc. Natl. Acad. Sci. USA* 92:7951–7955.

Friedberg, E.C., Walker, G.C. and Siede, W. (1995) *DNA Repair and Mutagenesis.* ASM Press, Washington, DC.

Gigliani, F., Ciotta, C., Del Grosso, M.F. and Battaglia, P.A. (1993) pR plasmid replication provides evidence that single-stranded DNA induces the SOS system *in vivo. Mol. Gen. Genet.* 238:333–338.

Gross, M.D. and Siegel, E.C. (1981) Incidence of mutator strains in *Escherichia coli* and coliforms in nature. *Mutat. Res.* 91:107–110.

Higashitani, N., Higashitani, A., Roth, A. and Horiuchi, K. (1992) SOS induction in *Escherichia coli* by infection with mutant filamentous phage that are defective in initiation of complementary-strand DNA synthesis. *J. Bact.* 174:1612–1618.

Higgins, N.P. (1992) Death and transfiguration among bacteria. *Trends Biochem Sci* 17:207–211.

Jyssum, K. (1960) Observation of two types of genetic instability in *Escherichia coli. Acta Pathol. Microbiol. Immunol. Scand.* 48:113–120.

Kannan, P. and Dharmalingam, K. (1990) Induction of the inhibitor of the RecBCD enzyme in *Escherichia coli* is a *lexA*-independent SOS response. *Current Microbiol.* 21:7–15.

Kimura, M. (1968) Evolutionary rate at the molecular level. *Nature* 217:624–626.

Kowalczykowski, S.C. (1994) *In vitro* reconstitution of homologous recombination reactions. *Experientia* 50:204–215.

Kuan, C.-T., Liu, S.-K. and Tessman, I. (1991) Excision and transposition of Tn5 as an SOS activity in *Escherichia coli. Genetics* 128:45–57.

Lane, D., Cavaille, J. and Chandler, M. (1994) Induction of the SOS response by IS1 transposase. *J. Mol. Biol.* 242:339–350.

Lenski, R.E. and Sniegowski, P.D. (1995) Adaptive mutation: The debate goes on. *Science* 269:285–288.

Levy, M.S., Balbinder, E. and Nagel, R. (1993) Effects of mutations in SOS genes on UV-induced precise excision of Tn10 in *E. coli. Mutation Res.* 293:241–247.

Li, W.-H. (1993) So, what about the molecular clock hypothesis? *Curr. Opin. Genet. Dev.* 3:896–901.

Lundblad, V. and Kleckner, N. (1985) Mismatch repair mutations. of *Escherichia coli* K12 enchance transposon excision. *Genetics* 109:3–19.

Maas, W.K., Wang, C., Lima, T., Zubay, G. and Lim, D. (1994) Multicopy single-stranded DNAs with mismatched base pairs are mutagenic in *Escherichia coli. Mol. Microbiol.* 14:437–441.

Maas, W.K., Wang, C., Lima, T., Hach, A. and Lim, D. (1996) Multicopy single-stranded DNA of *Escherichia coli* enhances mutation and recombination frequencies by titrating MutS protein. *Mol. Microbiol.* 19:505–509.

Matic, I., Rayssiguier, C. and Radman, M. (1995) Interspecies gene exchange in bacteria: The role of SOS and mismatch repair systems in evolution of species. *Cell* 80:507–515.

Matic, I., Taddei, F. and Radman, M. (1996) Genetic barriers among bacteria. *Trends Microbiol.* 4:69–73.

Modrich, P. (1991) Mechanisms and biological effects of mismatch repair. *Annu. Rev. Genet.* 25:229–253.

Modrich, P. and Lahue, R. (1996) Mismatch repair in replication fidelity, genetic recombination and cancer biology. *Annu. Rev. Biochem.* 65:101–133.

Morris, J.G. (1993) Bacterial shock response. *Endeavour* 17:2–6.

Moxon, E.R., Rainey, P.B., Nowak, M.A. and Lenski, R.E. (1994) Adaptive evolution of highly mutable loci in pathogenic bacteria. *Current Biology* 4:24–33.

Ninio, J. (1991) Transient mutators: A semiquantitative analysis of the influence of translation and transcription errors on mutation rates. *Genetics* 129:957–962.

Oishi, M. (1988) Induction of recombination-related functions (SOS functions) in response to DNA damage. *In*: K. Brooks Low (ed.): *The Recombination of Genetic Material*. Academic Press, San Diego, CA, pp. 445–491.

Pang, P.P., Lundberg, A.S. and Walker, G.C. (1985) Identification and characterization of the *mutL* and *mutS* gene products of *Salmonella typhimurium* LT2. *J. Bacteriol* 163:1007–1015.

Petit, M.A., Dimpfl, J., Radman, M. and Echols, H. (1991) Control of chromosomal rearrangements in *E. coli* by the mismatch repair system. *Genetics* 129:327–332.

Radman, M. and Wagner, R. (1993) Mismatch recognition in chromosomal interactions and speciation. *Chromosoma* 102:369–373.

Radman, M., Wagner, R. and Kricker, M.C. (1993) Homologous DNA interactions in the evolution of gene and chromosome structure. *In*: K.E. Davies (ed.): *Genome Analysis*, Volume 7. Cold Spring Harbor Laboratory Press, Cold Spring Harbor, pp. 139–152.

Radman, M., Matic, I., Halliday, J. and Taddei, F. (1995) Editing DNA replication and recombination by mismatch repair: From bacterial genetics to mechanisms of predisposition to cancer in humans. *Phil. Trans. R. Soc. Lond.* B 347:97–103.

Rayssiguier, C., Thaler, D.S. and Radman, M. (1989) The barrier to recombination between *Escherichia coli* and *Salmonella typhimurium* is disrupted in mismatch-repair mutants. *Nature* 342:396–401.

Ridley, M. (1993) The Red Queen: Sex and the evolution of human nature. *In*: *Penguin Science*. Penguin Books, London.

Rinken, R. and Wackernagel, W. (1992) Inhibition of the RecBCD-dependent activation of Chi recombinational hot spots in SOS-induced cells of *Escherichia coli. J. Bacteriol.* 174:1172–1178.

Roberts, D. and Kleckner, N. (1988) Tn10 transpositon promotes RecA-dependent induction of a λ prophage. *Proc. Natl. Acad. Sci. USA* 85:6037–6041.

Schaaper, R.M. and Radman, M. (1989) The extreme mutator effect of *Escherichia coli mutD5* results from saturation of mismatch repair by excessive DNA replication errors. *EMBO J.* 8:3511–3516.

Shapiro, J. (1995) Adaptive Mutation: Who's really in the garden? *Science* 268:373–374.

Swenson, P.A. and Schenley, R.L. (1974) Respiration, growth and viability of repair-deficient mutants of *Escherichia coli* after ultraviolet irradiation. *Int. J. Radiat. Biol. Relat. Study. Phys. Chem. Med.* 25:51–60.

Taddei, F., Matic, I. and Radman, M. (1995) Cyclic AMP-dependent SOS induction and mutagenesis in resting bacterial populations. *Proc. Natl. Acad. Sci. USA* 92:11736–11740.

Tessman, E.S. and Peterson, P. (1985) Plaque color method for rapid isolation of novel *recA* mutants of *Escherichia coli* K-12: New classes of protease-constitutive *recA* mutants. *J. Bact.* 163:677–687.

Thaler, D.S., Sampson, E., Siddiqi, I., Rosenberg, S.M., Stahl, F.W. and Stahl, M.M. (1988) A hypothesis: Chi-activation of recBCD enzyme involves removal of the recD subunit. *In*: E. Friedberg and P. Hanawalt (eds): *Mechanisms and Consequences of DNA Damage Processing*. Alan R. Liss, New York, pp. 413–422.

Tröbner, W. and Piechocki, R. (1984) Selection against hypermutability in *Escherichia coli* during long-term evolution. *Mol. Gen. Genet.* 198:177–178.

Vericat, J.A., Guerrero, R. and Barbé, J. (1988) Increase in plasmid transformation efficiency in SOS-induced cells. *Mol. Gen. Genet.* 211:526–530.

Weill, J.-C. and Reynaud, C.-A. (1996) Rearrangement/hypermutation/gene conversion: When, where and why. *Immunology Today* 17:92–97.

Woese, C.R. (1987) Bacterial evolution. *Microbiol. Rev.* 51:221–271.

Worth, L. Jr., Clark, S., Radman, M. and Modrich, P. (1994) Mismatch repair proteins MutS and MutL inhibit RecA-catalyzed strand transfer between diverged DNAs. *Proc. Natl. Acad. Sci. USA* 91:3238–3241.

Yasbin, R.E., Cheo, D.L. and Bayles, K.W. (1992) Inducible DNA repair and differentiation in *Bacillus subtilis*: Interactions between global regulons. *Mol. Microbiol.* 6:1263–1270.

Environmental Stress, Adaptation and Evolution
ed. by R. Bijlsma and V. Loeschcke
© 1997 Birkhäuser Verlag Basel/Switzerland

# Stress-resistance genotypes, metabolic efficiency and interpreting evolutionary change

Peter A. Parsons

*School of Genetics and Human Variation, La Trobe University, Bundoora, Victoria 3083, Australia*
*For correspondence: 21 Avenue Road, Glebe, NSW 2037, Australia*

*Summary*. Assuming stress levels to which free-living populations are normally exposed, an association between rapid development time, a long life, success in mating and size of sexual ornaments can be predicted. Fitness at one stage of the life cycle should therefore correlate with fitness at other stages under this environmental model. Assuming that stress targets energy carriers, high-energy efficiency underlain by stress-resistance genotypes that are likely to be heterozygous is the basis of this prediction.

Stress-resistance genotypes therefore have a role in promoting the energy efficiency required for organisms to accommodate a stressed world. Selection for energy efficiency to utilize heterogenous resources implies that the process of speciation should normally occur rapidly and be rarely observed. It follows that the ecological species concept is primary to other species concepts. The intensity of selection for stress resistance goes from an extreme in the highly disturbed and stressful environments of living fossils to relatively stable abiotic habitats, where specialist diversifications and adaptive radiations are likely. Between these extremes, a punctuated pattern of evolutionary change may occur in perturbed environments during a transient phase of increased resources. In abiotically benign tropical habitats where energy constraints are low, specialization of resource utilization by learning appears possible.

## The universality of stress

Natural environments have always been hostile for organisms. Exposure to climatic stress is the norm in nature, and hydrological changes, especially drought, are of major importance (Parsons, 1995a). Human interference can exacerbate these effects. Especially since 1980, trends in deforestation and atmospheric change are leading towards lowered precipitation, increased seasonality and more extreme weather conditions in the tropics (Hartshorn, 1992). Consequently, the slowest-growing shade-tolerant, and presumably stress-sensitive, tropical trees are particularly vulnerable to these increasingly extreme abiotic conditions (Phillips and Gentry, 1994).

The abundance of organisms is mainly determined by a relative shortage of resources for the young, including sources of nitrogen that animals need to produce protein (White, 1993). Such shortages are a feature of increasingly severe famines affecting human populations. Most animals have the capacity to produce far more offspring than survive, so mortality levels can be extreme. For instance, in male Soay sheep, *Ovis aries*, on the St. Kilda archipelago of Scotland, more than 90% of juvenile males may die in periodic over-winter crashes (Clutton-Brock et al., 1992). Together with

Table 1. Effect of shelter on lamb mortality

|                  | Exposed groups | Sheltered groups |
| ---------------- | -------------- | ---------------- |
| Born             | 287            | 278              |
| Dead within 48 h | 100            | 11               |

(Summarized from Rowley, 1970.)

adult and yearling mortalities, the population size can fall by around 65% during these die-offs.

Extreme weather increases the severity of nutritional deprivation and vulnerability to disease, especially in the young and the very old. For example, the major factor underlying neonatal lamb mortality in Australia is faulty nutrition of the ewe during pregnancy followed by climatic stress during and following parturition (Rowley, 1970). In one survey nearly half of the lambs were starving at death; lambs weakened by starvation became vulnerable to climatic stress, so predisposing conditions for predation and disease were established. Conversely, providing shelter substantially reduced mortality (Tab. 1).

Abiotic stress periods occur frequently, as shown by El Niño-Southern Oscillation events. During the 1982–83 episode, there were changes in primary and secondary productivity resulting in nutritional deprivation affecting many species (Glynn, 1988). In *Drosophila melanogaster*, high reproductive potential of these flies tends not to be expressed under natural conditions because of substantial and variable deficiencies in food availability (Boulétreau-Merle et al., 1987). In any case, adults of English *Drosophila* populations had a mean life expectancy of 1.3 to 6.2 days, which is at least an order of magnitude less than survival times under equable laboratory conditions (Rosewell and Shorrocks, 1987).

Many North American bird species have northern margins at around the mean minimum temperature at which the metabolic rate is 2.5 times the basal metabolic rate, regardless of habitat, body size or relatedness (Root, 1993). Therefore, the northern margins are determined by the amount of metabolic energy birds need to expend to keep warm. As conditions deviate from optimal, energy costs increase, so that physical conditions can limit organisms to particular habitats.

In modern times, climatic stress often occurs in combination with environmental pollutants and toxins. For instance, in polar bears, exposure to oil increased metabolic rate (Hurst et al., 1991), which would increase the demand for food. The consequence is higher vulnerability to death, especially during the stress of a hard winter, because there is insufficient energy from resources to counter the costs from abiotic stresses.

In feral rock doves, *Columbia livia*, lice reduce feather and host body mass, and increase thermal conductance and metabolic rate, indicating an energy cost which is exacerbated in a deteriorating abiotic environment during winter (Booth et al., 1993). More generally, increases in metabolic rate for small birds can be costly, especially during bad weather. In the colonially mating cliff swallow, *Hirundo pyrrhonata*, mark-recapture experiments over an 8-year period showed that the annual survivorship of birds parasitized with cimicid bugs, fleas and chewing lice was 0.38, compared with 0.57 for fumigated, non-parasitized birds (Brown et al., 1995).

Stress therefore has an energy cost threatening survival. Conversely, the more energy an organism can obtain from resources, the larger its reserves for accommodating stress. Since exposure to stress is the norm in natural populations, this implies a need for some energy to permit existence in any habitat. More often than not, the energy balance characteristic of free-living populations implies that organisms normally struggle to survive in an environment that is variably inadequate nutritionally and energetically. This conclusion is consistent with Kauffman (1993), who argues that organisms face an extremely perturbed world in which "life exists at the edge of chaos". Consequently, adaptive change occurs in the context of continuous environmental challenges. For change to be achievable, he proposes an underlying order in complex systems, under the control of a connected metabolism.

Following Lewontin (1970), the basic tenets of Darwin's (1859) theory of evolution are (1) organisms vary, (2) this variation is heritable and (3) this variation is selected. In selecting variation, Darwin emphasized biotic variables, especially complex interactions among species. Here, I assume a far more stressed world where abiotic variables dominate, and then proceed in the context of energy balances to develop an integrative reductionist approach to evolutionary change. Biotic variables such as competition can be incorporated in terms of energy costs, which appear second order in magnitude compared with abiotic stresses (Parsons, 1996a). Even if the model is too extreme, it provides a boundary condition which is being increasingly approached in many parts of the world today. In addition, the fossil record almost certainly reflects this boundary, since biotic factors are far more transient in historical and geological time than are abiotic factors.

## Adaptation under stress: Fitness and stress-resistance genotypes

Some organisms accommodate extreme aridity stress in nature by a reduction in the rate of metabolism, compared with closely related species from wetter regions. Hoffmann and Parsons (1991) developed a laboratory parallel from selection experiments for desiccation resistance in *D. melanogaster*. In these experiments, desiccation resistance rapidly increased, but metabolic rate fell. In addition, early fecundity and behavioural activity fell. Therefore, stress resistance can be readily acquired but is associated with reduced fitness in the absence of the stress.

High-stress resistance appears to be associated with the efficient use of metabolic resources underlying growth and reproduction, especially when resources are limited (Koehn and Bayne, 1989). This suggests that such genotypes can support growth under a wide range of conditions, i.e. there is selection for energy and metabolic efficiency. Quantitative shifts in life history traits should follow periods of strong selection under stress (Parsons, 1993). This assessment assumes that an important target of selection of desiccation and other generalized stresses is at the level of energy carriers, as implied by the fall in metabolic rate in the above experiments for desiccation resistance. More generally, Kohane (1994) provided the co-factor adenine nicotinamide dinucleotide (NAD) as a novel energy source which reduced development time in three isofemale strains of *D. melanogaster*. An energetic interpretation is supported, since larval development time is an important fitness component, and is partly under genetic control. Consequently, selection under stress, by targeting energy carriers, should directly affect this and other life history traits.

Taking into account progressively accumulating energy costs during ageing, an association between metabolic efficiency and stress resistance implies that a long life is likely to be underlain by genes for stress resistance (Parsons, 1995b). Indeed, a correlated response to selection for desiccation resistance in *D. melanogaster* is increased longevity (Hoffmann and Parsons, 1993). Similarly, the *age-1* gene of the nematode *Caenorhabditis elegans* increased life span by 65%, together with a correlated increase in resistance to high temperature (Lithgow et al., 1995).

Oxidative stress causing damage from free radicals makes a major contribution to the rate of ageing. Accordingly, mutants with extended life-span tend to show high resistance to oxidative stress. Conversely, *D. melanogaster* mutants in which specific components of the oxygen defense mechanism were disrupted have a short life-span and are stress-sensitive (Hilliker et al., 1992).

Because metamorphosis in *Drosophila* imposes a crisis of oxygen stress on the developing embryo, and an ability to cope with the high energy cost of development is at a premium, this is a time when metabolic efficiency is important. Stress-resistant individuals that develop rapidly should be favoured. Interestingly, and in agreement, Loeschcke and Krebs (1996) found that temperature-resistant strains of *D. buzzatii* often had a relatively short development time. Rapid emergence thus appears to be a forerunner of a long life in free-living populations. Therefore, following Sohal and Allen (1990), oxidative stress is a major component of the stress theory of ageing which incorporates the developmental phase (Parsons, 1995b, 1996b). Under this theory, high vitality, vigour, resilience and physiological homeostasis enable survival to an old age in the face of the internal (mainly oxidative) stresses to which all organisms are exposed, and to the external (mainly abiotic from climatic extremes) stresses to which free-living organisms are additionally exposed.

With reference to two major and extensively discussed evolutionary theories of ageing, the mutation-accumulation and the antagonistic pleiotropy mechanisms, Curtsinger et al. (1995) concluded that experimental support for them is not strong, and they favour a model correlating old and young fitness components. This conclusion agrees with a major prediction of the stress theory of ageing that fitness at various developmental stages should be correlated in free-living populations.

Secondary sexual traits are extremely exaggerated forms of ordinary traits, close to their limits of production and maintenance, implying substantial energy costs. Ultimately, under extreme circumstances, even a minor increase in size could be deleterious. Further, these are times when ornaments are very sensitive to environmental stresses, especially abiotic (climatic) perturbations, inadequate nutrition and parasites (e.g. Halliday, 1978; Zuk et al., 1990). This suggests that a high level of metabolic efficiency is a prerequisite for carrying extreme sexual ornaments, which should therefore be underlain by genes for stress resistance (Parsons, 1995 c). A similar argument applies for mating success, with respect to fast mating individuals.

Assuming that fast mating individuals that may have extreme ornaments carry stress-resistance genes, such males should transfer these genes to their offspring, who should develop rapidly and live longest. Examples of such associations are discussed by Parsons (1996b). They include (1) the African cockroach *Nauphoeta cinera*, which show a substantial correlation between offspring development rate and attractiveness of males in mating (Moore, 1994), (2) the damsel fly, *Ischnura graellsii*, for which the best predictor of male lifetime mating success is mature life span (Cordero, 1995), and (3) barn swallows, *H. rustica*, whose offspring longevity is correlated with ornament size of male parent (Møller, 1994).

Collectively, these and a few other examples suggest that rapid development, long life span, extremes of sexual ornament and success in mating can be considered as a suite of characters conferring high fitness, assuming the stressful environments of free-living populations. The paucity of empirical observations for this relationship may relate to the unlikeliness of studies carried out under relatively benign laboratory conditions to be efficient in revealing such associations. On the other hand, irrespective of the background environment, associations of development time and life span with mating success and the size of ornaments should be detectable, because mating and the development and maintenance of sexual ornaments are normally energetically expensive processes.

The genetical situation is, however, more complex, since there is a tendency for heterozygosity levels in natural populations to correlate with fitness, expecially for enzyme loci influencing metabolism and contributing to the amount of energy available for development and growth. The correlation is clearest under stressful conditions, when energy demands from the environment are high (Mitton, 1993). Heterozygotes tend to have lower energy requirements than homozygotes, suggesting greater metabolic efficiency, and they should have the potential to develop and reproduce under a wider range of environmental conditions than do homozygotes (Koehn and Bayne, 1989). Based upon a substantial and diffuse literature, the collective evidence indicates that rapid development, long life span, extremes of sexual ornaments and success in mating tend to be correlated with high heterozygosity. Furthermore, these associations extend to low fluctuating morphological asymmetry, which is a general indicator of high fitness (see Parsons, 1997, and Møller, this volume).

Two approaches suggest parallel associations for a range of fitness traits. The first approach commences at the whole organism level and leads to genes for stress resistance, while the second approach commences at the gene level using electrophoretic variants and leads to generalized heterozygous advantage. These approaches have developed largely independently in the past, but can be unified by the need for metabolic and energy efficiency to adapt to the stresses to which free-living populations are normally exposed. The stresses are quite general, and appear not to be specific to a particular metabolic pathway. The generalized advantage of heterozygotes under stress does, however, suggest that many interacting loci could be involved, although this does not preclude some genes of major effect, e.g. isozymes at the phosphoglucose isomerase locus, which appear targeted by abiotic stress since the locus regulates flux through glycolysis (Watt, 1985; Riddoch, 1993). Even so, the "good genes" hypothesis in interpreting sexual selection appears genetically simplistic for natural populations, and should be replaced by "good genotypes" hypothesis (Parsons, 1997).

The above considerations directly translate into an expectation of positive correlations among life history traits. It is impossible to review the enormous literature on this topic (but see Stearns 1992; Roff, 1996). Much of life history theory assumes limited energy availability of resources which gives negative associations among fitness characters (e.g. Sibley and Calow, 1986). Rollo (1994) makes the point that negative correlations based upon the principle of resource allocation have been most consistently obtained in laboratory comparisons utilizing experimental manipulations of phenotypes and the consequences of artificial selection regimes on key fitness components. Studies based on unmanipulated populations yielded a mixture of results, but positive correlations predominated. Unmanipulated populations are likely to reflect the situation in nature more than laboratory populations, expecially those that have been in such "artificial" or "domesticated" environments for many generations.

The heterogeneity of results from unmanipulated populations is reasonable, since not all populations would be in a state of energetic stress. However, the overall shift towards positive correlations is to be expected. At the extreme is the model analyzed in this paper, which, if nothing else, provides a boundary condition for comparisons with more benign circumstances. In this context, long-term ecological studies on stress levels in key natural populations, including *Drosophila*, are a high priority.

In summary, under the stressful scenario pertaining in free-living populations, and taking into the account the characteristics of those individuals surviving to an extremely old age (Parsons, 1996c, d), it appears that stress resistance and metabolic efficiency in free-living populations should be associated with

(1)   Stress resistance genes and genotypes
(2)   High (electrophoretic) heterozygosity
(3)   High vitality, vigour and resilience
(4)   High homeostasis in response to external stresses
(5)   Low fluctuating asymmetry
(6)   Rapid development
(7)   High male mating success
(8)   Extremes of sexual ornaments
(9)   Long life-span
(10) Positive correlations among fitness traits

This generalization is based on the assessment that stress acts as an environmental probe targeting energy balances, and hence various fitness traits. Fitness in nature is therefore maximized by having stress-resistance "good genotypes". This conclusion provides a background for considering patterns of evolutionary change.

## Conditions favouring evolutionary change

*The punctuated evolutionary pattern versus stasis*

Evolutionary change at the morphological level is arrested in "living fossils" over long time periods that may encompass one or more mass extinction events. Living fossils include stromatolites in supersaline regions of searing heat and low precipitation. These habitats are so stressful that predators and competitors do not survive, so stresses must be totally abiotic. In these energetically costly environments little obvious evolutionary change occurs, and morphological stasis prevails (Ward, 1992; Parsons, 1994).

Following the great Palaeozoic extinction crisis, when abiotic stresses were probably less than during the actual extinction event, early Triassic sequences were characterized by abundant stromatolites (Schubert and Bottjer, 1992; Erwin, 1993). This implies an increase in resource availability compared with the time of the crisis, and this may have underlain the diversification commencing in the early Triassic. Similarly, it is likely that the explosive period of diversification in the early Cambrian was associated with rapidly changing abiotic stress combined with an increase in nutrient availability and type (Bowring et al., 1993). Furthermore, the first occurrence of post-Palaeozoic marine benthic invertebrate higher taxa occurred in resource-rich estuarine habitats, where there were large fluctuations in abiotic stress (Jablonski and Bottjer, 1990).

The increase or change in resources would be patch in populations that were probably subdivided into isolated and semi-isolated habitats, interspersed with inhospitable areas. A diversity of forms is then possible following adaptation of organisms to these varying habitats. Even though many forms would not survive, rapid diversification could occur among the survivors.

A possible example of this scenario in the living biota comes from succulent plants of the Mesembryanthemaceae from the Succulent Karoo region, an arid area of southern Africa of heterogeneous soils, local climatic conditions, and varying levels of soil moisture. Rapid evolutionary divergence has occurred within the last 5 million years. In spite of generally arid conditions, the wide heterogeneity of fragmented microhabitats appears to underlie high species diversity, approaching that of rain forests (Ihlenfeldt, 1994). There is a parallel in southwestern Australia, where there is an exceedingly diverse flora in a fragmented unstable transitional zone between a humid, high rainfall zone and a low rainfall, extremely arid zone (Hopper, 1979).

Accordingly, disturbed conditions appear to be conducive to major diversifications, assuming that the energy from resources is sufficient to overcome the costs from abiotic stresses. Under these circumstances, there could be periodic relaxations of stress intensity, permitting the development of morphologically diverse forms. However, on a reversion to more stressed circumstances, especially during mass extinction events, phylogenetically bizarre forms including carriers of sexual ornaments would be vulnerable (Eldredge, 1991; McLain, 1993). Small, common and generalist species that are genetically stress-resistant should survive such crises best, and would be living fossils under the most extreme of conditions.

Two situations are apparent above: (1) stasis, where survival is at a premium, and (2) evolutionary change under energetically more permissive conditions, where there is increased resource availability in abiotically disturbed habitats. In combination, these situations give a pattern of stasis punctuated by brief periods of diversification. This is the punctuated evolutionary pattern of Eldredge and Gould (1972). Sheldon (1993 and this volume) argues for this situation in abiotically unstable environments, based on his work on Builth trilobites.

## Specialist diversification

Many tropical habitats are relatively stable abiotically, and the premium on the energy cost to counter stress is thus less than in temperate zones.

Similarly, some marine and lake habitats are relatively stable abiotically. In both habitat categories specialist diversifications, and in some cases adaptive radiations, occur.

Recent work on fish in lake habitats provides useful models. In the Arctic charr, *Salvelinus albinus*, in Iceland, four morphs adapt to a range of resource categories with considerable precision or fine-tuning (Skúlason et al., 1989; Malmquist et al., 1992). Consequently, the energy returns to charr appear to be maximized by genetic divergence among morphs, since this reduces competitive interactions and enables the efficient use of a range of resources (Parsons, 1996a). Such energy efficiency would appear compatible with ecological speciation based upon niche-specific fitnesses in habitats showing considerable stability in total nutrient availability.

Schluter and McPhail (1993) record a number of examples where similar morphological and ecological shifts have occurred multiple times under similar circumstances to evolve sympatric species involving limnetic and benthic forms. The limnetic species are tyically smaller, with a smaller mouth and larger, more numerous gill rakers than benthic species. The morphological differences therefore correlate with differing feeding performances. Schluter and McPhail (1993) conclude that the benthic-limnetic split could be a predictable first step in the diversification of many fish taxa in low-diversity post-glacial lakes, where the efficiency of utilization of restricted available resources is at a premium.

In contrast, in the Arctic charr the resource polymorphism was swamped when excess resources were added artificially (Skúlason et al., 1989), presumably breaking down the energetic fine-tuning under more restrictive conditions. Similarly, in many habitats the energy cost of increased climatic instability would appear sufficient to swamp the progressive selection for energy efficiency implied by the continual selection for increasing fine-tuning for adaptation to resource heterogeneity. This situation may be occurring in fragmented rain forests today (Phillips and Gentry, 1994).

During rather benign periods, when there is a relaxation of selection for stress resistance, specialist diversifications are likely. However, these ecologically specialized species are vulnerable to environmental disturbance, and would be expected to exist on a shorter time frame than generalist species (Vrba, 1987). Rapid habitat change would therefore differentially eliminate stress-sensitive species. Vrba (1987), however, argues that such habitat disruption would lead to disjunct and outlier populations from which new species could occasionally evolve, differing from those gone extinct. Events of rapid environmental change are therefore likely to be associated with both extinction and speciation. The few resource-use generalists would carry stress-resistance genes enabling persistence in the face of such environmental perturbations.

Since specialist diversification tends to occur under rather restricted conditions, fragility in these habitats is implied. The reduced premium on selection for stress resistance in these habitats means that genes for stress sensitivity could accumulate as found under relatively benign laboratory and domesticated circumstances (Parsons, 1996c, d). Similarly, tropical resource-specialist *Drosophila* species tend to be sensitive to stress compared with those from temperate regions (Parsons, 1982).

An extreme version of specialization in an environment where energy constraints are apparently minimal comes from the finch, *Pinaroloxias inornata*, on Cocos Island, Costa Rica. These birds are exposed to an aseasonal tropical environment with very few competing species, and where there is high availability, variety and predictability of food resources. Here, energy costs from the abiotic and biotic environments appear low. Under these circumstances, the finches show behavioural specialization, which is transmitted at least partly culturally, for the utilization of an extremely heterogeneous range of resources (Werner and Sherry, 1987). In other words, these tropical birds may have been selected for learning ability enabling the efficient exploitation of heterogeneous resources, in an environment where the energy cost from abiotic perturbations is minimal.

## Evolutionary patterns

Based on the magnitude of abiotic perturbations, the above examples can be placed in a continuum with living fossils at one extreme and diversification following learning at the other. At a reductionist level, this is a gradient of environmental stability in the face of stress levels ranging from extreme to benign, as follows:

(1) Stasis under extreme perturbations represented by living fossils;
(2) An ameliorated version of (1) so that rapid evolutionary change can briefly occur, followed by stasis giving a punctuated evolutionary pattern;
(3) Relatively stable abiotic environments where there is a premium on energy and metabolic efficiency, enabling adaptation to heterogeneous resources leading to specialist diversifications; and
(4) An aseasonal unstressed tropical environment with adequate to abundant resources, where diversification can occur by learning without requiring the energy cost of genetic change.

There is a descending intensity of selection for genes for stress resistance from (1) to (4), so there would be a tendency towards stress sensitivity as the usage of energy becomes progressively more fine-tuned for the exploitation of diverse resources.

The extremes of (1) and (4) are at the opposite ends of a continuum of increasing efficiency of the use of metabolic energy in excess of that needed for maintenance and survival, since the energy needed to counter abiotic stress decreases along the same continuum. This situation is shown by the

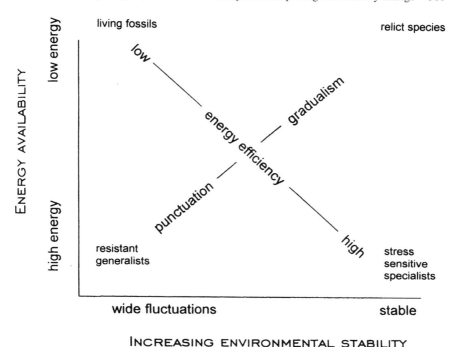

Figure 1. The interaction between the stability of environmental fluctuations and energy availability in excess of maintenance and survival. The four corners represent limiting extremes. The leading diagonal indicates increasing energy efficiency from the extreme of living fossils to stress-sensitive specialists, while the reverse diagonal indicates patterns of evolution ranging from punctuational change to phyletic gradualism. (Adapted from an initial presentation in Parsons, 1993.)

leading diagonal in Figure 1, which is a plot of the level of environmental stability and available metabolic energy.

The other diagonal represents a continuum from generalist species at one extreme to relict species at the other. Stress-resistant generalists occur in widely fluctuating abiotic environments, where there is sufficient energy for rapid punctuational change to occur at favourable times. Relict species are found in stable, low-energy situations, for example some deep-sea habitats. Sheldon (1993) gives an example of a limited relaxation of this stable situation in the Builth trilobites from Wales that thrived in a marine environment several hundred metres deep. Long-term environmental fluctuations were assumed normally to be low in this environment, in association with low energy availability from low primary productivity. Sheldon (1993 and this volume) argued that such a relatively invariant physical environment is consistent with an evolutionary pattern of the gradualist type that he observed.

## Conclusion

The unifiying premise in this chapter is that there is a tendency for organisms to maximize energy and metabolic efficiency in their habitats. This is most difficult in unstable habitats, hence stress-resistant generalist species evolve that can use a range of resources. However, as the energy cost from abiotic stress falls, increasingly precise fine-tuning becomes possible for adaptation to resource heterogeneity, which in some situations could lead to sympatric speciation.

Assuming strong selection for the maximization of energy and metabolic efficiency, the direct observation of speciation events would be rare. Successful species should exist in a state of equilibrium between the physiology of component organisms and their habitats. It would be uncommon to observe the progression towards this situation, which is likely to be underlain primarily by ecological shifts. For instance, Kambysellis et al. (1995) propose that much of the adaptive radiation into alternative breeding sites in Hawaiian *Drosophila* occurred rapidly, so the evolving species would quickly develop behavioural, physiological and anatomical adaptations to the varying breeding sites.

Viewed in this way, the process of evolutionary change as manifested by speciation as the end product becomes compatible with the model of organism-environment interactions assumed in this chapter. Simplistically, this model is the selection for energy efficiency to accommodate a stressed world, where stress-resistance genotypes play a pivotal role in determining the survival and potential of organisms for evolutionary change. In other words, following van Valen (1976), the ecological species concept is fundamental. While other species concepts are more derivative, all should be correlated. Assuming that the initiating event in speciation is ecological, it should occur before visible modifications of the karyotype appear, which are often important in making permanent primary ecological divergences.

*Acknowledgments*
I thank the organizers of the Symposium in Budapest upon which this volume is based for the opportunity to present this approach to evolutionary change.

## References

Booth, D.T., Clayton, D.H. and Block, B.A. (1993) Experimental demonstration of the energetic cost of parasitism in free-ranging hosts. *Proc. Roy. Soc. London B* 223:125–129.
Boulétreau-Merle, J., Fouillet, P. and Terrier, O. (1987) Seasonal variations and balanced polymorphisms in the reproductive potential of temperate *D. melanogaster* populations. *Entomol. Exp. Appl.* 43:39–48.
Bowring, S.A., Grotzinger, J.P., Isachsen, C.E., Knoll, A.H., Pelechaty, S.W. and Koslov, P. (1993) Calibrating rates of early Cambrian evolution. *Science* 261:1293–1298.
Brown, C.R., Brown, M.B. and Rannala, B. (1995) Ectoparasites reduce long-term survival of their avian host. *Proc. Roy. Soc. London B* 262:313–319.
Clutton-Brock, T.W., Price, O.F., Albon, S.D. and Jewell, P.A. (1992) Early development and population fluctuations in Soay sheep. *J. Anim. Ecol.* 61:381–396.

Cordero, A. (1995) Correlates of male mating success in two natural populations of the damselfly *Ischnura graellsii* (Odonata, Coenagrionidae). *Ecol. Entomol.* 20:213–222.

Curtsinger, J.W., Fukui, H.H., Khazaeli, A.A., Kirscher, A., Pletcher, S.D., Promislow, D.E.L. and Tatar, J. (1995) Genetic variation and aging. *Annu. Rev. Genet.* 29:553–575.

Darwin, C. (1859) *On the Origin of Species by Natural Selection*. Murray, London.

Eldredge, N. (1991) *The Miner's Canary: Unravelling the Mysteries of Evolution*. Prentice-Hall, New York.

Eldredge, N. and Gould, S.J. (1972) Punctuated equilibria: An alternative to phyletic gradualism. *In*: T.J.M. Schopf (ed.): *Models in Paleobiology*. Freeman and Cooper, San Francisco, pp. 82–115.

Erwin, D.H. (1993) *The Great Paleozoic Crisis: Life and Death in the Permian*. Columbia University Press, New York.

Glynn, P.W. (1988) El-Niño-Southern Oscillation 1982–1983: Nearshore population, community and ecosystem responses. *Annu. Rev. Ecol. Syst.* 192:309–345.

Halliday, T.R. (1978) Sexual selection and mate choice. *In*: J.R. Krebs and N.B. Davies (eds): *Behavioural Ecology: An Evolutionary Approach*. Blackwell, Oxford, pp. 180–213.

Hartshorn, G. (1992) Possible effects of global warming on the biological diversity in tropical forests. *In*: R.L. Peters and T.E. Lovejoy (eds): *Global Warming and Biodiversity*. Yale University Press, New Haven, pp. 137–146.

Hilliker, A.J., Duyf, B., Evans, D. and Phillips, J.P. (1992) Urate-null rosy mutants of *Drosophila melanogaster* are hypersensitive to oxygen stress. *Proc. Natl. Acad. Sci. USA* 89: 4343–4347.

Hoffmann, A.A. and Parsons, P.A. (1991) *Evolutionary Genetics and Environmental Stress*. Oxford University Press, Oxford.

Hoffmann, A.A. and Parsons, P.A. (1993) Selection for adult desiccation resistance in *Drosophila melanogaster*: Fitness components, larval resistance and stress correlations. *Biol. J. Linn. Soc.* 48:43–54

Hopper, S.D. (1979) Biogeographical effects of speciation in the southwest Australian fauna. *Annu. Rev. Ecol. Syst.* 10:399–402.

Hurst, R.J., Watts, P.D. and Oritsland, N.A. (1991) Metabolic compensation in oil-exposed polar bears. *J. Therm. Biol.* 16:53–57.

Ihlenfeld, H.-D. (1994) Diversification in an arid world: The Mesembryanthemaceae. *Annu. Rev. Ecol. Syst.* 25:521–546.

Jablonski, D. and Bottjer, D.J. (1990) Onshore-offshore trends in marine invertebrate evolution. *In*: R.M. Ross and W.P. Allman (eds): *Causes of Evolution: A Paleontological Perspective*. University of Chicago Press, Chicago, pp. 21–75.

Kambysellis, M.P., Ho, K.-F., Craddock, E.M., Piano, F., Parisi, M. and Cohen, J. (1995) Pattern of ecological shifts in the diversification of Hawaiian *Drosophila* inferred from a molecular phylogeny. *Curr. Biol.* 10:1129–1139.

Kauffman, S.A. (1993) *The Origins of Order: Self-Organization and Selection in Evolution*. Oxford University Press, New York.

Koehn, R.K. and Bayne, B.L. (1989) Towards a physiological and genetical understanding of the energetics of the stress response. *Biol. J. Linn. Soc.* 37:157–171.

Kohane, M.J. (1994) Energy, development and fitness in *Drosophila melanogaster*. *Proc. Roy. Soc. London B* 257:185–191.

Lewontin, R.C. (1970) The units of selection. *Annu. Rev. Ecol. Syst.* 1:1–18.

Lithgow, G.J., White, T.M., Melov, S. and Johnson, T.E. (1995) Thermotolerance and extended life-span conferred by single-gene mutations and induced by thermal stress. *Proc. Natl. Acad. Sci. USA* 92:7540–7544.

Loeschcke, V. and Krebs, R.A. (1996) Selection for heat-shock resistance in larval and in adult *Drosophila buzzatii*: Comparing direct and indirect responses. *Evolution* 50:2354–2359.

Malmquist, H.J., Snorrason, S.S., Skúlason, S., Jonsson, B., Sandlund, O.T. and Jónasson, P.M. (1992) Diet differentiation in polymorphic Arctic charr in Thingvallavatn, Iceland. *J. Anim. Ecol.* 61:21–35.

McLain, D.K. (1993) Cope's rules, sexual selection and the loss of ecological plasticity. *Oikos* 88:490–500.

Mitton, S.B. (1993) Enzyme heterozygosity, metabolism and developmental variability. *Genetica* 89:47–63.

Møller, A.P. (1994) Male ornament size as a reliable cue to enhanced offspring viability in the barn swallow. *Proc. Natl. Acad. Sci. USA* 91:6929–6932.

Moore, A.J. (1994) Genetic evidence for the "good genes" process of sexual selection. *Behav. Ecol. Sociobiol.* 35:235–241.

Parsons, P.A. (1982) Evolutionary ecology of Australian *Drosophila*: A species analysis. *Evol. Biol.* 14:297–350.

Parsons, P.A. (1993) The importance and consequences of stress in living and fossil populations: from life-history variation to evolutionary change. *Am. Nat.* 142:S5–S20.

Parsons, P.A. (1994) Morphological stasis: An energetic and ecological perspective incorporating stress. *J. Theor. Biol.* 171:409–414.

Parsons, P.A. (1995a) Evolutionary response to drought stress: Conservation implications. *Biol. Conserv.* 74:21–27.

Parsons, P.A. (1995b) Inherited stress resistance and longevity: A stress theory of ageing. *Heredity* 75:216–221.

Parsons, P.A. (1995c) Stress and limits to adaptation: Sexual ornaments. *J. Evol. Biol.* 8:455–461.

Parsons, P.A. (1996a) Competition versus abiotic factors in variably stressful environments: Evolutionary implications. *Oikos* 75:129–132.

Parsons, P.A. (1996b) Rapid development and a long life: An association expected under a stress theory of aging. *Experientia* 52:643–647.

Parsons, P.A. (1996c) The limit to human longevity: An approach through a stress theory of ageing. *Mech. Age. Dev.* 87:211–218.

Parsons, P.A. (1996d) Stress, resources, energy balances and evolutionary change. *Evol. Biol.* 29:39–72.

Parsons, P.A. (1997) Success in mating: A co-ordinated approach to fitness through genotypes incorporating genes for stress resistance and heterozygous advantage under stress. *Behav. Genet.* 27:75–81.

Phillips, O.L. and Gentry, A.H. (1994) Increasing turnover through time in tropical forests. *Science* 263:954–958.

Riddoch, B.J. (1993) The adaptive significance of electrophoretic mobility in phosphoglucose isomerase (PGI). *Biol. J. Linn. Soc.* 50:1–17.

Roff, D.A. (1996) The evolution of genetic correlations: An analysis of patterns. *Evolution* 50:1392–1403.

Rollo, C.D. (1994) *Phenotypes: Their Epigenetics, Ecology and Evolution*. Chapman & Hall, London.

Root, T. (1993) Effects of global climate change on North American birds. *In*: P.M. Kareiva, J.G. Kingsolver and R.B. Huey (eds): *Biotic Interactions and Global Change*. Sinauer, Sunderland, MA, pp. 280–292.

Rosewell, J. and Shorrocks, B. (1987) The implication of survival rates in natural populations of *Drosophila*: Capture-recapture experiments in domestic species. *Biol. J. Linn. Soc.* 32:373–384.

Rowley, I. (1970) Lamb predation in Australia: Incidence, predisposing conditions and the identification of wounds. *CSIRO Wildlife Research* 15:79–123.

Schluter, D. and McPhail, J.D. (1993) Character displacement and replicate adaptive radiation. *Trends Ecol. Evol.* 8:197–200.

Schubert, J.K. and Bottjer, D.J. (1992) Early Triassic stromatolites as post-mass extinction disaster forms. *Geology* 20:883–886.

Sheldon, P.R. (1993) Making sense of microevolutionary patterns. *In*: D.R. Lees and D. Edwards (eds): *Evolutionary Patterns and Processes*. Academic Press, London, pp. 20–31.

Sibley, R.M. and Calow, P. (1986) *Physiological Ecology of Animals: An Evolutionary Approach*. Blackwell, Oxford.

Skúlason, S., Noakes, L.G. and Snorrason, S.S. (1989) Ontogeny of trophic morphology in four sympatric morphs of arctic charr *Salvelinus alpinus* in Thingvallavatn, Iceland. *Biol. J. Linn. Soc.* 38:281–301.

Sohal, R.S. and Allen, R.G. (1990) Oxidative stress as a causal factor in differentiation and aging: A unifying hypothesis. *Exp. Gerontol.* 25:499–522.

Stearns, S.C. (1992) *The Evolution of life Histories*. Oxford University Press, Oxford.

Van Valen, L. (1976) Ecological species, multispecies and oaks. *Taxon* 25:233–239.

Vrba, E.S. (1987) Ecology in relation to speciation rates: Some case histories of Miocene-Recent animal clades. *Evol. Ecol.* 1:283–290.

Ward, P.D. (1992) *On Methuselah's Trail: Living Fossils and Great Extinctions.* W.H. Freeman, New York.

Watt, W.B. (1985) Bioenergetics and evolutionary genetics: Opportunities for new synthesis *Am. Nat.* 125:118–143.

Werner, T.K. and Sherry, T.W. (1987) Behavioral feeding specialization in *Pinaroloxias inornata*, the "Darwin's Finch" of Cocos Island, Costa Rica. *Proc. Natl. Acad. Sci. USA* 84:5506–5510.

White, T.C.R. (1993) *The Inadequate Environment: Nitrogen and Abundance of Animals.* Springer-Verlag, Berlin.

Zuk, M., Thornhill, R. Ligon, J.D. and Johnson, K. (1990) Parasites and mate choice in red jungle fowl. *Amer. Zool.* 30:235–244.

Environmental Stress, Adaptation and Evolution
ed. by R. Bijlsma and V. Loeschcke
© 1997 Birkhäuser Verlag Basel/Switzerland

# The Plus ça change model: Explaining stasis and evolution in response to abiotic stress over geological timescales

Peter R. Sheldon

*Department of Earth Sciences, The Open University, Milton Keynes MK7 6AA, UK*

*Summary.* The Plus ça change model addresses the surprising finding that morphological stasis seems to be the usual response to widely fluctuating physical stresses over geological timescales (until thresholds are reached). The lineages that survive in an environment of fluctuating sea levels, climate, substrate, salinity and so on appear to be those that are relatively inert to each environmental twist and turn, in contrast to more sensitive lineages in less changing environments. Generalists in this long-term sense are species with properties that enable them to survive throughout wide fluctuations in the physical environment for millions of years. The model predicts a tendency for continuous, gradualistic evolution on land in the tropics and in the deep sea, and for more stasis (and occasional punctuations) in shallow waters and temperate zones. Punctuated equilibrium may be mistakenly perceived as the usual pattern in the history of life because the environments in which gradualism predominates are rarely preserved in the fossil record. Given the Quaternary climate upheavals, relatively little evolution may be occurring worldwide at present compared with, say, 3 million years ago (except for evolution associated with human activity).

## Introduction

*The problem of relating phenomena across scales is the central problem in biology* (Levin, 1995, p. 311).

*Plus ça change, plus c'est la même chose* (The more that changes, the more it's the same thing).

Evidence emerging from the fossil record suggests that over long timescales (e.g. a million years) the relationship between environmental stress and evolution is far from straightforward and is even counterintuitive. The fossil record can offer a perspective that cannot be gained from a study of living organisms alone. The fact that palaeontologists' data are time-averaged can be a distinct advantage, smoothing out short-term responses that may be a wholly misleading indicator of long-term change. Indeed, there seem to be major discrepancies between the nature of evolutionary patterns observed in the short term (over usual biological timescales) and those that emerge over geological timescales.

Of course, "geological timescales" are not, in reality, separate from biological or ecological timescales – they encompass them. Many important biological changes occur over long time intervals, such as the "Cambrian Explosion", and, by contrast, some "geological" events are very rapid: a flash flood, an earthquake, a meteorite impact. The net outcome of individual events depends so much on the time interval being considered. For example, an ecologist's "unpredictable environment" can, as far as incumbent species are concerned, be highly predictable over geological timescales. Levin (1995, p. 295) commented that in locally disturbed systems "local unpredictability is globally the most predictable feature of the system.... As the scale of description is increased beyond the scale of individual disturbances, variability declines, and predictability correspondingly increases." It is genuinely rare, unpredictable events that cause thresholds to be crossed. As an extreme case, large meteorite impacts, such as that which occurred at the Cretaceous-Tertiary boundary and which probably contributed to mass extinction, are unpredictable at any biological level.

The recent fashion for fractals – patterns that are self-similar over many spatial scales – may be unwittingly reinforcing the tendency to extrapolate from short-term observations to long-term outcomes. But evolutionary systems cannot be predicted so easily. For example, the relationship between population extinction (i.e. local extinction) and species extinction (i.e. global extinction) remains little understood. Lawton (1994) reviewed population dynamics principles and discussed the factors that make some species more extinction-prone than others. There are many subtle and unresolved issues; for example, a species that an ecologist reports as susceptible to local extinction may be relatively resistant to global extinction.

Here I present a general model about how species respond to abiotic stresses over geological timescales. Such a model is bound to have many exceptions; it is only about tendency and relative frequency. Nor can any single model explain pattern on all scales; the key to interrelating phenomena at different scales is discovering what fine detail is relevant at the higher levels and what is noise (Levin, 1995). The French saying above conveys an important concept in the bridging of timescales. It is often used when events are no longer a surprise; people just say "Plus ça change..." to indicate an ironic "What's new?" to, for example, yet another political blunder, or divorce in the British royal family.

## The Plus ça change model of long-term response to environmental stress

Eldredge and Gould's model of punctuated equilibrium has exerted a strong influence on the investigation and interpretation of evolutionary patterns (see reviews by Gould and Eldredge, 1993; Erwin and Anstey, 1995). According to punctuated equilibrium, most evolutionary change in the history of life is concentrated in brief periods of rapid transformation. A new species originates rapidly and then remains virtually unchanged – in stasis – for the rest of its existence. Gradualism, according to Eldredge and Gould, is very rarely found in the fossil record.

The starting point for the Plus ça change model was a study of about 15,000 Ordovician trilobites from central Wales that revealed a pattern of broadly parallel, gradualistic evolution in eight lineages (Sheldon, 1987). Over a period of about 2 million years, a variety of changes took place at different times in different lineages, with patterns far closer to gradualism than punctuated equilibrium. The "missing links" between previously described, successive trilobite species were no longer missing: intermediate horizons yielded specimens of intermediate morphology. These trilobites thrived in a relatively stable, narrowly fluctuating, low-energy, low-oxygen (dysaerobic) environment. They were rather specialized, benthic forms, at least as adults, and probably lived in a basin several hundred metres deep (Sheldon, 1987, 1988). They are found in fairly monotonous dark shales, and except for occasional inarticulate brachiopods, there is little or no associated macrobenthos. Despite this quiet environmental setting, the trilobites underwent a pattern of long-term phyletic evolution that resembles the gradualism model far more than stasis.

Evidence from this and other studies led to the proposal that continuous phyletic evolution is characteristic of relatively stable environments, whereas stasis tends to prevail in unstable environments (Sheldon, 1987, 1990). The kind of environmental stresses for which this generalization is proposed are physical (abiotic) variables that can be studied over geological timescales, such as changing sea level, substrate, salinity and climate (e.g. mean temperature).

Biologists tend to expect major changes in the physical environment to result either in evolution or extinction. There is, however, much evidence against this generalization. For example, it is highly significant that the intense physical stresses of the Quaternary ice age have not (at least, not yet) generated many new invertebrate species (see, for example, Valentine and Jablonski, 1991). According to Williams (1992, p. 131): "The current facts and understandings of population genetics would be thoroughly compatible with major changes in the adaptations of most lineages of animals and plants during the last million years. Instead, a large proportion of Recent species are essentially identical to their Pliocene ancestors."

Jackson (1994) reviewed comprehensive evidence that communities of molluscs, reef corals and planktonic foraminiferans have changed very little since the end of the Pliocene about 2 million years ago. The turnover of Late Pliocene faunas in apparent response to the onset of glaciation required only a few hundred thousand years, in contrast to at least 8 million years of previous relative faunal stability. Both the extinction rate and the origination rate of molluscs increased at this time. Once past the initial thermal filter of glacial cooling, rapid rises and falls in sea level and in temperature often had negligible effect on the surviving species. But Jackson warns that although integrated communities may persist for millions of years, often in the face of a fair degree of environmental change, *once a threshold is exceeded*, collapse is abrupt and recovery impossible. For example, a rapid rise in tropical sea surface temperatures of only 2–3°C above present values is likely to be catastrophic for reef corals, whose upper thermal tolerance is typically only a few degrees above present seasonal maxima. Higher temperatures are beyond the evolutionary experience of modern corals (see references in Jackson, 1994).

There is overwhelming evidence of approximate morphological stasis lasting millions of years in shallow marine shelly organisms such as bivalves (Stanley and Yang, 1987); bryozoans (Cheetham, 1987; Cheetham and Jackson, 1990, 1995), and in "living fossils" such as lingulid brachiopods and some limulids (though well-documented cases of persistent species-level identity in "living fossils" are rarer than often supposed). Several species of living bivalves studied by Stanley and Yang (1987) amazingly show no more evolutionary change over the past 17 million years than there is geographic variation at the present day. Cenozoic marine ostracods (Cronin, 1985) exhibit the most stasis during major high-frequency climatic oscillations. Quaternary beetles (Coope, 1990, 1994) are well known for showing stasis throughout glacial-interglacial cycles.

Brett and Baird (1995), studying shallow marine Devonian faunas, found a pattern of species appearing in discrete packages that appear and disappear more or less in synchrony, a phenomenon they coined "coordinated stasis". Once originated, new biotas persisted more or less intact in both morphologies of constituent species and ecological associations for several million years, despite environmental fluctuations. Only the occasional, more extreme perturbations produced change when thresholds were exceeded; for further discussion see Morris et al. (1995).

Further evidence supporting the model, obtained from a wide range of fossil groups, is given in Sheldon (1996). Despite the various descriptive biases that obscure evidence of gradualism (Sheldon, 1993, 1996), approximate stasis does seem to be a widespread feature of the shallow marine species which dominate the fossil record; and the fact that Linnean taxonomy can often be applied successfully to fossils as well as living organisms is itself strong evidence for the prevalence of stasis.

Intuitively, one might perhaps expect a changing environment to lead to changing morphology, and a stable environment to stable morphology. However, over long timescales, the saying "Plus ça change, plus c'est la même chose" may be especially apt, as far as the relationship between physical change and phyletic evolution is concerned. Perhaps more widely fluctuating environments (seen over geological rather than ecological timescales) maintain their own kind of stability *within wide reflecting boundaries*, and selection soon tends to favour lineages with "all-purpose" hard-part morphologies that are relatively inert to each environmental twist and turn (Fig. 1). The fluctuations, in effect, put the brakes on further evolution, and the long-term "generalist" lineages that emerge then persist with no net change until thresholds are exceeded.

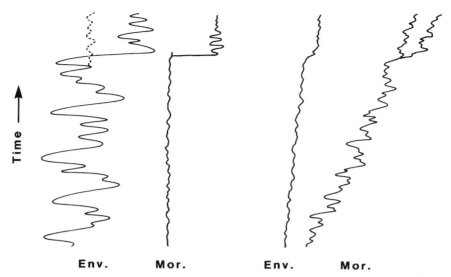

Figure 1. The Plus ça change model proposes that, over geological timescales (e.g. a million years), gradualism (right) is characteristic of narrowly fluctuating, relatively stable environments such as the terrestrial tropics and the deep sea. By contrast, net stasis (with occasional punctuations) (left) is expected to prevail in the more unstable environments that dominate the fossil record, especially shallow seas. In a widely fluctuating environment (left), a threshold might occur not only when some environmental variable exceeds wide reflecting boundaries, but also when it contracts to become narrowly fluctuating (dotted line). In the relatively stable environment (right), there is a slight net environmental shift (to the right) but this is not necessary to produce the net morphological change depicted. Env. = some long-term physical aspect of environment, e.g. sea level, substrate or mean temperature. Mor. = some aspect (or aspects) or morphology. High-frequency environmental oscillations such as annual cycles are not shown. See text for discussion. (From Sheldon, 1993; modified from Sheldon, 1990.)

Dobzhansky (1950, p. 216), discussing the adaptations of species living in temperate climates, wrote: "To survive and reproduce, any species must be at least tolerably well adapted to every one of the environments which it regularly meets.... Changeable environments put the highest premium on versatility rather than on perfection in adaptation." This argument can be scaled up to geological intervals: the species that survive early on after the onset of widely fluctuating conditions have to be generalists in a very long-term sense. Whether by selection for habitat tracking (as invoked by Coope, 1990, 1994, for Quaternary insects), or selection for widely tolerant morphotypes, or by the mechanisms of lineage selection discussed below, lineages might, *from their perspective*, experience widely fluctuating stresses as stable environments until thresholds are reached. As Coope (1990, p. 214) put it for insects, "Those that survived the first onslaughts of the [Pleistocene] climatic changes were well fitted to withstand subsequent ones."

There may be fruitful analogies between species that live in a fire-adapted ecosystem (Allen and Hoekstra, 1992) and the shallow marine species that often persist in approximate stasis over millions of years through major changes in sea level, substrate and temperature. To such species, the stresses may be equivalent to all but the least frequent and most severe forest fires. Perhaps those major (but not extreme) "geological" perturbations somehow effectively isolate some species from major selection pressures until a threshold is reached.

Invert that French saying, and we get the other side of the coin: *the less that changes, the less it's the same thing.* In quieter, less dynamic environments, perhaps organisms need not be so generalist in the above sense; the brakes are off and finely tuned, less time-averaged adaptations will come and go more frequently (see Fig. 1). Despite deeply-engrained predispositions, evolutionary reversals can be expected to occur very often, though numerous biases hinder their detection in the fossil record; for detailed discussion, see Sheldon (1993, 1996). Even when the narrowly fluctuating environment changes relatively little, so perhaps must its lineages. Evolutionary change presumably has to occur more continuously in fine-tuned, specialized lineages to avoid extinction. Such lineages would be more sensitive to minor environmental nudges, but the nudges themselves would be hard to detect in the geological record. When preserved, such lineages would tend to show successive intergrading chronospecies, each having, on average, a shorter duration than static species in widely fluctuating environments.

The Plus ça change model (Sheldon, 1990, 1993, 1996) predicts a tendency for more continuous phyletic evolution in deep waters, and in the tropics generally, and for more stasis (and occasional punctuations) in shallow waters and in temperate zones. A key point is that as shallow marine sediments provide the vast majority of the macrofossil record, it is not surprising that many macrofossil lineages show stasis and occasional punctuated change. According to the model, documented cases of gradualism are rare primarily because the environments in which gradualism predominates (such as on land in the tropics and in the deeper sea) are rarely preserved in the fossil record.

On the left of Figure 1, a threshold effect occurs when environmental stresses exceed the widely reflecting boundaries that they have remained within for perhaps a million years or more. Thus abiotic environmental stress is certainly a motor of evolution, and no doubt many major environmental events in Earth history have elicited strong evolutionary responses. However, in the model it is change *relative to pre-existing conditions* that is important: the impact of an event is entirely contingent on history. If a system has been subjected many times to large disturbances, yet another one may yield no response at all. After a while, nothing is perceived to change, just as in human society rebellion can become conformity. In contrast, a medium magnitude disturbance in a system that has experienced only low magnitude events for a long time may have a large effect, like a loud sneeze in a hushed library compared with in a football crowd.

In the case of widely fluctuating stresses, a threshold might occur not only when some environmental variable exceeds wide reflecting boundaries, but also when it *contracts* to become narrowly fluctuating (see Fig. 1). Speciation and rapid phyletic evolution might therefore be expected, for example, when the Quaternary upheavals are finally over (see also below).

In less dynamic environments, stresses exceeding rather narrow limits of speed and extent might tend to cause speciation or extinction (or, in some cases, simply speed up directional change). If cladogenesis tends to occur when certain environmental variables exceed "reflecting boundaries", another factor contributing to high tropical and deep-sea diversities could be a higher frequency of such events there, albeit physical events with excursions of far lower amplitude than those elsewhere, so we would rarely detect them. For simplicity, only one such event is shown for the narrowly fluctuating environment on the right of Figure 1; note that to emphasise the difference in species diversities expected in the different settings, no net increase in diversity has been shown for the widely fluctuating environment on the left.

The tropics (and the deep sea) are, of course, far from being stable; for example, Huston (1994, p. 527) states: "Tropical forests are extremely dynamic, subject to a variety of disturbances and environmental fluctuations." However, in terms of some of the physical variables to which the Plus ça change model relates, especially climate change, the amplitudes of oscillations in the tropics are usually thought to be small. Organisms living in a relatively stable, narrowly fluctuating physical environment such as the tropics may be expecially sensitive to biotic interactions (Dobzhansky, 1950). Higher-diversity, smaller, more isolated populations, and a greater tendency for specialization, are all features consistent with relatively continuous evolution in the tropics compared with temperate zones. Similar arguments may apply to the deep sea as opposed to shallow marine settings.

In the Plus ça change model, the concepts "long-term generalist" and "long-term specialist" are related to but distinct from the ecological division into eurytopes and stenotopes. A generalist in the long-term, geological sense is a species (not an individual) that can survive throughout wide environmental fluctuations over geological timescales; it is sufficiently adapted to "perceive" such an environment as little changing. A specialist in a long-term sense would not be able to cope with such wide fluctuations over geological timescales but would thrive in more stable settings. In the simplistic terms of this model, the tropics and deeper sea tend to be dominated by long-term (albeit continuously evolving) specialists, and shallow marine shelves by long-term generalists.

In any one environment (wherever it is) in which there is a mixture of ecological eurytopes and stenotopes, there may be a tendency for eurytopes to show more stasis and stenotopes to show more gradualism. Some very ancient lakes (e.g. Lake Tanganyika) illustrate some of these aspects. Here there is evidence of very rapid speciation followed by stasis, especially in littoral communities, and of long-term gradualism in some pelagic communities (Martens et al., 1994). Where benthos alone are concerned, changeability fostered eurytopes and stasis, whereas stability and permanence produced stenotopes and gradual phyletic change. According to Coulter (1994, p. 132), "Fishes of the benthic zones of Lakes Malawi and Victoria are speciose and many are stenotopic, perhaps because of greater environmental stability at the bottom", whereas benthic fishes of Lake Tanganyika are typically eurytopic and live in an unstable environment.

An important feature of the model (Fig. 1) is that net long-term stasis in more widely fluctuating environments is not simply a case of major morphological shifts that keep getting cancelled out by broad, zigzagging reversals in the main population (though there will be frequent abortions of short-lived peripheral offshoots). Instead, the morphological response has become damped compared with the environmental shifts, and of less amplitude than typical responses in a narrowly fluctuating environment. In Figure 1, for simplicity, the amplitude of morphological reversals in the widely fluctuating environment on the left is drawn about as small as the amplitude of environmental reversals in the narrowly fluctuating environment on the right. The bryozoans described by Cheetham (1987) and the bivalves described by Stanley and Yang (1987), for example, apparently show narrow oscillations about their mean stasis, despite living in widely fluctuating environments.

A greater amplitude of morphological responses in a narrowly fluctuating environment can be expected, because selection probably has the chance to act on a greater range of variation tolerated (and perhaps produced) in a stable environment under relaxed selection pressures (e.g. see Williamson, 1987). Presumably lineages that produce a wider range of variation "when the going is good" are more likely to survive severe new environmental stresses. If so, this would be one mechanism of lineage selection, i.e. the preferential sorting of species. The evidence, however, concerning the extent of morphological variation in different types of environments is rather thin.

Direct support for the model has come from Parsons (1991, 1993, 1994a, b), who argues that it is consistent with the evolutionary genetics of populations under varying degrees of environmental stress. Maximum evolutionary rates are expected in habitats characterised by narrowly fluctuating (but not invariant) environments implying moderate stress, moderate genetic variability of ecologically important traits and not unduly restrictive metabolic costs.

Parsons (1994b and references therein) developed a model to explain morphological stasis or change in terms of energetic costs to organisms living in varying degrees of stressful environments. Assuming that preferred habitats are regions of relatively low energetic costs, such costs would increase towards the limits of distributions, ultimately becoming restrictive for evolutionary change and thereby promoting stasis. He argues that, although a variety of local abiotic and biotic factors influence species ranges, as the timescale increases towards geological time, climatic factors dominate as the remaining factors become increasingly transient. On the basis of experiments with *Drosophila*, Parsons proposes that morphological stasis may not be accompanied by stasis for physiological traits, such as metabolic rate, which might be changing much more often. Most physiological traits would not, however, be preserved in the fossil record.

Parsons (1994b) suggests that living fossils, e.g. horseshoe crabs, stromatolites and lingulid brachiopods, which show remarkably stable morphology (at least as genera if not species) over tens or even several hundreds of millions of years, tend to live in harsh shoreline environments with widely fluctuating stresses and high energetic costs that preclude major evolutionary change.

The predicted evolutionary pattern in a single lineage is shown in Figure 2: stasis when there is greater environmental fluctuation; net directional change and/or more widely fluctuating morphologies when stresses are more narrowly fluctuating. This can be tested by following one or more fossil lineages through strata showing varying degrees of environmental fluctuation.

A case that supports this prediction is that of the *Melanopsis impressa* to *M. fossilis* Miocene gastropod lineage summarised in Geary (1995). *M. impressa* underwent stasis for 7 million years, followed by directional change over about 2 million years. According to Geary (personal communication, 1993), during the interval of stasis the melanopsids were living on the margins of a basin in transient, fresh-brackish water bodies – i.e. in a widely fluctuating environment. When the salinity of the basin reduced, the gastropods invaded the basin. Thereafter, during the interval of directional change in morphology, the salinity of the basin was stable.

The most likely causal mechanisms of stasis in the context of the model remain to be established. Perhaps the most important mechanism may be a type of lineage selection with two stages: (1) if an established or an incipient species experiences and survives a widely fluctuating environment on geological timescales, the evolutionary response (morphological change) tends to become damped with time, and (2) those species that are least sensitive to environmental change (the most generalised in a long-term sense) are the ones that tend to persist, remaining in morphological stasis until a threshold is reached. In prolonged, widely fluctuating environments, most new, rapidly diverging sidelines may soon be wiped out by return to different (repeated) conditions that the generalist ancestral lineages are better able to thrive in.

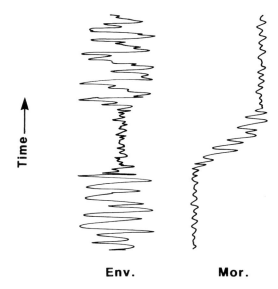

**Env.**              **Mor.**

Figure 2. One of the testable predictions of the Plus ça change model shown in Figure 1. A single lineage should show more stasis when there is greater environmental fluctuation, and net directional change and more widely fluctuating morphology when conditions are relatively stable. In this case, a new equilibrium (with the species in stasis) is attained with the return of widely fluctuating environments. Alternatively (not shown), the lineage might return to the same equilibrium state after exhibiting more widely fluctuating morphology during an interval of narrowly fluctuating environments. The total range of morphological variation at any one time (not shown) may vary inversely with the width of environmental fluctuations over time. Env. = some long-term physical aspect of environment, e.g. sea level, substrate or mean temperature. Mor. = some aspect (or aspects) of morphology, as in Figure 1. The time span represented is at least half a million years. (From Sheldon, 1996.)

It must be emphasised that such a model is bound to have many exceptions; it is only a hypothesis about tendency. An immensely complex interplay of factors will determine the details of patterns in individual cases. More detailed discussion can be found in Sheldon (1996).

## Implications for evolutionary responses to stress today and in the future

Evolution today should be easier to observe in the tropics than in temperate zones. However, given the Quaternary climate upheavals, many of today's species across the globe may be relatively inert to environmental twists and turns – i.e. short-term evolution may be genuinely hard to detect because little is going on generally compared with quieter times. Although the Recent climate of the last 10,000 years seems to have been very stable relative to the Pleistocene, 10,000 years may not be a sufficiently long interval of stability to relax the promotion of stasis.

Bradshaw (1991) perceives a general lack of evolutionary change – what he calls "evolutionary failure" – in most Recent species, especially plants. He attributes lack of evolution mostly to lack of appropriate genetic variation, i.e. "genostasis". Recent species have, however, made it over the climatic hurdles of the Quaternary assault course, so presumably they should have the appropriate genetic attributes for survival, and stasis has been the norm for some time. (As Bradshaw points out (1991, p. 297), most cases in which rapid evolution has been reported are in new environments associated with human activity.) Irrespective of human influences, increased evolution, including speciation, can be expected when the huge fluctuations finally cease.

Recent concern about global warming has provided a focus for considering evolutionary responses of extant species to changes in the physical environment. The Plus ça change model would predict that, other things being equal, human-induced global climate change will make little difference to the evolution of most species *because* there has been so much global change in the last 2 million years. But other things are far from being equal: thresholds are now being exceeded. Stresses induced by human activity such as tropical habitat destruction and chemical pollution are much more threatening – not for fear of evolution but of extinction.

## Conclusion

Somewhat surprisingly, morphological stasis seems to be the usual response to widely fluctuating physical stresses over geological timescales of $\sim 10^5 – 10^6$ years (until thresholds are reached). The lineages that survive in an environment of fluctuating sea levels, climate, substrate, salinity and so on appear to be relatively inert to each environmental twist and turn, in contrast to more sensitive lineages in less changing environments. Generalists in this long-term sense are species with properties that enable them to survive throughout wide fluctuations in the physical environment for millions of years; they are thus related to but distinct from ecological eurytopes. The Plus ça change model predicts a tendency for continuous, gradualistic evolution on land in the tropics and in the deep sea, and for more stasis (and occasional punctuations) in shallow waters and temperate zones. Such a model is bound to have many exceptions; it is only about tendency and relative frequency.

Punctuated equilibrium could be being mistakenly perceived as the usual pattern in the history of life because the environments in which gradualism predominates are rarely preserved: the vast majority of the fossil record comes from dynamic shallow marine settings where lineages are likely to show approximate stasis and occasional punctuations.

The fossil record enables us to estimate better the stresses that living species and their ecosystems can be expected to tolerate, and beyond which extinction may result. Given the Quaternary climate upheavals, relatively little evolution may be occurring worldwide at present compared with quieter times, say, 3 million years ago; most reported cases of rapid evolution today are in new environments associated with human activity.

## References

Allen, T.F.H. and Hoekstra, T.W. (1992) *Toward a Unified Ecology*. Columbia University Press, New York.

Bradshaw, A.D. (1991) Genostasis and the limits to evolution. *Phil. Trans. R. Soc. Lond. B* 333:289–305.

Brett, C.E. and Baird, G.C. (1995) Coordinated stasis and evolutionary ecology of Silurian to Middle Devonian faunas in the Appalachian Basin. *In*: D.H. Erwin and R.L. Anstey (eds): *New Approaches to Speciation in the Fossil Record*. Columbia University Press, New York, pp. 285–315.

Cheetham, A.H. (1987) Tempo of evolution in a Neogene bryozoan: Are trends in single morphologic characters misleading? *Paleobiology* 13:286–296.

Cheetham, A.H. and Jackson, J.B.C. (1990) Evolutionary significance of morphospecies: A test with cheilostome bryozoa. *Science* 248:579–583.

Cheetham, A.H. and Jackson, J.B.C. (1995) Process from pattern: Tests for selection versus random change in punctuated bryozoan speciation. *In*: D.H. Erwin and R.L. Anstey (eds): *New Approaches to Speciation in the Fossil Record*. Columbia University Press, New York, pp. 184–207.

Coope, G.R. (1990) The invasion of Northern Europe during the Pleistocene by Mediterranean species of Coleoptera. *In*: F. Di Castri, A.J. Hansen and M. Debussche (eds): *Biological Invasions in Europe and the Mediterranean Basin*. Kluwer Academic Publishers, Dordrecht, pp. 203–215.

Coope, G.R. (1994) The response of insect faunas to glacial-interglacial climatic fluctuations. *Phil. Trans. R. Soc. Lond. B* 344:19–26.

Coulter, G. (1994) Speciation and fluctuating environments, with reference to ancient East African lakes. *In*: K. Martens, B. Goddeeris and G. Coulter (eds): *Speciation in Ancient Lakes*. Arch. Hydrobiol. Beih. Ergebn. Limnol. 44:127–137.

Cronin, T.M. (1985) Speciation and stasis in marine Ostracoda: Climatic modulation of evolution. *Science* 227:60–63.

Dobzhansky, T. (1950) Evolution in the tropics. *Am. Sci.* 38:209–221.

Erwin, D.H. and Anstey, R.L. (1995) Speciation in the fossil record. *In*: D.H. Erwin and R.L. Anstey (eds): *New Approaches to Speciation in the Fossil Record*. Columbia University Press, New York, pp. 11–38.

Geary, D.H. (1995) The importance of gradual change in species-level transitions. *In*: D.H. Erwin and R.L. Anstey (eds): *New Approaches to Speciation in the Fossil Record*. Columbia University Press, New York, pp. 67–86.

Gould, S.J. and Eldredge, N. (1993) Punctuated equilibrium comes of age. *Nature* 366:223–227.

Huston, M.A. (1994) *Biological Diversity*. Cambridge University Press, Cambridge.

Jackson, J.B.C. (1994) Constancy and change of life in the sea. *Phil. Trans. R. Soc. Lond. B* 344:55–60.

Lawton, J.H. (1994) Population dynamic principles. *Phil. Trans. R. Soc. Lond. B* 344:61–68.

Levin, S.A. (1995) The problem of pattern and scale in ecology. *In*: T.M. Powell and J.H. Steele (eds): *Ecological Time Series*. Chapman and Hall, New York, pp. 277–326.

Martens, K., Coulter, G. and Goddeeris, B. (1994) Speciation in ancient lakes – 40 years after Brooks. *In*: K. Martens, B. Goddeeris and G. Coulter (eds): *Speciation in Ancient Lakes*. Arch. Hydrobiol. Beih. Ergebn. Limnol. 44:75–96.

Morris, P.J., Ivany, L.C., Schopf, K.M. and Brett, C.E. (1995) The challenge of paleoecological stasis: Reassessing sources of evolutionary stability. *Proc. Natl. Acad. Sci. USA* 92:11 269– 11 273.

Parsons, P.A. (1991) Stress and evolution. *Nature* 351:356–357.

Parsons, P.A. (1993) Stress, extinctions and evolutionary change: From living organisms to fossils. *Biol. Rev.* 68:313–333.

Parsons, P.A. (1994a) Habitats, stress and evolutionary rates. *J. Evol. Biol.* 7:387–397.

Parsons, P.A. (1994b) Morphological stasis: An energetic and ecological perspective incorporating stress. *J. Theor. Biol.* 171:409–414.

Sheldon, P.R. (1987) Parallel gradualistic evolution of Ordovician trilobites. *Nature* 330: 561–563.

Sheldon, P.R. (1988) Trilobite size-frequency distributions, recognition of instars and phyletic size changes. *Lethaia* 21:293–306.

Sheldon, P.R. (1990) Shaking up evolutionary patterns. *Nature* 345:772.

Sheldon, P.R. (1993) Making sense of microevolutionary patterns. *In*: D.R. Lees and D. Edwards (eds): *Evolutionary Patterns and Processes*, Volume 14. Linnean Society Symposium. Academic Press, London, pp. 19–31.

Sheldon, P.R. (1996) Plus ça change – a model for stasis and evolution in different environments. *Palaeogeogr. Palaeoclimatol. Palaeoecol.* 127:209–227.

Stanley, S.M. and Yang, X. (1987) Approximate evolutionary stasis for bivalve morphology over millions of years: A multivariate, multilineage study. *Paleobiology* 13:113–139.

Valentine, J.W. and Jablonski, D. (1991) Biotic effects of sea level change: The Pleistocene test. *J. Geophys. Res.* 96:6873–6878.

Williams, G.C. (1992) *Natural Selection: Domains, Levels and Challenges*. Oxford University Press, Oxford.

Williamson, P.G. (1987) Selection or constraint? A proposal on the mechanism of stasis. *In*: K.S.W. Campbell and M.F. Day (eds): *Rates of Evolution*. Allen and Unwin, London, pp. 129–142.

# Subject Index

322

324

**B. Streit / T. Städler,** *Univ. of Frankfurt, Germany*
**C.M. Lively,** *Indiana Univ., Bloomington, IN, USA (Eds)*

# Evolutionary Ecology of Freshwater Animals

## Concepts and Case Studies

1997. 384 pages, Hardcover • ISBN 3-7643-5694-4 (EXS 82)

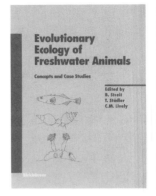

Evolutionary ecology includes aspects of community structure, trophic interactions, life-history tactics, and reproductive modes, analyzed from an evolutionary perspective. Freshwater environments often impose spatial structure on populations, e.g. within large lakes or among habitat patches, facilitating genetic and phenotypic divergence. Traditionally, freshwater systems have featured prominently in ecological research and population biology.

This book brings together information on diverse freshwater taxa, with a mix of critical review, synthesis, and case studies. Using examples from bryozoans, rotifers, cladocerans, molluscs, teleosts and others, the authors cover current conceptual issues of evolutionary ecology in considerable depth.

The book can serve as a source of critically evaluated ideas, detailed case studies, and open problems in the field of evolutionary ecology. It is recommended for students and researchers in ecology, limnology, population biology, and evolutionary biology.

A book of interest to:
Scientists, libraries, students

# Birkhäuser Verlag • Basel • Boston • Berlin

## DATE DUE

DEMCO. INC. 38-2971